An Introduction to Island Studies

An Introduction to Island Studies

James E. Randall

ISLAND STUDIES PRESS AT UPEI / ROWMAN & LITTLEFIELD
Lanham • Boulder • New York • London

Published in cooperation with Island Studies Press at UPEI
550 University Avenue, Charlottetown, PE C1A 4P3
www.upei.ca/isp

Published by Rowman & Littlefield
A wholly owned subsidiary of The Rowman & Littlefield Publishing Group, Inc.
4501 Forbes Boulevard, Suite 200, Lanham, Maryland 20706
Unit A, Whitacre Mews, 26-34 Stannary Street, London SE11 4AB
www.rowman.com

British Library Cataloguing in Publication Information Available

Library and Archives Canada Cataloging in Publication Available

ISBN 978-1-78661-545-9 (hardcover)
ISBN 978-1-78661-546-6 (paperback)
ISBN 978-1-78661-547-3 (epub)

For Barbara, Brenda,
Brian, Jen and Chris

Table of Contents

Acknowledgements

Although the "no man is an island" phrase by John Donne has often been taken out of context, the reason you hear it so often is that it seems to fit so many situations. So I will misuse it once again to thank those who have helped me during the journey to produce this book. I could never have completed this without the help of so many people. I would like to thank my friend and mentor Godfrey Baldacchino who has inspired me and so many students with his knowledge of islands. I only wish our island studies paths had crossed earlier. Thanks as well to my colleague and office neighbour Laurie Brinklow for picking up the slack when I was writing. Thanks to the following island studies scholars who reviewed various chapters along the way: John Connell, Lynda Harling Stalker, Andrew Jennings, and Patrick Nunn. The finished product is much better because of their time and patience. Thanks as well to the five anonymous reviewers who read the book proposal and provided excellent suggestions. I hope I have incorporated most of your ideas into this final product. I would like to thank the University of Prince Edward Island for giving me the opportunity to write this book. In particular, I took advantage of several Faculty Writing Workshops/Writing Retreats that allowed me to focus exclusively on writing. I was also fortunate enough to receive one of the UPEI Internal Publication grants that allowed for the co-publication between Island Studies Press and Rowman & Littlefield. Thanks for funding these programs must go to the Vice-President Academic & Research. Without graduate student Maggie Henry checking and correcting the many references, this book would not have been published until 2022! Bren Simmers, the Managing Editor of Island Studies Press, brought a "glass half full" optimism to copy-editing and page layout that I could never have matched if our roles had been reversed. Gurdeep Mattu and the folks at Rowman & Littlefield took a chance on this book and are responsible for putting it in the hands of so many readers across this world of islands. Thanks as well to my family who had the patience to put up with me during this process. Finally, I must acknowledge all of those islanders on my adopted Prince Edward Island home as well as islanders from so many other islands whose stories are represented on these pages.

Introduction

Island studies as an interdisciplinary field of inquiry is still in its infancy at many universities and colleges around the world. Despite this, a rich body of scholarship has emerged that considers islands as places that are critically important, not only for what they can tell us about the islands themselves and the estimated 600 million islanders who currently call them home, but also for the lessons they provide to our global society. Perhaps because of the absence of island studies programs at most universities and vocational colleges, even at those institutions of higher learning that play significant economic, social, and political roles on their islands, there are few introductory textbooks for undergraduate students and instructors who may wish to develop a more comprehensive understanding and appreciation of island issues. The purpose of this book is to fill that gap, by providing readers with an introduction to the many dimensions of island studies.

THEMES

This text employs several central themes in the form of contradictions or dichotomies as the framework for our discussion. At one level these dichotomies are fairly obvious. Islands have been described as sites of refuge and recreation (R. King 1993; Lockhart 1997), pilgrimage and punishment (Baldacchino 2005b; Gillis and Lowenthal 2007), and as places of value and neglect (Baldacchino 2015a; Danson and Burnett 2014). When we probe deeper into the contradictions and the questions they evoke, three general perspectives emerge. Are islands and islanders marginalized and vulnerable to global changes, or are they resilient and capable of responding quickly to external pressures? Are islands and island societies isolated and remote physically, culturally, and economically, or are they open and connected to the world around them? Finally, to what degree do islands share common features? Are islands and islanders diverse, heterogeneous, and unique, or do they share standard characteristics that may allow us to think of them as part of one or more relatively homogenous, coherent, groups?

As you may have already guessed by now, the three general themes of vulnerability and resilience, isolation and connectedness, and diversity and cohesion are not mutually exclusive.

As is often the case in our studies, there are no absolute right or wrong answers. For example, some island regions, such as the country of Haiti on the island of Hispaniola in the Caribbean Sea, have been decimated by natural disasters, economic despair, and social disorder. Islanders continue to survive by the most tenuous of margins. Other islands not far away, such as the British Virgin Islands, have prospered economically as centres of offshore banking and tourism. The residents of some islands, like those on Pitcairn Island in the South Pacific (of *Mutiny on the Bounty* fame) and the Saints of Saint Helena in the South Atlantic, have experienced high transport costs and inaccessibility. On the other hand, for thousands of years some peoples of the Pacific Ocean (or Oceania) have seen their islands as the centres of their universe and as historical hubs of trade and interaction across an interconnected marine "highway."

In order to set the stage for what this text is attempting to accomplish, and before providing a summary of the remaining chapters, it is important to take a moment to expand on these overarching conceptual themes.

Vulnerability and Resilience

ARE ISLANDS AND ISLANDERS MARGINALIZED AND VULNERABLE TO GLOBAL CHANGES, OR ARE THEY RESILIENT AND CAPABLE OF RESPONDING QUICKLY TO EXTERNAL PRESSURES?

Islands have been portrayed and judged throughout history, in Baldacchino's words, "by what they don't have" (2007, 14), including people, natural resources, or competitive advantages that limit their opportunities and possibilities. The words vulnerability and marginality have been applied as descriptors for islands by island studies scholars and laypersons alike (Briguglio 1999; Briguglio et al. 2009; Royle 2001). Often after the fact, vulnerability has been used simplistically to explain the failure of islands to follow a prescribed development trajectory, or as a convenient way to link the short-term consequences of natural disasters, loss of cultural identity, or outmigration. For small islands or atolls, vulnerability may be associated with an absence of options during periods of crisis. Unlike many continental regions where a natural or economic disaster may only affect part of the landmass, when a crisis occurs on a small island it is more likely to affect the entire island and islanders have nowhere to flee.

More recently, the terms vulnerability and marginality have been adopted by islanders themselves to garner global attention and resources to address the long-term impacts of sea-level rise. Perhaps Fischer (2012, 296) best captures this sentiment by stating, "Hurricanes, volcanoes, earthquakes, tsunamis—the Four Horsemen of Islands ... and a Fifth has now arrived, global warming." As such, islands and island societies are often seen as victims of circumstance. Human-induced climate change, globalization, challenges with governance, limited resources, and isolation are challenges that islanders and global development agencies seek to remedy by accessing external resources (Gough et al. 2010). Vulnerability has also been used in the context of islands to describe the state of island ecosystems, where endemic flora and fauna that have evolved in response to localized environmental circumstances are vulnerable to invasive plant and animal species introduced from the mainland. Many of these biological and zoological vulnerabilities are linked to the small space of the island and the degree of specialization of the species.

Môre recently, a countervailing narrative has emerged. This is one that sees many island societies as resilient, nimble, flexible, and adaptable to external events, ranging from the economic and cultural impacts of climate change (Lazrus 2012) to the consequences of natural disasters (J. Campbell 2009; Kelman and Khan 2013). Some case studies of prehistoric and historic island societies have suggested that mobility and interaction over long ocean distances were strongly linked to climatic conditions (Nunn et al. 2007). The broad expanses of ocean, seared into Western consciousness by historical tales of European explorers overcoming vast distances, are increasingly being viewed as routes and highways rather than as barriers. In fact, prior to European colonization of Oceanic islands, it was not uncommon for islanders facing crises to turn to other nearby islands for assistance of various kinds (J. Campbell 2009; Malinowski 1922).

Although many island cultures have seen their ways of life marginalized—with language, cultural practices, and symbols lost or distorted by exposure to Western cultural practices—island-based symbols of culture such as Reggae and Celtic music have developed a much larger continental audience, at least partly because of the island diasporas that introduce and reinforce these island cultural practices. A growing body of research is making a convincing case that small sovereign island states (SIS), as well as non-sovereign but subnational island jurisdictions (SNIJs), have developed substantial economic and political capabilities, including preferred trade deals, opportunities for labour mobility, access to domestic and/or foreign capital, and assistance in times of natural disasters (Baldacchino 2006b, 2006e; Dunn 2011; McElroy and Mahoney 2000). Rather than being poverty-stricken and destitute relative to many other regions of the world, island economies might be more accurately described as adaptable. For these reasons, this text uses the twin terms of vulnerability and resilience as one of its themes.

Isolation and Connectedness

ARE ISLANDS AND ISLAND SOCIETIES ISOLATED AND REMOTE PHYSICALLY, CULTURALLY, AND ECONOMICALLY, OR ARE THEY OPEN AND CONNECTED TO THE WORLD AROUND THEM?

Historically, the prevailing description of islands is one of isolation and peripherality. Islands have also been described as bounded, where the definitive boundary between land and water serves as a natural and perceptual barrier between what takes place on an island and what takes place elsewhere in the world (Hay 2003a; Royle 2001). In David Weale's (1992, 93) words, "What is it that gives an island its special charm ...? I think the main reason is that an island has clear physical limits, and the mind is able to grasp it and make a picture of it as a whole." This perception has led the general public and researchers to portray islands as closed systems or living laboratories. Rosemary Gillespie (2007) went so far as to call them "Nature's test tubes" (1), where physical and social experiments could be carried out separate and apart from the complications and messiness found on continents. For example, C. Michael Hall (2010) proposed that islands are natural laboratories for the study of tourism, and Evans (1973) used the same rationale to argue for the study of cultural practices on islands. Biologists, ecologists, and biogeographers have repeatedly pointed to the outcomes of endemism, including dwarfism and giantism, as a function of the isolation of islands and the closed nature of island environments (Carlquist 1974). DeLoughrey (2013) suggests that one of the best examples of the use of islands as closed ecosystems was the rationale used by the French, Americans, and British to test nuclear weapons on islands in the Pacific following the Second World War.

At the same time, islands have been conceptualized as extremely well-connected, with boundaries that are permeable (Baldacchino 2007) or porous (Stratford 2003). This dichotomy is identified by Epeli Hauʻofa (1994, 1998) to describe the historical role of islanders in Oceania. He suggests that the extreme isolation of many parts of this region may have shielded island societies from the cultural influences that raged across most of the continental world in the past. This isolation also led to the evolution of distinct island cultures without a continental influence. While it may have served as a shield, Hauʻofa (1998) makes the case that the distribution of islands within the Pacific Ocean has also provided waterways that have connected neighbouring islands into regional groups for cultural and economic exchange. Hay (2003b) suggests that the very "critical distance" of islands, or their relative isolation from the centres of the global economy, may enable them to generate real and radical alternatives in a more tightly controlled and centralized world. For all of these reasons, isolation and connectedness is a useful pair of concepts to better understand islands.

Diversity and Cohesion

> ARE ISLANDS AND ISLANDERS DIVERSE, HETEROGENEOUS, AND UNIQUE, OR DO THEY SHARE STANDARD CHARACTERISTICS THAT MAY ALLOW US TO THINK OF THEM AS PART OF ONE OR MORE RELATIVELY HOMOGENOUS, COHERENT, GROUPS?

As is often the case when we are introduced to new phenomena, focusing on unique features can be overwhelming. It is not surprising that we inevitably seek to better understand objects by distilling them down into more comprehensible categories, themes, or groups. This is sometimes referred to as the distinction between taking an idiographic or unique approach versus taking a nomothetic or generalizing approach. Such is the case with the study of islands. For example, Hay (2013, 209-210) makes reference to the "bewildering diversity among the planet's islands" while Gillis and Lowenthal (2007, iv) note that the essays in their special edition of the *Geographical Review* reflect the physical and cultural diversity of the hundreds of thousands of islands of the world and their inhabitants. Although it is hard to fathom how one set of essays might be able to reflect the diversity inherent in hundreds of thousands of places, the point being made in both cases is that islands are more different than they are alike.

There is also a countervailing tendency to seek order or structure to the islands of the world. For example, in the same article by Gillis and Lowenthal, they go on to say that islands are clearly defined or confined by the water surrounding them and are wrapped up in mythic metaphors. In other words, despite their obvious differences, islands share certain innate features on the basis of their geography or the common experiences and beliefs of islanders. Some might call this "islandness." Conkling (2007) states that the boundedness and isolation of island life create conditions that transcend local culture, time, and space. Much of island literature and poetry takes the same approach. For example, David Weale (1991, 6) writes that "Islandness becomes part of your being, a part as deep as marrow, and as natural and unself-conscious as breathing." Recognizing both the diversity of islands and the role they may play as a group in answering bigger world questions, Godfrey Baldacchino (2007) views islands as the repository of new things and as sites of agency. In fact, in an attempt to get readers to think of islands in totality as being central to larger issues, and following Quammen (1996) and Hauʻofa (1994), Baldacchino (2007) titled his edited collection of ground-breaking island scholarship *A World of Islands* rather than the more conventional "islands of the world."

This apparent contradiction between the unique and the general is important because it frames how islands and islanders are often understood. Premdas (1996) provides an excellent example of this complexity in reference to the "Caribbean region." He notes that although we may have a mental image of the Caribbean as a unified region of places, peoples, and cultures, likely created for us by tourism marketing, the area is one of the most culturally diverse regions in the world. This dichotomy between diversity and cohesion or homogeneity forms the third theme in this book.

BOUNDARIES AND LIMITATIONS TO THIS STUDY OF ISLANDS

The lower threshold to be defined as an island as suggested by Christian Depraetere and Arthur Dahl (2007, 2018) is ten square kilometres and the upper boundary is one million square kilometres. Based solely on total land area this means there would be 5,675 islands in the world. At the lower end of this range, you would find islands such as Nauru in the Pacific Ocean, while the upper end would include Madagascar, off the east coast of Africa. This definition of islands excludes almost nine million islets that are smaller than ten square kilometres, as well as the millions of other sparsely or uninhabited rocks. It also excludes many atolls and archipelagos that consist of small, individual islands, but collectively are much larger than ten square kilometres, especially when you include all of the marine space between and around the islands.

This raises the question of scale. What are the boundaries or limits that define the term island and what constitutes the population of islands to be used in this book? The more difficult question of the definition of island is raised in the next chapter. Walker and Bellingham (2011) look closely at the influence of scale of islands as a fundamental dimension that shapes the unique characteristics and challenges facing islands. As to which islands should be included or excluded, I take a very liberal approach. For instance, the eight Senkaku/Diaoyu/Diaoyutai Islands in the East China Sea together total less than seven square kilometres in area and are uninhabited. However, their geopolitical significance in the surrounding region greatly outweighs their less significant size. At the other extreme, although many island studies scholars would not include Great Britain, the Japanese islands of Honshu or Hokkaido, or

> **THE CONNECTIONS THAT INDIVIDUALS HAVE WITH THEIR ISLANDS ARE COMPLEX; MUCH MORE SO THAN A LOCATION THAT MIGHT APPEAR ON YOUR BIRTH CERTIFICATE. THERE IS A LOT MORE MESSINESS TO THE LABELS "ISLANDER" AND "MAINLANDER."**

the island of Java in Indonesia to illustrate island-oriented subjects, it would be shortsighted to exclude them out of hand simply because of their larger populations.

Taking this issue of the relevance of scale as a limiting factor in the study of islands one step further, much attention has been paid to the adjective "small" as a defining characteristic of island studies scholarship and instruction. Take, as examples, the Small Island Developing States (SIDS) organization, the International Small Islands Studies Association (ISISA), the Small Island Cultures Research Initiative (SICRI), the Small Island Tourist Economies (SITES) development model, not to mention all of the literature that focuses on small islands (Brinklow, Ledwell, and Ledwell 2000), sustainable development (Kakazu 1994), and "the political economy of small islands" (Baldacchino and Milne 2000). Some researchers go through considerable methodological gymnastics to set parameters around scale and "smallness." For example, in examining the hydrology and water resources of small islands, Falkland (1991) defines the upper limit of a small island as being no more than 2,000 square kilometres in total and no wider than ten kilometres. Bass and Dalal-Clayton (1995) define a small island state as having fewer than one million inhabitants and a landmass smaller than 1,000 square kilometres, while Péron's (2004, 328) study of islands off the coast of France was limited to those "large enough to support permanent residents, but small enough to render to their inhabitants the permanent consciousness of being on an island." The inference from this body of work is that islands greater than a certain size, as defined by population, or areal extent or identity or some other feature, lose the right to belong to this group. There are perfectly reasonable reasons for limiting the scale and scope of island studies in these ways. However, as suggested by Depraetere and Dahl (2007, 2018) and as implied above, drawing a line between something too large or too small to be studied as an island is ultimately an arbitrary decision.

In order to avoid these methodological conundrums, this text does not set formal exclusionary boundaries on the range of islands used as examples. As such, it takes a broad survey of concepts applied to islands in their many forms. Granted, most of the examples used in the following chapters fall within the general consensus of what constitutes a small island. However, if the best example of a concept or idea just happens to be associated with Great Britain or Japan at one extreme or Pitcairn at the other extreme, then so be it. In the same vein, even though they are not "completely surrounded by water," we are not going to eliminate Prince Edward Island, Canada, from discussion just because it is now connected to the mainland by a bridge, or Manhattan, New York, because it is connected to the rest of continental North America by twenty-one bridges and four tunnels, or Singapore because it is linked by a causeway to the Malay Peninsula across the narrow Straits of Johor. Finally, although most of the examples that are discussed in island studies are surrounded by saltwater, there is no reason to exclude islands in freshwater lakes or those formed as the alluvial fans of river deltas. Geographically, all of these are islands.

Before proceeding, I must make a confession regarding my own personal background. I was not born on an island and have only called an island (Prince Edward Island in the Gulf of St. Lawrence, Canada) my home since 2010. In the jargon of Atlantic Canada, I am therefore a "come from away" or CFA. If I was living on Nantucket Island in the northeastern seaboard of the United States, I might be called a "wash ashore," or in the British Virgin Islands, a "non-belonger." As such, I may be viewed by some as a bit of an outsider looking in on island issues. In fact, some might argue that my birth on a continent gives me less of a right to speak on the subject matter of islands (i.e., "islands for islanders" in Nunn, 2004b). I am the first to admit that I may never be able to understand the deep connection that some islanders express for their island homes. However, the connections that individuals have with their islands are complex, much more so than a location that might appear on your birth certificate. There is a lot more messiness to the labels "islander" and "mainlander."

As is the case with many island studies scholars in a field with few doctoral-level university programs, I have come to the study of islands from another disciplinary home. In my case, my academic training was as a geographer. With due deference to my island studies colleagues who were trained as economists, sociologists, anthropologists, biologists, political scientists, etc., being a geographer and an island studies scholar is not that unusual. Perhaps it is because of the inherent interdisciplinarity of geography and the love and appreciation that many geographers have for space and place. However, all of us perceive and understand phenomena based on our own personal experiences and professional training. As I have done so far in my island studies journey, I would encourage you to use your own experiences and knowledge to question prevailing assumptions and beliefs.

OUTLINE OF CHAPTERS

This introduction has already provided a hint to some of the challenges surrounding the definition of islands. In **Chapter One** (Definitions and Classifications of Islands), we expand on some of these issues. For example, how do functional and perceptual definitions of islands differ from what you might find listed under the word "island" in a dictionary? What is the difference between a continent and an island, or is there any difference? The word "island" itself has been used metaphorically in so many ways, such as "a heat island" or "a traffic island" or "a kitchen island." What do these and other appropriations of the word island tell us about our shared understanding of islands? We also review several of the common typologies or classifications of islands, as seen through geomorphology, political science, and international relations.

Much has been written regarding the physical processes that have formed islands, from subduction and the movement of the Earth's tectonic plates to volcanic upheaval and the deposition of riverine alluvial sediments. **Chapter Two** (Physical Processes and Islands) examines the processes that have created and modified the islands of the world. In addition to geomorphological processes, the evolution of island flora and fauna is one of the most fascinating aspects of island studies. Island locations offer some of the best examples of endemism and species biodiversity and, as such, are a cultural and economic treasure for islanders and the larger global society. Whether from the perspective of Charles Darwin's analysis of finches on the Galápagos Islands (now a subnational island jurisdiction of Ecuador) or Alfred

Wallace's description of species differentiation across the islands of the Malay Archipelago, the examination of island biogeography has important implications for the evolution of the human species and life on our planet. Chapter Two also examines the climatology of islands, from the microclimates that create lush rainforests amid desert-like regions to the longer-term impacts of climate change on islands.

Popular images and stereotypes of islands have been created through a variety of mediums from the sagas told by returning European explorers and merchants, to children's adventure stories, to cinema, television, music, literature, and even advertising. **Chapter Three** (Images of Islands from Literature and the Popular Media) looks at these popular expressions of islands from the perspective of how they have shaped the images of islands and islanders and how these representations have influenced how islanders see themselves.

Chapter Four (The Settling of Islands and Indigenous-Outsider Interactions) examines the settling of islands, from the perspective of Indigenous peoples to the interactions between Indigenous island societies and non-Indigenous societies. The islander-outsider relationship is one that has spanned thousands of years. Historic examples of these relationships and the perceptions that have been generated by islander and non-islander interactions will help us to better understand the current social, economic, and political life of islanders. The more recent relationships between islanders and mainlanders informs **Chapter Five** (Islands, Islandness, and Culture). The concept of islandness, not to be confused with insularity, is one of the most prominent and significant terms in island studies. This chapter provides a comprehensive discussion of the interpretations and applications of islandness, primarily from the perspective of islanders, but also from those who have not been raised in island environments.

In spite of an assumed and imputed association with marginalization, many islands continue to hold a prominent political place in the world. **Chapter Six** (Geopolitics and Island Governance) describes the geopolitical roles played by the world's islands and the level of power and autonomy they may or may not exercise. The past fifty years have seen the creation of many more sovereign island states, the recognition of exclusive economic zones extending from coastlines, and the consequences of climate change. As a result islands increasingly find themselves at the centre of some aspects of international relations. This chapter discusses the importance of these trends, but also introduces us to the much larger group of islands referred to as subnational island jurisdictions (SNIJs) that remain tied historically, politically, economically, and culturally to mainland jurisdictions. Some of these SNIJs make the conscious choice to forego nationhood in favour of an ongoing political, economically dependent, and often ambiguous relationship with a larger host nation or metropole (e.g., the Falkland Islands or Martinique). Internally, islanders have adopted a variety of governance systems that reflect their island circumstances. Chapter Six describes these systems of governance and the roles that these structures have played in the ongoing political evolution of islands.

Chapter Seven (Islands, Population, and the Movement of People) examines the roles that islands have played in the movement of people. Some of the most prominent cultural perceptions of islands are as disembarkation points for migrants and as sites for refugees. Therefore, this chapter examines islands as entry and exit points for humanity. It also looks at the exodus of populations from islands, the factors that have prompted these emigrations, and the consequences of these diasporas, both for the islands themselves and for the displaced islanders.

Islands have been useful in advancing our understanding of health, wellness, and epidemiology. Everything from our understanding of the evolution of epidemics (Cliff, Haggett, and

Smallman-Raynor 2000) to the genetics of colour blindness (Sacks 1996) has benefited from exploring these health problems on islands. **Chapter Eight** (Island Health and Epidemiology) describes the lessons that have been learned from examining population health on islands. Unfortunately, some island peoples suffer from more severe health problems, including obesity and the consequences of obesity, such as diabetes. This chapter also outlines some of the health challenges facing islanders and the role that living on an island may have played in contributing to these health problems.

Arguably the largest body of research on islands has focused on their economies. **Chapter Nine** (Economic Change, Development, and Islands) looks at the economic fortunes of islands and islanders using several models. Some island economies are almost entirely reliant on foreign aid and remittances from those islanders who have moved away, while others, like offshore financial centres, have sustained their economy based on their ability to negotiate financial relationships with other nations and organizations. This chapter takes a look both conceptually and empirically at the many ways that islands have developed their economies.

Increasingly, islands in temperate locations are using tourism to shape their image and attract a share of the global tourism market. The stereotypical island as a tropical paradise, framed by crystal blue waters, white sands, and palm trees, may be one of the most well-known elements of travel industry imagery. **Chapter Ten** (Island Tourism) looks at the ways in which tourism has shaped many islands and island peoples, economically, politically, culturally, and ecologically. Given the impacts being placed on island ecosystems and societies as a result of increasing numbers of world travellers, sustainable tourism management is now emerging as an important component of island tourism. This chapter discusses the role and potential for this model of island-based tourism.

Since sustainability encompasses so many aspects of the issues discussed earlier, a broader discussion of this term forms the central theme of **Chapter Eleven** (Islands in the Age of Sustainability and Sustainable Development). Sustainability may be one of the most overused (and misused) terms in modern languages. However, it is still a powerful device to better understand the relationships between humans and their social and natural environments. In this chapter, island sustainability is discussed from both historical and current perspectives.

Islands hold a special place in the world, not just for those who have experienced island life, but also for those whose relationships with islands have been formed indirectly. Therefore, the **Conclusion** (Conclusions and Future Directions in Island Studies) will provide insights on the future of islands and island peoples. It will also address the institutional structures of island studies, where it started, and where it might be heading. Throughout the book, and especially in this concluding chapter, the many examples and concepts are woven together using the three central themes of vulnerability/resilience, isolation/connectedness, and diversity/cohesion.

One of the goals of this text is to show readers that island studies is relevant to both islanders and mainlanders. Island phenomena play an important role in allowing us to better understand issues in mainland communities, regions, and mainland society. In the chapters that follow, issues that are of broader relevance will be highlighted in dialogue boxes. The content of these dialogue boxes will emphasize, for example, the importance of islands for endemism and biodiversity, sea-level rise and climate change adaptation, the evolution of species, how adventure novels have shaped our perceptions of islands, and why islands have been so successful as sites for financial transactions.

Key Readings

Baldacchino, Godfrey, ed. 2018. *The Routledge International Handbook of Island Studies*. New York: Routledge.

Gillespie, Rosemary G, and David A. Clague, eds. 2009. *Encyclopedia of Islands*. Berkeley: University of California Press.

Hauʻofa, Epeli. 1994. "Our Sea of Islands." *The Contemporary Pacific* 6 (1): 148-161. http://hdl.handle.net/10125/12960.

Hay, Pete. 2013. "What the Sea Portends: A Reconsideration of Contested Island Tropes." *Island Studies Journal* 8 (2): 209-232.

Chapter One

Definitions and Classifications of Islands

Islands and continents are but names we give to different parts of one interconnected world.

John Gillis, "Island Sojourns," 277

DEFINITIONS OF THE TERM "ISLAND"

Dictionary Definitions

Defining an island would seem to be a simple exercise. As a noun, almost all dictionaries provide a variation on the following, "A tract of land completely surrounded by water, and not large enough to be called a continent" (Dictionary.com 2020a) or "A landmass, especially one smaller than a continent, entirely surrounded by water" (TheFreeDictionary.com n.d.). However, when you start to scratch the surface of these definitions, you encounter more questions than answers. For example, is the difference between an island and a continent based solely on size? And, if an island has to be completely surrounded by water, then how do you define those places such as Eilean Tioram ("Dry Island") on the west coast of Scotland that are surrounded by water at high tide yet are connected to the mainland at low tide (see Figure 1.1). Taken to an extreme, what about those many islands that are now linked to the mainland by a bridge or a tunnel or a causeway, like Prince Edward Island or Great Britain or Cape Breton Island? Is Manhattan still an island despite being connected to the mainland by twenty-one bridges and four tunnels? At what point does the degree of connection, or distance to the mainland, change the status of an island, if not functionally then perhaps in the minds of those living on them? And what about all those uninhabited rock outcroppings and sandbars? Does the "tract of land" have to be a certain size, exist for a certain length of time, have a minimum number of residents, or be able to sustain human life of some kind, to be called an island? These are just some of the questions that will be addressed in this chapter.

In addition to its physical or "real" existence, the term "island" has also been described as one of the most powerful metaphors in Western culture (Hay 2006). Therefore, this chapter will examine the idea of the island: how it has been understood functionally, institutionally,

Figure 1.1   Eilean Tioram ("dry island"). *Source:* **Chris Allan**

and psychologically, and the difference between what might be called a "real" island and an "imagined" or a metaphorical island. This chapter will also show several of the many ways in which islands have been classified and categorized.

Historical and Linguistic Origins of the Word "Island"

There seems to be general agreement among linguists and etymologists that the current English word "island" first appeared in the fifteenth century, but there is less agreement regarding its precise origin (Ronström 2009). One theory is that it came from the Norse/ Vikings, a people with a long seafaring history. For them, the word *is* meant water, hence the word "island" would indicate the mixture or interface of land and water at the coast (Shell 2014). Old English also provides clues to the origin of the word "island." Accordingly, the letter combinations *ieg, ig,* and *eg* are all Anglo-Saxon/Old English in reference to a small island (Ronström 2009). Also, the term *ey* means island or a low-lying land by water and may have evolved to the present-day island as shown on old maps as Eyland Madagascar. This is why it would be redundant to add the word "island" after the names of many of the islands around Great Britain, such as Jersey, Guernsey, and Anglesey, since the word is already contained in the suffix of their names. Second, the proto-Indo-

Figure 1.2　Manhattan Suspension Bridge. *Source:* **Natalia Bratslavsky**

> **IS MANHATTAN STILL AN ISLAND DESPITE BEING CONNECTED TO THE MAINLAND BY TWENTY-ONE BRIDGES AND FOUR TUNNELS? AT WHAT POINT DOES THE DEGREE OF CONNECTION, OR DISTANCE TO THE MAINLAND, CHANGE THE STATUS OF AN ISLAND?**

European *ea* means river, or more broadly any body of water, so ea-land or island literally means river-land (Holm 2000). Finally, the French *yle* (as in the Yle of Skie/Skye in Scotland) or *ile*, the Spanish *isla*, the Italian *isola*, and the Portugese *ilha* all stem from the Latin *insula*. The French derivation may have contributed to the naming of places as *ile-land,* such as Isle de France, renamed Mauritius (Figure 1.3), which is not too different from the current English term "island."

In Japanese the word for island is *shima*. Many of the islands that are part of the Japanese archipelago have *shima* as part of their name, as in Tori-shima, meaning "Bird Island," or Nishi-no-shima meaning "western island" (see Figure 1.4). However, *shima* can also mean an area or community on an island. For example, the capital and largest city on the southwestern tip of Kyushu is Kagoshima. Perhaps more so than when the word "island" is used in English, *shima* in Japanese has

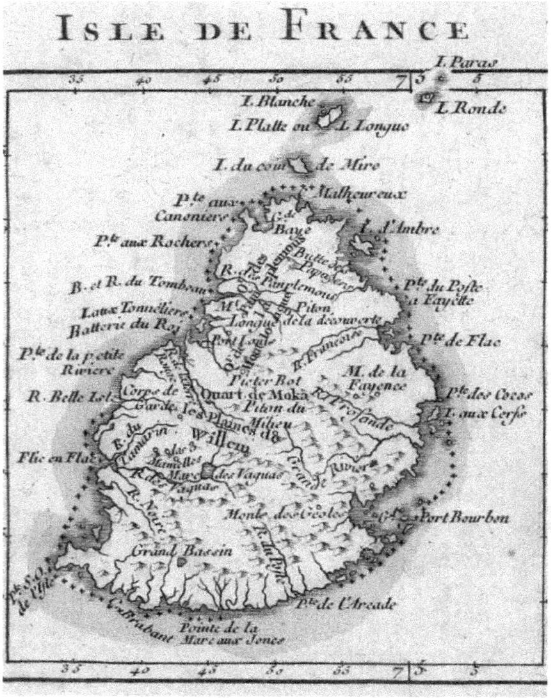

Figure 1.3 Isle de France by Rigobert Bonne, 1791. Currently known as Mauritius. *Source:* **public domain**

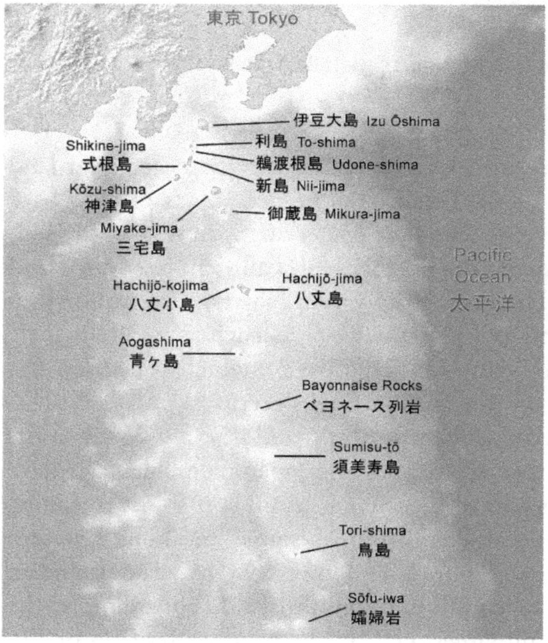

Figure 1.4 Izu island chain in Japan. *Source:* **public domain**

a deeper cultural meaning and can refer to a place of sanctuary and shelter, often bounded by waters, coastline, and hills, to a set of rituals, music, and beliefs, and finally to the productivity of the land and marine resources. Suwa (2007) provides a much more complete explanation of the word *shima*.

In Greek, the words *nesos*, *nes*, and *nesia* are all versions of the word "island" in English. One of the largest islands in Greece is known as "Nisos Kriti" or the "Island of Crete." Several islands and island regions of the world incorporate this as a suffix in their names, as in Indonesia. One of the more common categories of islands in the Pacific Ocean also incorporates this Greek word fragment in their names, as in Polynesia (or "place of many islands"), Micronesia (or "place of small islands"), and Melanesia (meaning "black islands" or literally, "islands of the black-skinned people"). It should be noted that some researchers critique this three-part division of Pacific Island zones as being arbitrary and a legacy of colonialism, preferring instead to use the more inclusive term Oceania to describe the region (Hau'ofa 1993, 1998). Moreover, some local peoples believe that even being referred to as Pacific Islanders carries with it a colonial stigma. The terms Pacifikans and Pacific Peoples (especially in New Zealand/Aotearoa) have emerged as alternative labels of self-identification that exclude all of the descendants of immigrants from Europe and elsewhere outside of the Pacific.

Functional and Institutional Definitions of an Island

The earlier discussion suggests that scale may not be the only criterion that needs to be taken into account when defining an island. As Shell (2014, 28) states, "Privileging size … no matter how attractive, rules out the realities of human experience." This is not a new idea. Ronström (2009) shows us that the existence of an island, historically and linguistically, has been based on many different features. For example, the Swedes have different words for islands based on the topography and composition (e.g., forested or sandy). Russell King (1993, 16) notes that the Vikings defined an island as a place that could be reached from the mainland by a ship with a rudder in place, while the 1861 Scottish census defined an island as "an area of land inhabited by man where at least one sheep could graze." Anders Källgård (2005) notes that estimates of the number of islands in Sweden have ranged from 24 to 221,800. The higher number used satellite imagery and digital maps—in other words, the "dictionary" definition of an island. The lower threshold number was arrived at by using the more complicated European Union's (EU) criteria for an island, including that it has to have a land area of at least one square kilometre, has to be at least one kilometre from the mainland, must not be permanently connected to the mainland, must have at least fifty inhabitants, and must not house the capital city of an EU member state. All of these examples point to the importance of characteristics other than scale or population size in defining an island. In this latter case, the degree of isolation, the connectivity, and the political role were also taken into account in determining what constituted an island.

In case you're starting to think that these kinds of definitions are only quirks or oddities that have no current relevance to islands, the United Nations definition of the "regime of islands" through the Convention on the Law of the Sea (United Nations 2018) provides an excellent example of how the definition of an island can be critically important to the lives and livelihoods of islanders. Article 121 of this Convention suggests that the main difference between rocks and islands is that the former "cannot sustain human habitation or economic life of their own." As such, the territorial sea surrounding uninhabited rocks would not include any exclusive economic zone (EEZ) or continental shelf. Exclusive economic zones, as defined by the Convention, can extend up to 200 nautical miles from an island and some

HOW MANY ISLANDS ARE THERE IN THE WORLD?

It is not surprising that different websites will give you different answers to this question, largely because of the ambiguity in defining an island as suggested above. Depraetere and Dahl (2007, 58) have probably been the most systematic in addressing this question from a statistical and positivist approach. Using a standard dictionary-based definition of "any piece of land surrounded by water, whatever its size or distance to the closest mainland," and employing the tools of remote sensing satellite imagery, they estimated that there are 5,675 islands that range in size from ten to a million square kilometres. Since these are islands found only in oceans and seas, this number would not include the thousands of islands within this size range that are located in freshwater lakes and rivers. Using a process of mathematical extrapolation, they then go on to suggest that there are also approximately 8.8 million islets, ranging from .001 to 10 square kilometres, and about 672 million rocks ranging from about one square foot to .001 square kilometres.

continental shelves can extend much farther than this. For example, the Grand Banks off Canada's Atlantic coast extend for 730 kilometres and are estimated to be 280,000 square kilometres, an area just slightly smaller than the size of Italy. Historically, this region has been a rich source of commercial marine life and is increasingly important to the local economy of the Canadian province of Newfoundland and Labrador for its oil and natural gas reserves. If an area is within the exclusive economic zone, it means that the coastal state has the right and responsibility to exploit, manage, and conserve the natural resources in the waters, on the seabed, and in the subsoil of the seabed. This distinction of a piece of land being either a rock or an island can be very important for nations, especially in disputes regarding who controls these areas and their current and future resources. Several centuries ago, many small tropical islands of the world were claimed by colonial powers, largely to harvest the phosphate-rich guano from seabirds for use as a natural agricultural fertilizer (Van Dyke and Brooks 1983). With the ratification of the United Nations Convention, these small oceanic islands took on much greater geopolitical significance.

The dispute among Japan, China, and Taiwan over control of the Senkaku (Japan)/Diaoyu (China)/Diaoyutai (Taiwan) Islands, located southwest of Okinawa in the East China Sea and as seen in Figure 1.5, is a case in point. These eight, formerly privately owned islands and uninhabited rocks are now at the heart of rising tensions among these nations, not coincidentally because the sea around these specks of land includes fishing grounds, potentially large reserves of oil and gas, and military significance at a time when China is flexing its geopolitical muscles in the region.

Consider also the example of Rockall in Figure 1.6, a seventy-four-square-metre outcropping of land almost 400 kilometres west of Scotland. Once described by Basil Hall (1831, as cited in F. MacDonald 2006, 631) as "the smallest point of a pencil could scarcely give it a place on any map which should not exaggerate its proportions," this piece of land has been claimed by Iceland, Denmark (via the Faroe Islands), Ireland, and the United Kingdom (Royle 2001, 2007; Royle and Brinklow 2018). In order to bolster its claim as an island controlled by the United Kingdom, a former British soldier spent forty days on Rockall in 1975. Since then, this record for continuous occupation of Rockall has been reset twice, first by three Greenpeace members in 1997 and then by a Scotsman in 2014. Of course, none of these adventures have constituted a change in Rockall's status from a rock to an island.

Another current example where the legal definition of the word "island" is linked to territorial disputes is in the High Arctic. With global warming and sea-level rise, the debate

Figure 1.5 Disputed islands in the East China Sea. *Source:* **Jackopoid**

Figure 1.6   Rockall "island" in the North Atlantic. *Source:* **Andy Strangeway**

regarding the feasibility of a northern sea passage and the resources that might be contained underneath these waters has once again become important. Canada's claim to the islands of the Arctic and control of the sea lanes between these islands is based at least partly on human occupation, although given the severity of the climate, very few of these islands see regular human activity or habitation. More importantly, Canada has not totally abandoned this region of islands, and historically other countries have either implicitly or explicitly recognized Canada's territorial claim. The question that is more likely in dispute is whether Canada can claim the areas surrounding these islands as part of its exclusive economic zone and whether the extension of the continental shelf in the area allows Canada to claim that its EEZ extends much farther in places than the standard 200 miles (Lasserre 2011).

This discussion of land areas and exclusive economic zones should also make us reconsider how we think about the presence of island territories in the world. If measured in terms of the total area encompassed by exclusive economic zones, island jurisdictions would constitute a much larger share of the Earth's surface, much more than 7 percent of the land area and 10.5 percent of the world's population as described by Baldacchino (2007) and others. To put this into perspective, the sum total of the land area of twenty-one island states in the Pacific Ocean is just over 91,000 square kilometres. However, when their EEZs are included, the total area expands to 27.5 million square kilometres, or just less than the size of the continent of Africa (World Bank.org 2000). As we will discuss later, these much larger island "territories" create an entirely new set of opportunities and challenges for island governments.

Indigenous and Spiritual Definitions of an Island

It is well understood that many Indigenous peoples understand land very differently than Anglo-American and other Western peoples. In reference to Australian aboriginals, Hill (1995) notes that white Australians think of land primarily through a material/economic lens, while Australian Aboriginals take a spiritual/cultural approach. As such, "the meaning of land goes beyond legal possession to include a personal and physical relationship that is tied to the Dreaming as well as recent ancestors" (Hill 1995, 314-315). Anthropologists often point out that Indigenous peoples, including those living on islands, are much more likely to incorporate informal systems such as kinship, friendship, and spiritual beliefs to understand the formal world around them (Skinner 2002). Reliance on myth and folklore is important to many Indigenous islander societies in understanding the origin and disappearance of islands as well. As Nunn (2001) notes, myths and oral traditions have long been used to help describe island flooding and the rise or subsidence of islands in the Pacific. Nunn (2003, 2008) also reminds us that many Pacific Island peoples believe that their islands were created by either being "fished up" (i.e., a [demi]god dropped a hook and line in the sea where it caught on submerged land and was hauled to the surface) or "thrown down" (i.e., where the earth was spilled from a basket carried by a god, as in the Marshall Islands, or leaves that fell from a tall tree, as in the islands of western Kiribati). As we will see in chapter four, not only do Indigenous people attach different meanings to their own islands, they also have a different understanding of the relationships between islands. Thus, while early European explorers viewed water as a formidable barrier and islands as specks of rock in a vast sea, Polynesians viewed the water between islands as something that connected island peoples, especially in times of crisis.

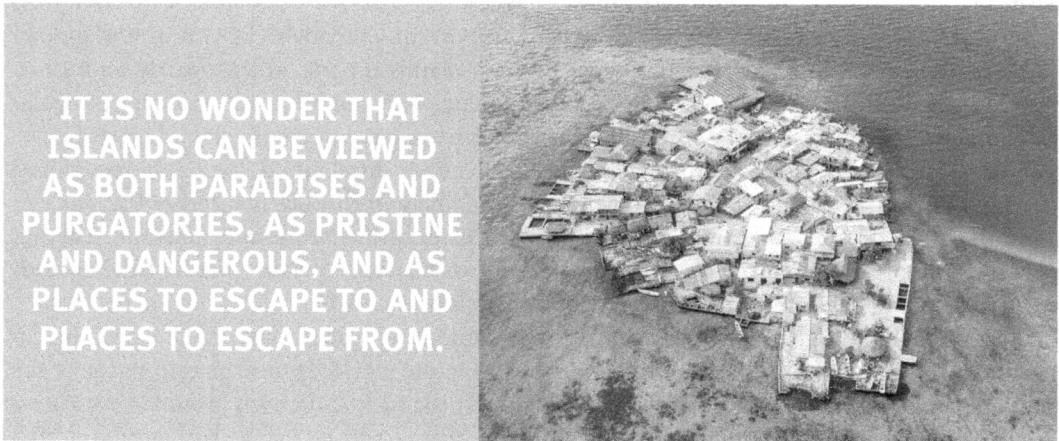

IT IS NO WONDER THAT ISLANDS CAN BE VIEWED AS BOTH PARADISES AND PURGATORIES, AS PRISTINE AND DANGEROUS, AND AS PLACES TO ESCAPE TO AND PLACES TO ESCAPE FROM.

Perceptual/Cultural/Psychological Definitions of an Island

There seems to be a growing consensus that adopting some numerical threshold for defining an island based on area, population size, or even physical features, is arbitrary and perhaps even irrelevant (Kerr 2005). Marc Shell (2014) forces us to think about this level of island "subjectivity" by asking us to imagine a scenario wherein Greenland's global ice sheet disappeared, a not so farfetched scenario given global temperature increases. What you would have then is not the largest single island in the world but an archipelago comprised of several smaller islands. Patrick Nunn (1994) suggests that it is not absolutely necessary to arrive at a general, precise definition of an island, and especially one that uses population size or areal extent as parameters in that definition, in much the same way that you cannot precisely define where a river ends and the sea begins. Rather, "we know instinctively what an island or a river is when we see one, and that is the important thing ... for anything that reduces the amount of rote learning in a subject is desirable" (Nunn 1994, 1). In trying to answer the question about the distinction between a continent and an island based on relative size, Holm (2000, 3) comes to a similar conclusion, "Then what is the size of a continent you ask? How many galaxies between here and infinity? How many thimbles of water in the Pacific? How many boxelder bugs in your own tree? An island is whatever we call an island. Or whatever I call an island. I am one and thus have rights and prerogatives in this matter."

At the start of this chapter, I alluded to some of the ambiguities surrounding this concept of "island" when I asked, only partly in jest, whether bridges or tunnels or causeways connecting islands to mainlands might mean they should no longer be considered islands. Steinberg (2005, 254) takes this one step further in asking the following not so hypothetical questions: (1) When an island is bisected by a canal, are two new islands created?, (2) When a small island is located near a large island, does the large island take on the properties of a mainland?, and (3) If a canal is built at the base of a peninsula, does this create an island?

What Steinberg is leading us to with these somewhat rhetorical questions is that islands are social constructions; they are created and evolve through time and space based on the characteristics, values, cultures, and experiences of individuals and societies. As such, it is no wonder that islands can be viewed as both paradises and purgatories, as pristine and dangerous, and as places to escape to and places to escape from. A good example of islands as social constructs again comes from Marc Shell's 2014 book *Islandology*. In it, and as illustrated in Figure 1.7, he reminds us that John Venn often used the British Isles to illustrate set theory, i.e., mathematical Venn Diagrams, to an audience familiar with the Victorian British empire. A more personal account of islands as social constructs is provided by Connell (2013a) who described spending three months on the tiny Micronesian atoll of Woleai. After walking around it in less than an hour on his first day on the island, he wondered how he was going to cope with being on such a small place. Months later, after getting to know the several hundred local islanders, he became resentful at being asked to walk 300 metres. In other words, the island had grown in size as his own personal social construction.

One way to understand islands as social constructions better is to think of the countless times in poetry and literature where islands and islanders are described, defined, and characterized in the words, songs, and art of the islanders themselves. For example, in his novel

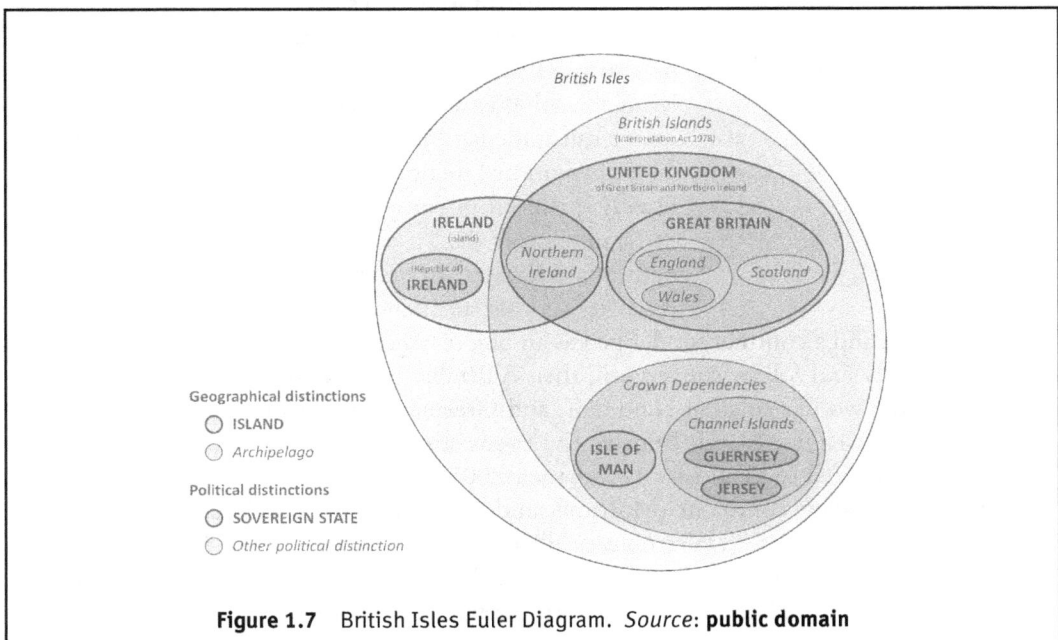

Figure 1.7 British Isles Euler Diagram. *Source:* **public domain**

The Magus, John Fowles (1965) describes the protagonist's thoughts of the fictional Greek island of "Phraxos" as "the boundedness of the smaller island, encompassable [sic] in a glance, walkable in one day, that relates it to the human body closer than any other geographical conformation of land." In reference to Prince Edward Island, David Weale (1991, 7) notes that "The topography and landscape of this province—that is to say, its Islandness—is the source and reference point for the imagination of Islanders. It is the primal source of our communal insight and wisdom ..." What these and many other poetic descriptions of islands are telling us is that islands are defined as much by how they are understood and perceived by those experiencing them, whether for a lifetime or for a short visit, as they are by their physical or functional characteristics. Lisa Fletcher (2011) calls this the real versus the imagined interpretation of islands and encourages us to see islands in a "performative" dimension.

As the scholarship and understanding of islands grows, especially among mainlanders, we need to be careful that we do not perceive islands using a "mainlander gaze" (Baldacchino 2008; McCusker and Soares 2011). In other words, we must be careful that we do not think of islands and islanders solely on the basis of our own preconceived notions. Today we would label the 1863 descriptions of Sandwich Islanders as ludicrous and insulting. These ancestors of modern-day Hawaiians were described in a grade school geography textbook as "ignorant of God, and addicted to some revolting customs, among which were infanticide, promiscuous concubinage, and the eating of human flesh" (Morse 1863, 65). Who is to say that our own descriptions of islanders will be perceived to be any less bizarre and ignorant of islanders' everyday lives when they are read in the twenty-second century?

AN ISLAND OR A CONTINENT

IS THE DIFFERENCE BETWEEN AN ISLAND AND A CONTINENT BASED SOLELY ON SIZE?

We have now seen that there is a great deal of ambiguity regarding the definition of a small island at the lower end of the size threshold. Therefore, it should not come as much of a surprise to realize that there is just as much ambiguity at the other end of the spectrum. In other words, why do we call some landmasses continents and others islands? The conventional dictionary definitions of the word island usually qualify it in comparison to the word continent, as in, "but smaller than a continent." In fact, this arbitrary nature is evidenced by Jedrusik (2011, 202) who goes so far as to say that an island is defined as an object smaller than a continent and a continent is defined as an object larger than an island. If we accept this definition literally and follow convention, then Australia, by total area at 7.6 million square kilometres, is the world's smallest continent, and Greenland, at 2.13 million square kilometres, is the world's largest island. Prior to the twentieth century, it was not uncommon for Australia to be referred to as an island (Ruhanen 2005). More recently, the consensus is that Australia is the smallest continent (Mortimer and Campbell 2014), or even, having it both ways, an "island continent" (Dalrymple, Williamson, and Wallace 2003; McMahon 2010; McMahon and Perera 2009; T. Ward and Butler 2006). Most researchers have deferred to geological processes and plate tectonics to distinguish an island from a continent. However, as is shown in Figure 1.8, there is still ambiguity, since the major continental and oceanic plates are all floating across the surface of the lithosphere and asthenosphere. A simplified explanation of the processes associated with plate tectonics and islands is provided in the next chapter.

A more comprehensive assessment of the continent/island conundrum might incorporate the relative distinctiveness of the flora and fauna of the areas, the unique cultures and languages represented, the orientation of their respective economies, or even whether the people themselves consider that they live on an island or a continent. In the case of Australia and Greenland, Australia is more likely than Greenland to have distinct animal and plant species and a distinctive culture. While Australians have a substantial connection to the land through agriculture and mining, Greenlanders are almost completely oriented towards the sea for fishing, hunting, and travel. As to whether they perceive themselves to be islanders, there is no definitive research that proves this, but anecdotal evidence suggests that Greenlanders are more likely to consider themselves islanders while Australians have a more mixed opinion that includes thinking of their country as an "island continent" (McMahon 2010).

There appears to be agreement, and especially among those concerned with how human activity has degraded the planet, that this discussion of islands versus continents is merely a distraction to the more important point: that Planet Earth is an island in a vast universal "sea" of deep space. The first colour images of Earth from space, taken almost half a century ago, have seared into our consciousness that, as with our stereotypes of an island, the Planet is fragile, vulnerable, and must be self-sustaining. Astronomer Carl Sagan (1994) alluded to this sea and island metaphor by referring to the Earth as both a pale blue dot and a very small stage in a vast cosmic arena.

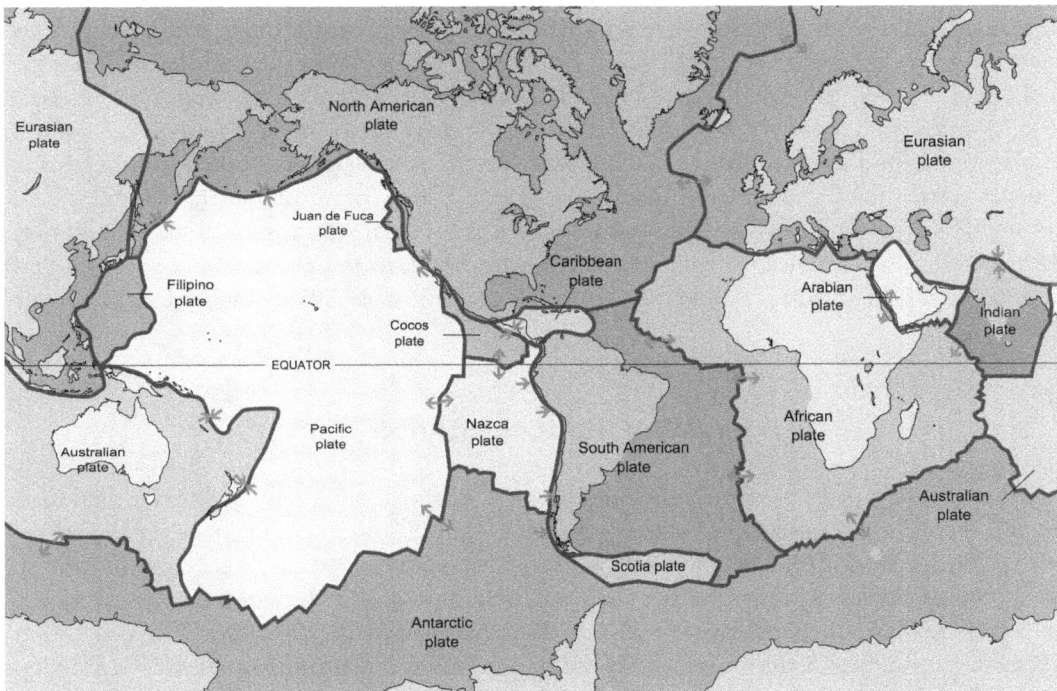

Figure 1.8 Major tectonic plates of the world, as mapped in the second half of the twentieth century. *Source:* **public domain**

INSULARITY, ISOLATION, AND ISLANDNESS

Students of island studies will quickly come across three "I" words when it comes to describing islands: insularity, isolation, and islandness. Starting once again with what we might find in dictionaries, there appears to be a great degree of confusion regarding these terms. The definition of insularity is sometimes synonymous with islands, as in "of or relating to an island or islands" (Dictionary.com 2020b). This doesn't help us very much. As is the case with the term "isolation," insularity is also used as an adjective, not only to describe the location of islands, but also to describe many other phenomena that are completely unrelated to islands, as in an insular language, or an insular cell, or even as a characteristic of islanders themselves, i.e., the islander community was "socially insular." In this way, insularity has also been used pejoratively to suggest that a person or a group is small-minded, inward-looking, or has a narrow view of the world. As such, most island studies scholars now avoid the term "insular" when describing islands and islanders.

Islandness is an important word and will be taken up in much more detail, and with more examples, later in this book. However, for now, think of islandness in two ways. First, it was described by Stephen Royle (2001, 42) as a set of constraints, including remoteness, isolation, smallness, and peripherality imposed upon small islands as a function of their insularity, where insularity is used as a synonym for island. Although this may have been the case during the state of island studies at the time Royle was writing, this definition is less prevalent today. In fact, there seems to be a growing consensus that isolation is purely a mechanistic measure of distance or time and that this distance shapes outcomes on islands and islanders in various ways (Baldacchino 2004; Hepburn 2012). For example, the cost of transporting small batches of goods to and from an island that is a great distance from world markets is undoubtedly higher than it would be to transport goods from places that are more centrally located to these markets. However, this relative remoteness may also serve to develop a strong sense of community and shared identity, characteristics that might serve the island well in response to both internal crises and external shocks. Taking this perspective, islandness, therefore, constitutes the characteristics of islanders and island societies. In Elaine Stratford's work, islandness "is a complex expression of identity that attaches to places [islands]" (Stratford 2008, 160). It may be associated with a sense of belonging and a worldview, to the point that it becomes part of your being (Platt 2004; Randall 2014; Royle and Brinklow 2018; Taglioni 2011; Weale 1991).

METAPHORICAL ISLANDS

Another complication surrounding the definition of island arises when you look at dictionaries and find multiple entries provided for the word "island." A sample includes (1) something resembling an island, especially in being isolated or having little or no direct communication with others (Dictionary.com 2020a), (2) a superstructure on the deck of a ship, as in an aircraft carrier (Merriam-webster.com n.d.), or (3) an island of peace, calm, sanity … a particular place that is peaceful, calm … when other surrounding places are the opposite (Dictionary.cambridge.org n.d.).

It is also not unusual for biologists to refer to a habitat or ecozone surrounded by another ecozone as an island, as in mountaintops separated by forested valleys, oases within a desert, hillocks of mixed forest in a bog or swamp, or even lakes in a forested environment (R. Gillespie and Clague 2009; Whittaker and Fernandez-Palacios 2007). Whittaker and Fernandez-Palacios refer to these as "habitat" islands to differentiate them from "true" islands. This is just the beginning of the use of islands as metaphors. As John Gillis (2007, 276) notes, "We see islands everywhere." There are "urban heat islands," islands of syntax (Boeckx 2012), traffic or highway islands, kitchen islands, and islands in the pancreas (i.e., the Islets of Langerhans) and the brain (Island of Reil). David Pitt (1980) coins the term "social islands" to describe groups of people who are cut off from the rest of society by social boundaries and are inward-looking. One of the most famous island metaphors is coined by the British clergyman and poet John Donne (1959) who, in 1624, wrote,

> *No man is an Island,*[italics added] entire of itself; every man is a piece of the Continent, a part of the main; if a clod be washed away by the sea, Europe is the less, as well as if a promontory were, as well as if a manor of thy friends or of thine own were; any man's death diminishes me, because I am involved in Mankind; And therefore never send to know for whom the bell tolls; It tolls for thee. (108-9)

One interpretation of Donne's words is that the relationship between individuals and the larger society is analogous to the relationship between one island and the larger mainland. Conversely, Steven Fischer (2012, 7) ends the preface to his book, *Islands: From Atlantis to Zanzibar*, with the quote, "Every man is an island." Although an exaggeration, by alluding to the Donne quote he seems to be implying that every one of the millions of islands and islets out there tells a story. Those who have lived on islands find these stories irresistible, whether they are about their own islands or other islands.

CLASSIFICATIONS AND TYPOLOGIES OF ISLANDS

From a physical geography perspective, most authors have followed Alfred Wallace's (1902) classification of "true" islands in distinguishing between two kinds of islands: those that are found fairly close to a continent and are really part of the same geomorphologic structure as the mainland, and those oceanic islands that are separated from the continental landmasses. The National Geographic website takes this one step further (Nationalgeographic.com 2012b). Recognizing that these are not necessarily mutually exclusive, they classify islands into six types: continental, tidal, barrier, oceanic, coral, and artificial. We will go into more detail on the physical formation of islands in Chapter Two. However, at this stage, it is just important to know that this classification is based on where they are located, for example their distance from a continent, and how they were formed, e.g., through volcanic activity, siltation, subduction and the movement of tectonic plates or, in the case of artificial islands, human engineering.

Other research has provided classifications of islands based on their proximity to other islands and their geological composition. For example, Lewis (1999) distinguishes single islands such as Nauru and Barbados from groupings of several islands in close proximity to each other (i.e., archipelagos). These archipelagos can consist of very few islands, as is the case with Malta or Tuvalu, or they can number in the hundreds (e.g., Ireland, Tonga) or even in the thousands (e.g., Indonesia, the Maldives, the Swedish coastline, and the Canadian Arctic). Archipelagos can be extremely vast and complex, physically and politically. For example, the Indonesian archipelago consists of more than 17,000 islands (6,000 inhabited) with an incredible diversity of cultures and languages. Islands can also be distinguished on the basis of their elevation, ranging from those dominated by mountainous landscapes such as Dominica in the Caribbean to atolls and reef islands that are composed almost entirely of coral and are rarely higher than a few metres above sea level (e.g., Kiribati, Tuvalu).

A different distinction can be made between those islands formed by the forces of nature and those created, either intentionally or unintentionally, by human beings. These artificial or anthropomorphic islands can take many forms. For example, the siltation that occurs at the mouths of most large rivers creates islands in the delta as sediment is deposited when the water from the river slows as it meets the sea. Although these islands are usually formed by natural processes, deforestation upstream by humans has the unintended consequence of increasing the amount of erosion and therefore washes even more silt and sand down to the delta. It creates examples such as the ephemeral char "floating islands" in Bangladesh's Bay of Bengal, which are formed and reformed by the silt washed down the Ganges River during the annual cyclones and monsoons. Interestingly, the Bangladeshi government wants to relocate the refugee Rohingya people to one of these islands (Bhasan Char), yet another example of islands being used as places of detention.

There are many examples of islands being constructed in harbours near built-up areas to provide space for airport landing strips, especially in large metropolises where undeveloped land is both scarce and expensive. These fabricated islands include Kansai Airport in Osaka Bay, Japan, Chūbu Centrair Airport near Nagoya, Japan, and Chek Lap Kok (i.e., Hong Kong International Airport). Perhaps more common than islands intentionally created for airport runways are islands that are created or expanded with the garbage of millions of nearby residents or tourists, including Semakau Island near Singapore (itself an island), and Thilafushi Island in the Maldives, created from the garbage of the millions of international tourists. On a smaller scale, there are thousands of examples of offshore oil drilling platforms or rigs, islands created specifically to serve as bird sanctuaries, and islands built as residential developments.

Probably the grandest and most ostentatious artificial island projects are found off the coast of Dubai, United Arab Emirates, on the shore of the Persian Gulf, where a large number of artificial islands have been constructed or are planned. The most ambitious of these projects is the partially completed World Islands archipelago, where islands have been created from sand dredged from rivers and surrounding coastlines and positioned to roughly conform to the world continents. The ultimate goal of this project is to sell these "luxury private islands" to individuals or corporations as resorts or island residences.

Several of the most prominent artificial islands are associated with the impacts of global pollution. They consist of large accumulations of human debris, primarily microscopic pieces of plastic, which have coalesced in various locations in the Pacific Ocean as a result of the prevailing ocean currents. Since the boundaries of these "gyres" are not definite and are difficult to detect on the surface or by satellite imagery, estimates of their size vary widely. However, as accidental artificial islands, they are just one of the reminders of the impact humans have on the natural environment.

A DIFFERENT DISTINCTION CAN BE MADE BETWEEN THOSE ISLANDS FORMED BY THE FORCES OF NATURE AND THOSE CREATED, EITHER INTENTIONALLY OR UNINTENTIONALLY, BY HUMAN BEINGS.

Complementing a classification based on physical features and geographical location, islands have also been classified using a political lens. For example, a large number of islands or island archipelagos are sovereign states, including Cuba, New Zealand, and Madagascar. The United Nations recognizes that many of these island states share similar challenges and has therefore created the label of Small Island Developing States (SIDS) to recognize their common features and to give them a stronger and more coherent international voice.

Some sovereign nations share an island. These "divided" islands or, as Shell (2014) calls them, "condominium" islands, are relatively rare. There are only ten inhabited islands today that are shared by more than one country. The most prominent of these are Hispaniola (Haiti and Dominican Republic), Ireland (the Republic of Ireland and the United Kingdom), Timor (East Timor or Timor-Leste and Indonesia) and New Guinea (Indonesia and Papua New Guinea). The website WorldIslandInfo.com also provides lists of islands by various categories, including those islands divided by international borders. Many more islands have ambiguous political relationships with other countries, often through a former colonial link. These include the Falkland Islands/Malvinas (United Kingdom, but also claimed by Argentina), Greenland (Denmark), Puerto Rico (US), and French Polynesia (France). There are many labels that have been given to these semi-autonomous island entities, including territories, protectorates, principalities, associated states, and autonomous regions. Still other islands are formal states or provinces within a larger continental nation, sharing the same or similar jurisdictional powers as other states or provinces in the federation. Examples include the Hawaiian Islands (US), Prince Edward Island (Canada), Tasmania (Australia), and Hainan (China). Given the complexity and often unique political and economic relationship that many of these islands have with their affiliated countries, it is almost impossible to articulate a coherent, mutually exclusive typology (see, however, Stuart 2008 and Watts 2009). Nonetheless, the catch-all phrase "subnational island jurisdiction" (or SNIJ) has emerged in an attempt to encompass all of the islands that have some form of politically dependent relationship with a metropole (Baldacchino 2006b; Baldacchino and Milne 2006, 2009; Grydehøj 2011).

Edward Warrington and David Milne (2007) are critical of the standard typologies of islands as being ahistorical, conflating smallness with islandness, and having too many contradictions or exceptions. Instead, they take a political economy approach that views islands as being one of seven types: "civilization," "fief(dom)," "fortress," "refuge," "settlement," "plantation," or "entrepôt." One of the advantages of this approach is that it recognizes the evolution of the role of islands over time, based on their changing geostrategic role, world view, and political system (see Chapter Six). Finally, speaking to the theme of diversity versus cohesion, Baldacchino (2005, 247) reminds us to be careful when we attempt to categorize islands as if they can be slotted into a certain genre or type because it would "debase the myriad diversity which is one of the foremost characteristics of islands as laboratories for innovation …"

A more recent subgenre of island studies focuses on island cities or urban archipelagos. It is perhaps surprising that more attention has not been paid to these near-shore islands, considering how many major cities and ports have developed on small islands, including Singapore, Manhattan, Malé (in the Maldives), and Venice. Grydehøj et al. (2015) makes the distinction between cities that play a significant role on the islands on which they are located (e.g., Reykjavik, Iceland) and those highly urbanized islands or archipelagos such as the aforementioned Singapore, Manhattan, Montreal, etc.

ISLANDS AS DISTINCT FROM ARCHIPELAGOS, ATOLLS, ISLETS, CAYS, OR REEFS

> WHILE THE WORD "ISLAND" CONJURES UP A STEREOTYPE OF ISOLATION, "ARCHIPELAGO" BRINGS TO MIND CONNECTIVITY, AND ESPECIALLY ISLAND-TO-ISLAND INTERACTION.

As we have seen above, islands and groups of islands can have many other names attached to them. In the Caribbean, you might hear the word "key" or "cay" to describe a low island composed of sand or coral (Stoddart and Steers 1977). It is said to have originated from the Spanish cayo meaning shoal or reef. One of the more recognizable groups of keys is found at the southern tip of Florida. This string of islands, many connected by road, extends in an archipelagic arc for over 200 kilometres, ending at the island of Key West (see Figure 1.9).

The term "archipelago" has almost as rich a history as the word "island." Defined as "an extensive group of islands" in the Oxford English Dictionary, it is said to have originated with the Venetians, who stitched together the two Greek words of *arch* or *arche* (meaning original, principal, or prime) and *pelago* or *pelagos* (meaning deep, sea, or abyss) (see also Shell 2014). It was originally used to refer specifically to the many islands in the Aegean Sea, but has since been applied to any location with a large number of islands. Elaine Stratford and colleagues (2011) provide a description of the many applications of archipelago throughout history. One of the points that they and others (see Stratford 2013) try to make in this emergence of scholarship regarding the word "archipelago" is that, while the word "island" conjures up a stereotype of isolation, "archipelago" brings to mind connectivity, and especially island-to-island interaction. So, for example, many of the 25,000 islands in Oceania were settled through a progressive journey by early Polynesians moving from one island to another (Howe, Kiste,

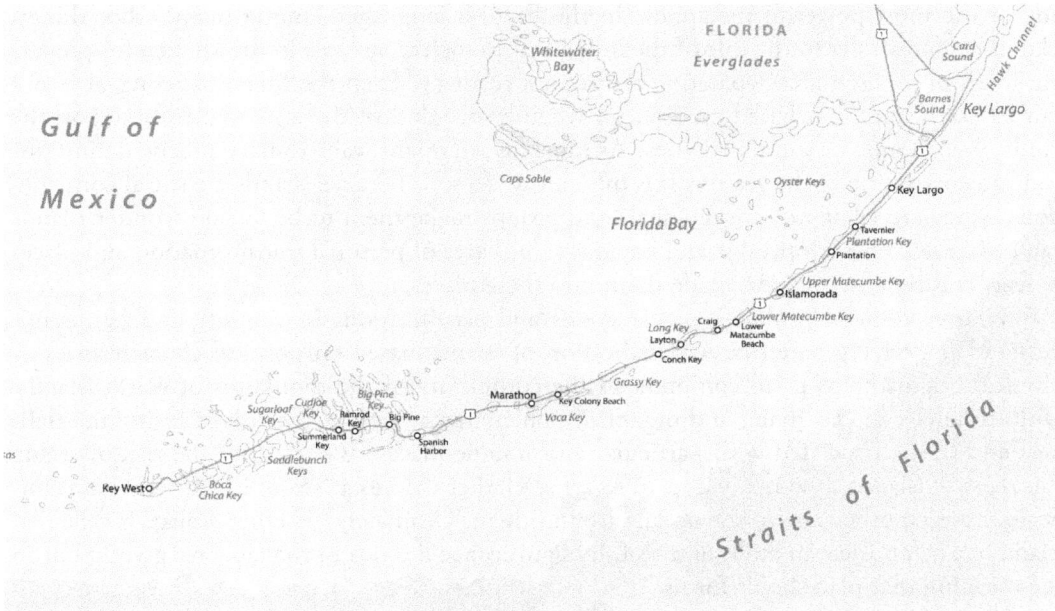

Figure 1.9 Map of Florida Keys. *Source:* **Rainer Lesniewski**

and Lal 1994). Also, prior to European arrival in the Caribbean, the relative locations of the Caribbean islands archipelago allowed the Taino culture to spread across the region, bringing with this mobility a shared language and trade relationships. Invasion and warfare with the South American Carib people was also made easier because of the close proximity of the islands (Ferguson 2008). Archipelagos have also been examined extensively by biogeographers to describe and explain the movement and evolution of animal and plant species through time and space (MacArthur 1967).

The processes that create atolls and reefs are described in the next chapter. However, at this point the easiest way to describe these formations is to say that they are all really subsets of islands, with an atoll encircling another island or lagoon, while a reef only sporadically encircles an island or lagoon. They are both distinguished from most other islands because they are formed as the result of the accumulation of the skeletons of living sea creatures and tend to not protrude above the surface of the sea to any great degree. National Geographic Education has an excellent series of images and captions that describe atolls and help explain their formation (NationalGeographic.com 2012a). An islet is normally considered to be a very small island. Depraetere and Dahl (2007, 2018) have estimated that there are 8.8 million islets in the world and they are distinguished from rocks and islands solely on the basis of size, with islets ranging from 100 square metres to 10 square kilometres (see Figure 1.10).

CONCLUSIONS

This chapter has shown that there is more to this idea of "island" than we might first have imagined. Beyond its literal definition as a piece of land surrounded by water, it has become one of the most powerful metaphors in the English language. Among many other things, islands can now refer to the role of the individual in society, an area in the kitchen for preparing food, or a raised area separating traffic in a roadway. Even the literal meaning of island can be confusing: is it based on size, on whether people live there permanently, or on the kinds of economic or social activities? As much as we might want to have precise definitions and categories, one of the reasons it is difficult to do so is because islands are social constructions—they are what we as individuals and groups make them to be. So no wonder islands can be considered both prisons and paradises, or places of personal transformation and places to fear. It is because we have made them into these places.

Similarly, we use typologies or categories and classifications to simplify and better understand underlying patterns. A classification of islands based on political characteristics or physical origins helps us to comprehend the complexity of the thousands of world islands. Unfortunately, in classifying and organizing phenomena, we run the risk of losing the richness and detail associated with particular and unique places. We also run the risk of seeing this mass of islands as so many dots in a vast sea rather than as a "sea of islands" that are connected to each other and to the oceans around them. Ultimately, whether a place is called an island, a cay, an islet, an atoll, or a reef, its significance lies not in the label we give it, but in the meaning that place holds for us.

Key Readings

Connell, John. 2013. *Islands at Risk?: Environments, Economies and Contemporary Change*. Cheltenham: Edward Elgar.

Depraetere, Christian, and Arthur Dahl. 2007. "Island Locations and Classifications." In *A World of Islands*, edited by Godfrey Baldacchino, 57-106. Charlottetown, Prince Edward Island: Island Studies Press.

Hay, Pete. 2006. "A Phenomenology of Islands." *Island Studies Journal* 1 (1): 19-42.

King, Russell. 1993. "The Geographical Fascination of Islands." In *The Development Process in Small Island States*, edited by Douglas Lockhart, David Drakakis-Smith, and John Schembri, 13-37. London: Routledge.

MacArthur, Robert H., and Edward O. Wilson. 1967. *The Theory of Island Biogeography*. Princeton, New Jersey: Princeton University Press.

Figure 1.10   New Eddystone Rock, Alaska, US. *Source:* **Jerzystrzelecki**

A lithograph of the island of Rakata before its disappearance in 1883. *Source:* **public domain**

Chapter Two

Physical Processes and Islands

They are generally made of rock and whatever garnishes nature or human ingenuity provide to decorate them. (Holm 2000, 4)

Despite the fact that islands make up only 7 percent of the world's land surface, they have played a key role in furthering our understanding of evolution, geology, biogeography, and, more recently, climate change. The main objective of this chapter is to give the reader a better understanding of the most important physical and natural processes at work on islands, including how they were formed, their climate, and why and how life has evolved on them. As an introduction to these broad processes, this chapter is not able to provide a comprehensive knowledge of all of the details of natural change on islands. The key readings section at the end of this chapter points you to sources that will help in this regard. However, by the end of this chapter, you should be able to understand and appreciate the physical forces affecting islands in the past, present, and in the future.

Perhaps it is human nature to think of the formation, number, and distribution of islands in terms of our own lifespan. In his book *Oceanic Islands* (1994) Patrick Nunn reminds us of the importance of considering the passage of time in a much more flexible way. As sea levels rise and fall, as plates collide and separate, and as volcanoes erupt and erode, the Earth's "island-scape" can appear very different throughout geologic time. For example, even as recently as 18,000 years ago, the sea level of the southwestern Pacific Ocean was 120 to 150 metres lower than it is today. If we could see it now, this region would have been dotted with hundreds of islands that are now no longer visible above the ocean's surface. Another way to think of this is to picture the eight main iconic Hawaiian islands that we often see on postcards, distributed in a diagonal line from the youngest "Big Island" of Hawai'i in the southeast to the oldest island of Kaho'olawe in the northwest. However, this island chain consists of many more proto-islands hidden beneath the surface of the ocean. These submarine islands or seamounts have either failed to break through the sea's surface or have sunk or been eroded over time. In fact, the entire Hawaiian island archipelago stretches a distance of over 5,100 kilometres from the Big Island to well past the Midway atoll (i.e., more than twice the distance from Kaho'olawe to Hawai'i and almost as far as the continental United States from east to west. The Hawaiian archipelago was created by the movement of the Pacific plate over a magma hotspot that has been spawning islands for over 80 million years and shows no signs of stopping any time soon.

This appearance and disappearance of islands in geologic time make islands appear to be fleeting and temporary, or in Steven Fischer's words, like "yesterday's snow" because they have either melted in the heat or are covered up by subsequent flakes (Fischer 2012, 8).

THE FORMATION AND STRUCTURE OF ISLANDS

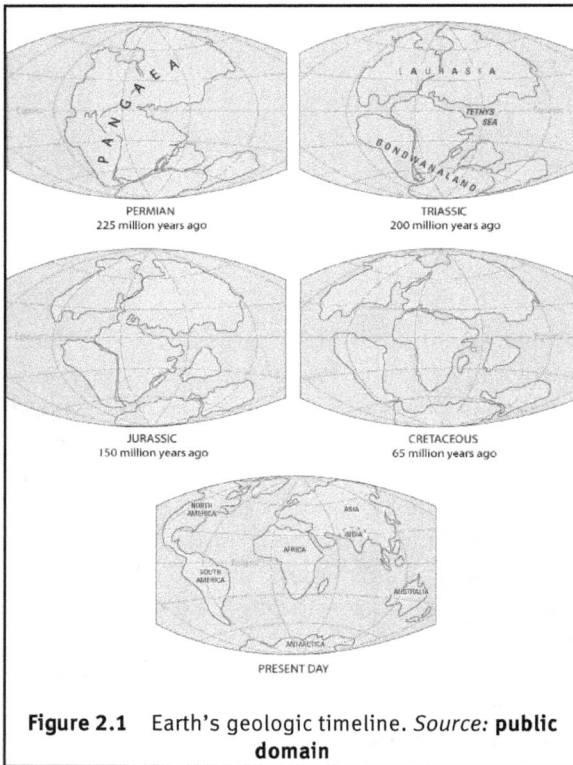

Figure 2.1   Earth's geologic timeline. *Source:* **public domain**

There are various points along Earth's geologic timeline where we could begin the journey to better understand the formation of islands. Perhaps the best starting point is more than 230 million years ago with our one "Earth Island," Pangaea, from the Greek pan meaning "all" and gaia meaning "earth." As the snapshot of time periods in Figure 2.1 shows, tectonic forces have acted to re-shape this supercontinent into the form we are familiar with today. The Earth's landmasses will continue shifting to a point where we may not recognize them in another 230 million years.

During this journey through and around the Earth's waters, islands have formed and disappeared in so many ways that it may be too simplistic to group the factors responsible for their creation into a few mutually exclusive categories. As Hopley (1994) suggests, islands can be bits of an ancient conti-nent, ocean sandbanks, the cooled tips of erupting volcanoes, and so much more. Despite this complexity, and as a way to begin to understand these many processes, it is still useful to think of islands as being formed in one of the following ways: 1) by volcanic eruption, 2) by subduction at the edges of plates, 3) as fragments of continents, 4) as a result of the extrusion and accumulation of living creatures, 5) from the uplifting of limestone and coral at plate boundaries, 6) from the deposition of silt and other sediments and, 7) as a com-bination of two or more of the major factors above, at the same time or sequentially. These major categories are described in more detail below, with island exemplars provided to better illustrate the processes involved for each type. It should also be noted that this categorization of island formation employs a Western science-based approach. There are many Indigenous islanders who embrace a very different world view to understand the formation of their is-lands. An introduction to some aspects of these belief systems is also described in this section.

Before outlining the first few categories of island formation, it may be helpful to provide a very brief introduction to the phenomenon of plate tectonics, a theory that is surprisingly still only fifty years old. If we were to take a cross section of the Earth as in Figure 2.2, we would find an outer lithosphere band approximately 100 kilometres thick encircling the globe and floating on a weaker asthenosphere. This lithosphere consists of layers of lighter, less dense continental crust and heavier, more dense oceanic crust divided into twelve major plates and a number of smaller plates. Being less dense, the continental plates tend to "float" an

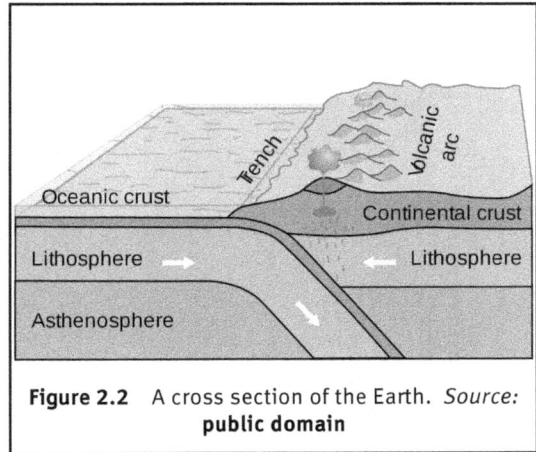

Figure 2.2   A cross section of the Earth. *Source:* **public domain**

average 4.6 kilometres higher than the oceanic crusts. As a result of gravity, water therefore tends to flow towards the lower oceanic crusts, collecting in what become the oceans and seas. These plates are constantly in motion through convection as heat is released from the core and moves through the asthenosphere. At the edges of the plates, there are three kinds of interactions or motions taking place: divergence, where two plates are moving apart (i.e., at mid-oceanic ridges), convergence, where two or more plates are moving towards each other (i.e., at subduction zones), and transform faults, where plate edges are moving horizontally past each other (for more on plate tectonics, see Searle (2009), A. Cox (2008), and Hamilton (1988)). With this brief introduction, let's turn our attention to the specific ways in which islands are formed.

Volcanic Origins and Hotspots

There are thousands of "oceanic" islands not connected to the plates or land masses of continents that have been created as a function of the eruption of magma at a break or a weak point in an oceanic plate. When the magma from these hotspots breaks through the mantle, it can reach the surface of the ocean, cool, and create a volcanic island. Of course, as was described earlier with the Hawaiian archipelago, there are also millions of submarine volcanoes or seamounts where the erupting magma does not reach the surface of the ocean. Most of the world's atoll islands also started out as volcanic islands.

Perhaps the best known example of a series of islands formed as a function of a volcanic hotspot are the Hawaiian Islands in the middle of the Pacific Ocean (see Figure 2.3). The hotspot, located at 18.92° N latitude and 155.27° W longitude, just southeast of the Big Island of Hawai'i, has been producing islands and submarine volcanoes for at least eighty million years. The next volcanic island in this steady procession is already being formed. *Lo'ihi*, the Hawaiian word for "long," is a seamount centred just thirty-five kilometres off the southeastern coast of the Big Island. It rises more than 3,000 metres above the seafloor and is within 1,000 metres of the ocean surface.

One of the most impressive geologic events is the rapid birth of an island as a result of an underwater volcanic eruption. The island of Surtsey, off the southwestern coast of Iceland,

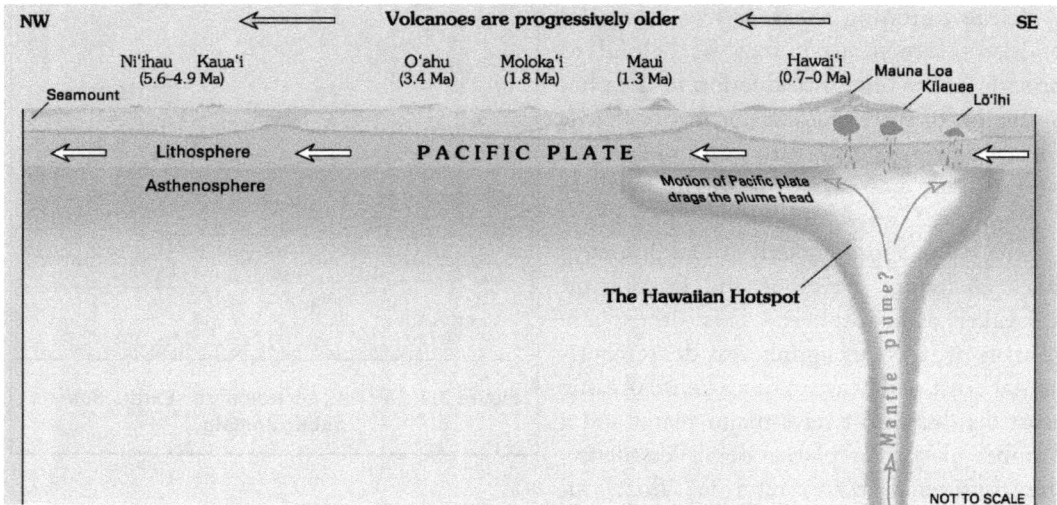

Figure 2.3   A cross section of the Hawaiian islands. *Source:* **public domain**

emerged in spectacular fashion in November 1963, and continued to add land mass until 1967. Named after a mythical Nordic fire giant, Surtsey grew to a size of 2.7 square kilometres and rose 175 metres above the surface of the Atlantic Ocean (Fridriksson 2009). Much to the credit of Icelandic authorities who have restricted island access to scientists only, Surtsey has served as a fascinating living laboratory for botanists and zoologists studying the building blocks and evolution of organic life on what started as a rock barren of life. Surtsey is also an incredible example of the power of wind and waves to diminish or destroy islands quickly. In the first forty-five years of its existence, almost half the island has already been eroded (Fridriksson 2009).

Nishino-shima is another example of how a new volcanic island has been used to better understand ecosystem change. One-thousand kilometres south of Tokyo, this islet was created following a 1973 volcanic eruption. Since then, scientists have been able to study its colonization with new species of flora and fauna. It has generally been much slower to adopt new species than Surtsey, largely because it is much more remote from existing landmasses (Abe et al. 2008).

Although hotspot volcanic eruptions can create new islands, they can also cause islands to disappear, sometimes in equally spectacular fashion. A good example occurred in August 1883 as a result of the eruption on Krakatau in what is now the Indonesian archipelago and was formerly known as the Dutch East Indies. Two-thirds of the island of Rakata disappeared beneath the surface of the sea, an estimated 36,000 people were killed, most from tsunamis, and it was reported that the eruption was heard as far away as Perth, Australia, a distance of over 4,500 kilometres from the eruption. As was the case with Surtsey, a new volcanic island has now emerged, in this case on the site of the original volcanic island. Called Anak Krakatau (meaning Child of Krakatau in Indonesian), this new island grew to a height of 300 metres and 3.7 square kilometres in area. Since complex life forms were wiped away on the former island of Rakata and the two surrounding islands as a result of the 1883 eruption, these islands have been the subject of considerable research on the regeneration of life (see Tagawa, Suzuki, Partomihardjo & Suriadarma 1985, Thornton 1996). Although very little new vegetative life has formed on the still active volcanic island of Anak Krakatau, on the older Rakata island there now exists over 500 vascular plant species and eighty-nine vertebrate species, including bats, reptiles, and snakes (Whittaker 2009). As continued evidence of the volatile nature of these islands, the volcano on Anak Krakatau erupted once again on December 22, 2018, causing much of the cone of the volcano to slide into the sea. This produced a tsunami that swept into seaside communities and killed an estimated 437 people (The Star Online 2018).

Subduction Zones: Islands at the Interstices of Plates

One of the other major causes of island formation occurs at the margins of plates on the Earth's surface. Plate boundaries can occur at the edges of existing continents such as those found along the west coasts of North and South America or in the middle of an ocean, far removed from any continent. For example, the Mid-Atlantic Trench in the Atlantic Ocean is an underwater mountain range between the Eurasian and North American plates in the northern hemisphere and the South American and African plates in the southern hemisphere. Islands like Iceland and the Portuguese territorial Azores are formed at the boundaries of these trenches. As the plates move apart (see Figure 2.4), you also find islands composed of older rock pushed farther away from the trenches, such as Cape Verde and Ascension Islands. Although it is not uncommon for volcanic activity to be associated with these boundaries, islands are also formed at the boundaries of plates as a result of the movement of one plate against, across, and sometimes underneath another plate. These subductions can force the lithosphere above the surface of the ocean, thus causing islands to form. The San Juan Islands located just off the west coast of Canada and the United States is a good example of an island chain formed as a result of the movement of two plates.

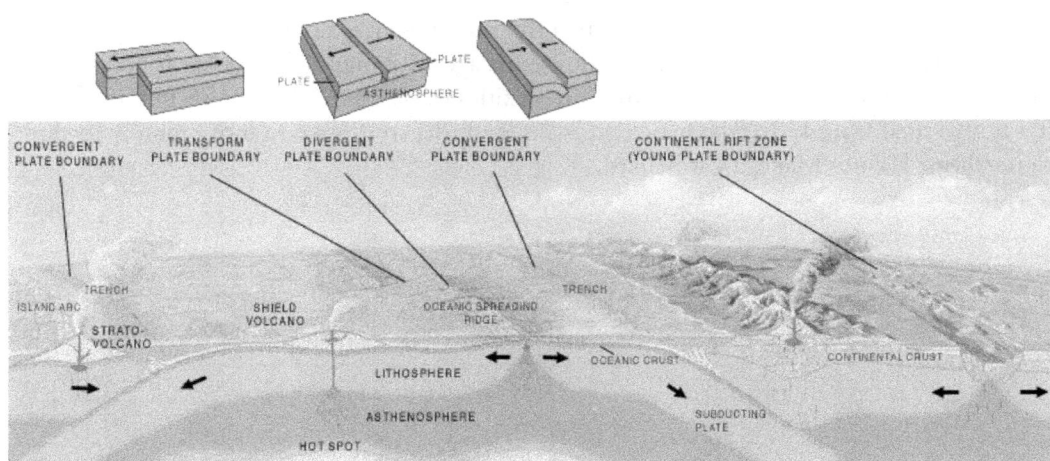

Figure 2.4   Movement of Plates. *Source:* **public domain**

Continental Island Fragment

Many islands located close to continents are really fragments of those continental land masses. They may have been formed as a result of sea-level rise, cutting them off from the rest of the continent, and/or by the subsidence of surrounding land resulting in the same outcome. For example, the British Isles have been a part of the European continental shelf for more than 550 million years. Along with the rest of Europe, these islands were part of the two supercontinents of Gondwana and Laurentia and have been on the move for more than half a billion years. Their status as islands has been more recent. They have alternately been connected to, and disconnected from, the rest of Europe in a region called Doggerland as a result of the rise and fall of the surrounding sea, particularly during periods of glaciation and deglaciation. The last land bridge connecting the British Isles to Europe disappeared beneath the sea as recently as 6,500 years ago.

Many of these continental island fragments are granitic islands, the most common rock type found in continental crusts, as opposed to basaltic rock, the most common form of rock found in oceanic crusts. Although both rock types are igneous formed originally by volcanic activity, granitic rocks are generally much older (greater than 350 million years) than most of the basaltic covering material on the ocean floors (less than 150 million years). Therefore, they have been eroded and compacted to a much greater extent over this longer time period, and now tend to be more stable compared to most oceanic islands. As we will see later in this chapter, if these older islands have remained separated from continents there is a greater likelihood that they will have developed a more diverse flora and fauna than oceanic islands (Coffin 2009, Nunn 1994).

Two other good examples of islands as fragments of continents are Tasmania, off the southern coast of Australia, and Madagascar, the fourth largest island in the world, located approximately 400 kilometres off the southeastern coast of Africa. In the case of Tasmania, the island was part of Australia as recently as the last ice age 10,000 to 12,000 years ago. It gained its current status as an island when sea levels rose by up to 120 metres and the connecting land bridge to Australia disappeared. In the case of Madagascar, although at the time it was part of the supercontinent of Gondwanaland, it broke off from Africa about 165 to 180 million years ago. It became its own island when the rest of what is now India and the Seychelle Islands broke off from it about 80–90 million years ago. Although Madagascar is occasionally referred to as a mini-continent given the complexities of plates in the western Indian Ocean, after it separated from India, the plate that included India continued to move rapidly towards the northeast leaving Madagascar behind.

ISLANDS ARE FORMED BY: 1) VOLCANIC ERUPTION, 2) SUBDUCTION AT THE EDGES OF PLATES, 3) FRAGMENTS OF CONTINENTS, 4) EXTRUSION AND ACCUMULATION OF LIVING CREATURES, 5) UPLIFTING OF LIMESTONE AND CORAL AT PLATE BOUNDARIES, 6) DEPOSITION OF SILT AND OTHER SEDIMENTS AND, 7) AS A COMBINATION OF FACTORS, AT THE SAME TIME OR SEQUENTIALLY.

Atolls and Reefs

Atolls and reefs are similar in origin, but they are often very different in shape. In both cases, they constitute the living homes and skeletal resting places for billions of sea creatures that have bonded to rocks, volcanoes, and each other. Over thousands of years, and combined with dropping sea levels, this accretion has enabled them to rise above the surface of the ocean. One of the features that sets atolls apart from other reefs is that inside an atoll no other land mass is visible, creating an idyllic and iconic image of a shallow lagoon partially surrounded by ring-shaped land (see Figure 2.5). There are many other examples of coral reefs that are closely associated with some other body of land, as in a fringing or barrier reef (see Hopley 2011; Spalding et al. 2001). In fact, most scientists would argue that an atoll is really just one type of coral reef (Guilcher 1988; Spalding et al. 2001).

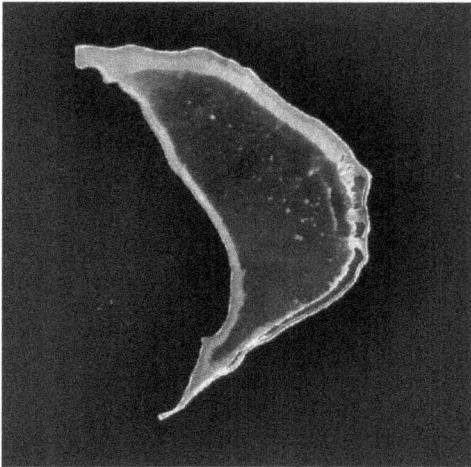

Figure 2.5   Bokak atoll. *Source:* **NASA/public domain**

Among his many other contributions to island-based knowledge, Charles Darwin is credited with providing the first accurate description of various types of reefs and atolls. Prior to his contributions, the prevailing theory was that atolls formed on relatively stable, submerged volcanic craters. By noting that atolls were often much more extensive than the volcanoes themselves, and that there was a link between the size of the atoll and the depth of the volcanoes associated with them, Darwin concluded that the process associated with the evolution of reefs and atolls was much more dynamic than previously suspected. Atolls are really just the last stage of coral reef accumulation that started in the form of a fringing reef, having been built up from the deposits on subsiding volcanoes, where the rate of growth of the coral reef occurs at approximately the same pace as the submergence of the volcano. As is often the case in science, this is not the complete story. It has now been shown that the subsidence or sinking of volcanoes is not the only factor leading to the formation of atolls. In addition, the rise and fall of sea levels through various glacial cycles may be just as important in creating and maintaining atolls.

The American territorial atoll of Midway (*Pihemanu* in Hawaiian), and named by the Americans as such because the archipelago is roughly midway between North America and Asia, is less than twenty-nine million years old and consists of three islands (Eastern, Sand, and Spit) surrounded by an elliptical reef. The highest point on these islands is less than twelve metres above sea level, but the average elevation above sea level on each of the main islands is only 2.6, 3.2, and 1.5 metres, respectively. The magnitude, richness, and significance of the coral reef is evidenced by the fact that while the three islands are collectively only six square kilometres in area, the coral reefs encompassing the islands above and below the surface spread out for 356 square kilometres. As is the case with many atolls, there are very few endemic species of flora and fauna represented on Midway. Although this lack of endemism is normally a function of the relative youthfulness and isolation of these oceanic atolls, in the case of Midway it is also a function of both the intentional and accidental introduction of non-indigenous species, including rats and the ironwood tree. Although unrelated to its formation, Midway has two other features that deserve mention. First, from the Secomd World War to the present, it has occupied a strategically important geopolitical role, and was the site of the Battle of Midway between Japan and the United States in 1942. More will be said about this aspect of islands in Chapter Six. Second, Midway was one of the first comprehensive examples of island-based conservation and restoration. From the early twentieth century when an American presidential decree was invoked to stop the slaughter of seabirds, to the establishment of the atoll and surrounding coral reefs as a protected marine conservation area that is closed to the public, to the current programme to eradicate non-indigenous species and reintroduce those species that existed on the islands prior to human occupation, Midway and the other islands in the northwestern Hawaiian island chain are good early examples of ecological sustainable planning (Craig 2006; Duffy 2010; Engilis and Naughton 2004; Flint 2009).

High Limestone Islands

Especially in Oceania, there is often a distinction made between "high" and "low" islands. Although this can be subjective, low islands are usually atolls and keys that have been exposed by the gradual subsidence of the surrounding ground or a decline in the sea level. On the other hand, high islands are often associated with dramatic topography, with steep coralline cliffs rising from the sea. A classic example is Rennell Island in the Solomon Islands archipelago where the cliffs can rise up to 150 metres above the surface of the sea. Although much of the exposed surface may have coral-like features, these islands have been shaped largely as a function of the tectonic forces associated with nearby convergent plate boundaries that force (or uplift) the surrounding land, including the coral reefs. You might hear the term "karst topography" associated with these islands. This means that the waves and water have dissolved

Figure 2.6   Togo Chasm on Niue Island.
Source: **fearlessRich**

material in the cracks of the surrounding limestone leaving behind caves, tubes, and columns that can extend for hundreds of metres and produce very unusual landforms (See Figure 2.6). Some of the best examples of this topography are Niue and the Henderson Islands in the South Pacific as well as some of the islands in the Philippines archipelago. Some islands, like Nauru and Kiribati, have been filled in by the guano of birds and, after reacting with the limestone, have produced deposits of phosphorus that were mined extensively in the past by colonial and postcolonial governments and companies for use as a rich source of natural fertilizer (Nunn 1999). In fact, much of the justification for colonial control of these islands was to allow companies associated with the major powers to strip the islands of their existing phosphorus (M. Williams and Macdonald 1985). For more information on the tragic exploitation of Nauru, see McDaniel and Gowdy (2000) and Connell (2006).

Islands Formed by Siltation or Sediment Deposits (Shoals, Sandbars, Delta Islands)

Earlier in this chapter, the emergence and disappearance of islands was described in geologic time. The lifespan of some islands makes them much more ephemeral (Fischer 2012; R. Gillespie and Clague 2009). Kano (2009) indicates that an ephemeral island is one that is "fated to disappear on a scale of hours to years after its appearance" (259). At one extreme, and as suggested in Chapter One, this could include a land mass that becomes an island at high tide and a peninsula at low tide. It might also include islands that are formed and disappear as a result of avalanches or landslides, the construction/destruction of a dam, or even cataclysmic volcanic eruptions or rapid erosion. This issue of the length of time for the existence of an island is a little misleading, since on a geologic time scale that spans millions of years, most of the world's islands could be characterized as ephemeral. Nonetheless, the process that has resulted in the creation and subsequent disappearance of many of these kinds of islands, especially those found in lakes, rivers, and alongside continents, is often closely linked to the deposition of sand, silt, or other matter by water or wind.

BECAUSE THEY ARE CONVENIENT TRANSSHIPMENT POINTS FOR HUMAN ECONOMIC ACTIVITIES, SOME OF THE LARGEST AND MOST WELL-KNOWN WORLD URBAN CENTRES HAVE BEEN BUILT ON THESE DEPOSITIONAL OR DELTA ISLANDS

Probably the most common kinds of islands in this category are those that have been created as a result of the deposition of sand, silt, or moraine at the mouth or delta of a river or where two or more rivers meet. As the current slows, heavier sediment being carried downstream can no longer be suspended by the moving waters and therefore sinks to the river bottom, eventually building up to form an island. Because they are convenient transshipment points for human economic activities, some of the largest and most well-known world urban centres have been built on these depositional or delta islands, including Mumbai, the most populous city in India at the mouth of the Thane River; many of the islands that make up the metropolis of Shanghai, China, at the mouth of the Yangtze River; Montreal, Canada, at the confluence of the St. Lawrence and Ottawa Rivers; and New Orleans, US, as part of the Mississippi River delta.

Even though it is far from the shores of the nearby continent, one of the most photographed examples of a depositional or barrier island is Sable Island, located in the Atlantic Ocean about 175 kilometres off the east coast of Nova Scotia, Canada. From the French word for sand (Île de Sable), Sable Island is really a forty-two-kilometre-long, narrow, crescent-shaped sandbar, known to most as home to a herd of feral or wild horses. Originally formed as a part of the terminal moraine during the last glacial ice age, it is now being added to by wind and waves in some places and subtracted by the same forces on other parts of the island.

Barrier island systems would also be part of this category. They are usually found parallel to shorelines and can extend for hundreds of kilometres. Some of the largest barrier islands are found along the eastern seaboard, such as those on the southeastern and Gulf coasts of the United States. Like Sable Island, their initial creation may be a result of glacial deposition followed by sea-level rise and/or the subsidence of surrounding land. Although this may explain their formation, they have been consequently shaped and modified through erosion and deposition by waves and tides. They serve multiple functions, including protecting the conti-

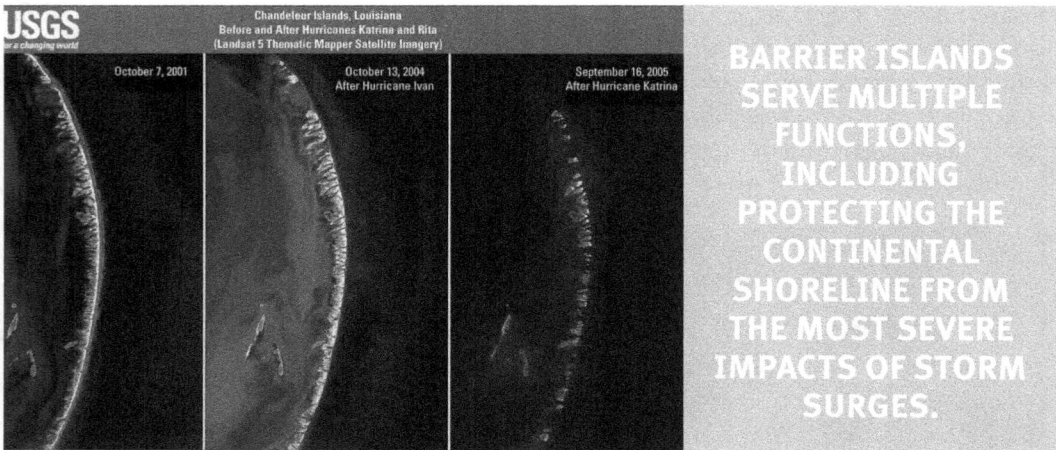

Figure 2.7   Barrier island system. *Source:* **National Park Service**

nental shoreline from the most severe impacts of storm surges, creating a sheltered waterway on the leeward side for the transportation of people and goods, and as a popular destination for recreational and tourist activities. Although these islands can be extensive and likely will be reshaped over time, they can disappear quickly when subjected to the power of hurricanes or typhoons, damaging the rich ecosystems associated with tidal flats and wetlands. The series of images in Figure 2.7, taken over only four years, shows the rapid change in the shape of the Chandeleur Islands off the coast of Louisiana in the Gulf of Mexico prior to, and then after, Hurricanes Ivan in 2004 and Katrina in 2005.

Formation by a Combination of Causes

More often than not, several major factors come together to create an island. This is especially so if we consider the multiple forces that have acted upon an island over its entire lifespan. We have already seen that many atolls are formed by several processes. One of the most fascinating examples is the island state of Nauru in the Central Pacific Ocean. As is the case with the Hawaiian Islands, Nauru was originally formed from volcanic eruptions at a hot spot in the Pacific Plate. Through subsidence and erosion the volcanic island diminished in size but was also transformed into a coral atoll with a thickness of about 500 metres. It then transformed once again to become a "phosphate" island. Many islands near the equator in the Pacific

Ocean were rich in phosphorus at least partly as a result of the accumulation of excrement from millions of seabirds over thousands of years and the chemical reaction this had with the underlying limestone (Hein 2009). Although these may have started out as volcanic islands that eventually became atolls, the combination of productive fishing grounds combined with a relatively dry climate has allowed for the accumulation of deep deposits of guano covering the surface of some of these islands. Once wave action has washed away the coral limestone deposits, we are left with considerable deposits of guano, rich in phosphorus and other minerals that were highly valued and mined as a source of fertilizer to enhance crop production. In fact, one of the initial factors leading to American expansionism on the islands in the Pacific Ocean was the presence of rich deposits of guano on many of these islands and the passing of the Guano Islands Act of 1856. This allowed US citizens to lay claim to islands containing guano deposits. By the end of the twentieth century, Nauru was the only island still mining and selling this phosphorous commercially. Although production on this island still takes place, it is at a much lower scale than earlier in the twentieth century. As we will see in later chapters, Nauru stands as a sad testament to some of the negative effects of globalization, political corruption, and unsustainable practices.

The origins and geology of Scotland's Shetland Islands, located in the North Atlantic Ocean slightly more than 200 kilometres northeast of Scotland and the island of Great Britain, is more varied than most other islands and may be as geologically diverse as any other group of islands existing today (A. Hall and Fraser 2004; Shetland Amenity Trust n.d.). During its 2.5 billion-year history, and thanks to plate tectonics, these islands have travelled from the South Pole to their present location south of the Arctic Circle. Parts of the Shetlands have been created as a result of volcanic eruptions, the layering of sand and other sediment

Figure 2.8 Cliffs at Bressay, Shetland. *Source:* **WRONAart**

when it was part of a lake bottom, and subduction forces that formed a mountain range that at one time rivalled that of the current Himalayas. Over the past sixty million years or so, the wind and waves in the North Sea have battered the shores of the islands, creating spectacular coastal cliffs as seen in Figure 2.8 and also influencing the culture and lifestyle of the islanders (McKirdy 2010).

Island Formation from an Indigenous Perspective

All of the island formation explanations described to this point are premised on a Western scientific world view. However, many islanders, past and present, hold belief systems that rely on legend, myth, and folklore to at least partially explain the origins of their islands. When we defined islands in the last chapter, we noted that Pacific societies in the past and present believed that gods or demigods created their islands by fishing them up on a hook and line, or by throwing them down in clumps of earth or leaves (Nunn 2008, 2003). Variations on the "fishing up" story have been used to explain the origins of Niue, the islands of French Polynesia, and the Hawaiian Islands (R. Williamson 2013). Other myths incorporate stamping stories to explain the origins of islands, as in the gods living under the sea were stamping their feet as they moved around and in so doing created islands. This has been used by Indigenous peoples to explain the origins of Tonga, Niue, the Cook Islands, and the Marquesas (Nunn 2004). Western science would interpret these stories as geologic events such as earthquakes or coastal landslides.

Geomorphological change on islands is also understood differently by Western science and Indigenous belief systems. For example, Vitaliano (2007) tells the story of the changing topography of Mangaia, one of the Cook Islands. In the myth, the island was originally smooth and regular. It developed its current eroded volcano as a result of a contest between the god of the sea and the god of rain. The former god attacked the island and eroded it, whereupon the latter god made it rain for five days, carving deep valleys into the slopes. We have to be cautious that we do not marginalize alternate world views just because they may not conform to our own perspectives.

ISLANDS, CLIMATE, AND CLIMATE CHANGE

In this chapter we have already seen examples of how climatic forces have created, modified, and even destroyed islands. The rise and fall of seas during periods of global cooling (glaciation) and global warming (deglaciation) and the effects of wind and water to erode islands has led to some spectacular island formations. In this section, we look more directly at the impact of climate on islands.

As is the case on continents, climate is a major factor affecting the physical geography of islands as well as the nature of life on those islands. One of the most important macro factors affecting an island's climate is its latitude. This factor affects both the temperature and the circulation of ocean currents and atmospheric systems. Islands near the tropics, sometimes referred to as "warm-water islands," are generally characterized by small seasonal differences in temperature and relatively large seasonal differences in precipitation. Islands in temperate or "cold-water" locations have distinct seasonal differences in temperature and relatively little seasonal variation in precipitation. Looking at two Atlantic Ocean islands illustrates how oceanic and atmospheric currents can have quite different impacts on temperature. Reykjavík, the capital of Iceland, and Iqaluit, the capital of the territory of Nunavut on Canada's Baffin Island, are both located at approximately the same latitude: 64° N of the equator. Warmed by a branch of the Gulf Stream current originating in the northern Caribbean Sea and off the coast of Florida, the average January high temperature in Reykjavik is +2°C, while Iqaluit experiences a much colder average January high temperature of -22.5°C. This example also points to the other major macro factor affecting the climate of islands: their distance to the nearest continental landmass. Relative to continents, large bodies of water are more likely to retain heat in winter and be cooler in summer. Therefore, those islands that are far removed from continents tend to experience less seasonal variation in temperatures.

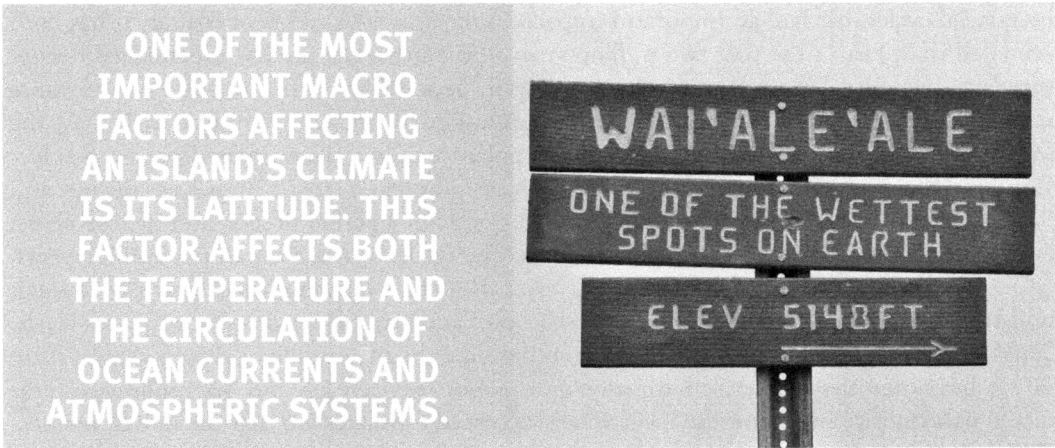

ONE OF THE MOST IMPORTANT MACRO FACTORS AFFECTING AN ISLAND'S CLIMATE IS ITS LATITUDE. THIS FACTOR AFFECTS BOTH THE TEMPERATURE AND THE CIRCULATION OF OCEAN CURRENTS AND ATMOSPHERIC SYSTEMS.

WAI'ALE'ALE
ONE OF THE WETTEST SPOTS ON EARTH
ELEV 5148FT

Some of the more significant climatic factors operate at the local level of individual islands and even parts of islands. The impact of the relative height and topography of an island is a case in point. Depending on their location relative to the prevailing trade winds, some small low-lying tropical islands are almost desert-like, making it difficult for plants and animals to survive and for human activity to be sustainable. Alternatively, islands nearby that are mountainous and exposed to moisture-laden air currents can experience significant differences in the amount of precipitation they receive on either side of the island. For example, on the mountainous Hawaiian island of Oahu, the mean annual rainfall on the eastern slope of the Hawaiian Island of Oahu can exceed seven metres (275 inches). Less than twenty-five kilometres west on the leeward or rain shadow side of the island, precipitation totals may average less than sixty centimetres (twenty-four inches) per year. Unlike the case with many tropical islands, on the high volcanic island of Pohnpei in the Federated States of Micronesia the wet season exists all year long. Rainfall can average between three and eight metres annually depending on where you are on the island. If you weren't saturated enough, on Kaua'i's (Hawai'i) Mt. Wai'ale'ale, in English roughly translated as "rippling or overflowing water," is believed to be one of the wettest places on Earth, with a reported 17.3 metres of rain falling in 1982 (Fischer 2012). As these examples suggest, microclimates can be very localized. The variation in topography on some islands, with steep valleys separating inlets, can create vastly different climates, each contributing to build distinct ecosystems, including many species of flora and fauna found nowhere else in the world.

Climate Change and Sea-Level Rise

There is sometimes ambiguity in the terms "climate variability" and "climate change." The popular media tends to use the phrase climate change loosely to refer to any variation in climate, regardless of time span. Most scientists reserve the term climate change for longer-term periods of at least thirty years. Although variations in climate from year to year or decade to decade can be as significant, these oscillations are more commonly referred to as climate variability (T. Schroeder 2009).

Long-term climate change, whether as a function of human activity or earlier glacial-interglacial cycles, has had an important impact on the ecology and life of islands. It has been estimated that, just in the past two million years, the Earth's climate has experienced twenty to twenty-five glacial-interglacial cycles, varying in periodicity from 100,000 years to more frequent cycles of 40,000 and 20,000 years in duration (Burney 2009; Nunn 2012). At the peak of glaciation, when more water is frozen in polar ice caps and continental glaciers, sea levels around the world are much lower, often decreasing by more than 120 metres (Burney and Flannery 2005). It just so happens that we are currently in a period of high and increasing sea levels, at least forty to fifty metres higher than the average level over the past two million years.

One of the outcomes of the most recent cycle of global warming, exacerbated by human industrial activity, has been a melting of these icecaps and glaciers, resulting in rising sea levels. Considerable research has been undertaken to describe the magnitude of this increase. Nunn (2012) has noted that sea levels have risen at a rate of between 1.8 and 3.2 millimetres/year for the past century or so and the most widely accepted estimates are that the rate of increase will be significantly greater over the next century. To be a little more precise about the historical record, using readings from twenty-seven stations in the Pacific Ocean over the past twenty-five years, Mitchell and colleagues (2001) reported an average annual sea-level rise of +0.77 millimetres/year. As you might expect from a shorter timespan and variation in ocean currents and sea-bottom topography, the rate of sea-level change is not the same across all oceans or even within the same oceanic region. For example, some climate stations in the Indian Ocean's Maldive Islands have shown sea levels rising by 3.2 to 6.5 millimetres/year in the 1990s while sea levels in the Canadian Beaufort Sea/Hudson Bay area have risen by an average of 3.0 millimetres/year (Mimura et al. 2007). Averaging out these geographic differences and the year-to-year variations due to periodic climate events such as El Niño and La Niña, the rate of sea-level change throughout the world's oceans has been about +1.8 millimetres/year from 1961 to 2003, for a total increase of 77.4 millimetres.

ONE OF THE OUTCOMES OF THE MOST RECENT CYCLE OF GLOBAL WARMING, EXACERBATED BY HUMAN INDUSTRIAL ACTIVITY, HAS BEEN A MELTING OF THESE ICECAPS AND GLACIERS, RESULTING IN RISING SEA LEVELS.

Projecting into the future, when the five most reliable climate modelling emission scenarios published by the Intergovernmental Panel on Climate Change (IPCC) in 2013 are compared, average sea levels are projected to rise by between 0.44 and 0.74 metres by the end of the twenty-first century (Church et al. 2013). It has been predicted that the upper estimate may still be too conservative. For example, Patrick Nunn (2012) suggests that a more probable upper limit for sea-level rise by the end of this century is closer to 1.9 metres. Either upper limit will result in significant loss of land for many low-lying atoll-based islands, including some in the states of Tuvalu, Kiribati, the Marshall Islands, Tonga, the Federated States of Micronesia, and the Cook Islands in the Pacific; Antigua and Nevis in the Caribbean Sea; the Maldive Islands in the Indian Ocean; and many other subnational island jurisdictions (Kelman and West 2009). It is not just sea-level rise and topography alone that combine to adversely affect small islands.

Human-induced global warming is also associated with an increasing frequency and severity of extreme climatic events, such as hurricanes, typhoons, and storm surges, that may be even more problematic in the short-term than the incremental increase in sea level. Even on mountainous islands, populations, public infrastructure, agricultural production, and tourism facilities tend to be concentrated in a thin strip along the coastlines, meaning the impact on the lives and livelihoods of islanders may be just as severe on these islands as for low-lying atolls.

Climate and Extreme Events

There are many catastrophic events we can point to that have devastated the lives and livelihoods of islands and islanders. For example, in September 2004 when Hurricane Ivan swept through the Caribbean with 200 kilometre/hour winds, it caused considerable damage to many islands in its path. The island of Grenada was hit particularly hard with estimated damage exceeding USD 900 million, twice the island's annual Gross Domestic Product (GDP) (Peters 2010). Ninety percent of hotel rooms and houses were destroyed or damaged, more than eighty percent of the nutmeg trees were lost, and agricultural sector losses were the equivalent of 10 percent of GDP (Mimura et al. 2007). Given the time it takes to bring new crops to maturity, it was estimated that nutmeg and cocoa would not be able to make a contribution to Grenada's GDP for another ten years. In September 2019, Hurricane Dorian devastated parts of the Bahamas causing extensive flooding, power outages, and killing over sixty people. Estimated to be the most powerful hurricane recorded in the Atlantic, Dorian will continue to have an enormous long-term impact on the economy of this archipelagic state.

These stories, and many others like them, point to an increasing frequency and intensity of extreme weather events and greater seasonal variability, two additional consequences of global warming that will have disproportionate impacts on small islands. Impacts include increased coastal erosion, greater likelihood of extinction of plant and animal species, inland flooding of productive agricultural land, changes in the volume and salinity of freshwater aquifers (thereby also affecting agricultural capacity), and coral reef deterioration because of warmer and more acidic ocean-surface waters (J. Smith et al. 2001). An increase in the number of extinctions may occur because, as is the case when invasive species were introduced onto small islands, indigenous plants and animals have nowhere to escape (Bramwell 2010). The potential impact on coral reefs is especially important on atolls because the societies living on these islands tend to have a greater economic and social dependence on the vast shallow-water resources associated with the surrounding coral reefs (Barnett and Adger 2003). These in turn may lead to food security issues, increased outmigration, social instability, and loss of income (Gaffin 1997; Nunn 2013; L. Nurse et al. 1997). Although many island societies are attempting to adopt strategies to meet the threats posed by climate change and extreme climate events, reliance on top-down governance models and non-local technologies that do not fit the cultural norms of the local populations may make them less effective (Nunn et al. 2014). Nunn (2013) summarizes the prospects of many oceanic island societies by stating that, "In many ways, the historical and modern Pacific will end within the next few decades" (143).

Although communicating the threats and vulnerabilities to small islands from climate change has undoubtedly led to a greater global awareness and more short-term political action through the United Nations and other organizations, it may also have had the unintended consequence of silencing alternative narratives, including those that speak to the resilience and resourcefulness of island societies to adapt to climate change (Farbotko 2005). So before we accept without question the rhetoric of vulnerability and marginalization, we should look to see if the devastation being predicted is inevitable. In the past, island peoples have been incredibly resilient to extreme events. Islands, and their societies, are not just small, isolated, and dependent, but rather are deeply connected internally and externally in ways that both constrain and facilitate their ability to meet the challenges of climate change (Lazrus 2012). Many researchers have made the case that adaptive strategies incorporating traditional ecological knowledge, and bottom-up, culturally appropriate consultation and decision-making are more likely to be successful in responding to climate change and extreme events (Barnett and Adger 2003; Lazrus 2012; Mimura et al. 2007; Nunn 2013; Nunn et al. 2014; Pelling and Uitto 2001).

One of the most compelling cases of supposed devastation on an island pertains to the story of the Pacific Ocean's Rapa Nui, better known as Easter Island or the home of the iconic Moai statues. At one point, Easter Island was the home to a thriving population. By the time Europeans arrived in the early 1700s, some researchers have suggested that the island had become almost completely deforested and population levels had declined precipitously. Because of this, the popular public explanation has been that this is an island laboratory version of the classic Malthusian Trap, where population increases at a rate faster rate than resources can support, resulting in "overshoot and collapse" outcomes (Ponting 1991; Weiskel 1989). In fact, Jared Diamond (2005) popularized this theory in his book *Collapse*, suggesting that Easter Islanders exploited their forest resource faster than it could regenerate, leading him to refer to this as one of the first times where a society had committed "ecocide." In fact, the factors that led to the loss of the forest and the subsequent collapse of civilization on Easter Island may be much more complex than these early theories suggested. For example, Terry Hunt (2007) makes the case the more likely contributing causes for deforestation and population decline were fire and the introduction of rats to the island by Polynesian colonizers, or diseases brought by the first Europeans (T. Hunt and Lipo 2009). Peiser (2005) suggests that the slave trade also was a major contributor to a rapid decline in the population of Indigenous islanders throughout the 1800s. Finally, researchers have also made a compelling case that around the year 1300 AD, climate change may have been a more significant factor in reducing the population carrying capacity of Easter Island and many other islands in the region (see, for example, Nunn 2000; Nunn and Britton 2001; Nunn et al. 2007). Evidence seems to point to a relatively short period during which the region experienced a rapid temperature decrease, sea-level fall, and a more arid climate. These climatic conditions may have led the largely agrarian society to engage in a more intense competition for the limited supply of resources, resulting in greater deforestation and soil erosion, and greater social conflict. So, although the Rapa Nui society may have, indeed, sealed its own fate by overusing the existing resources, this may have occurred primarily because of a deterioration in growing conditions brought about by a combination of climate change, the introduction of European-based diseases, and the slave trade, and only indirectly as a function of increasing populations.

ISLAND BIOGEOGRAPHY

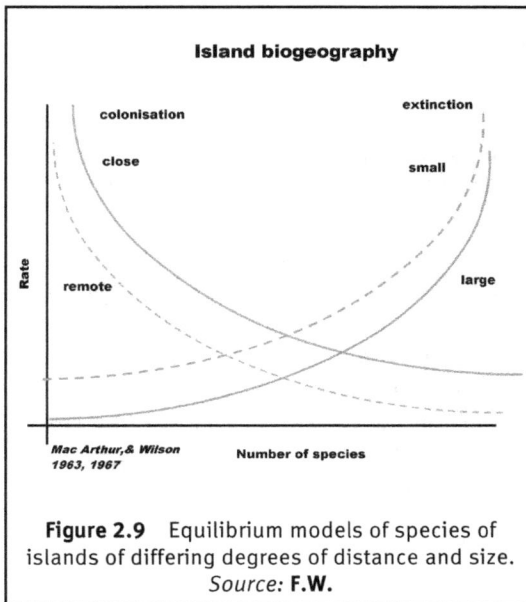

Figure 2.9 Equilibrium models of species of islands of differing degrees of distance and size. *Source:* **F.W.**

Arguably one of the most exciting contributions of islands to general scientific enquiry has been the field of island biogeography (Adsersen 1995; Bramwell 2011; Kostas 2011). This is not just the application of a theory to one of many different world contexts, but rather a means to better understand the fundamental relationships among ecology, conservation, population dynamics, evolution, and paleontology. For scientists, island biogeography has been a lynchpin in understanding evolution (Adsersen 1995).

The pioneers in this field were Edward Wilson and Robert MacArthur, with ideas initially conveyed in their seminal book, *The Theory of Island Biogeography* (MacArthur and Wilson 1967). As illustrated in Figure 2.9, the first of three basic components to their model was that the number of species in a taxonomic category increases with the size or area of the island. The second component was that the greater the distance (or isolation) of an island from a source area, the fewer the number of species will be present. Finally, the third component is a consistent inverse relationship between the rate of introduction of new species on an island and the rate of extinction of species, such that the number of species present would reach an equilibrium point. Although you would still see the introduction and extinction of new species over time, in general the total number of species present on an island would become relatively stable at this equilibrium point. One of the values of the theory of island biogeography is that it has allowed for the introduction of a predictive, quantitative model of species richness or diversity. In this model, the number of flora or fauna represented on an island is based on three population characteristics: 1) the number of species introduced or immigrated from another island or continent, 2) the speciation or evolution of species in situ, and 3) the extinction or turnover of species from an island.

In the simplest version of the model, the size of an island, sometimes used as a surrogate measure for the number and diversity of habitats, determines the rate of extinction, while the

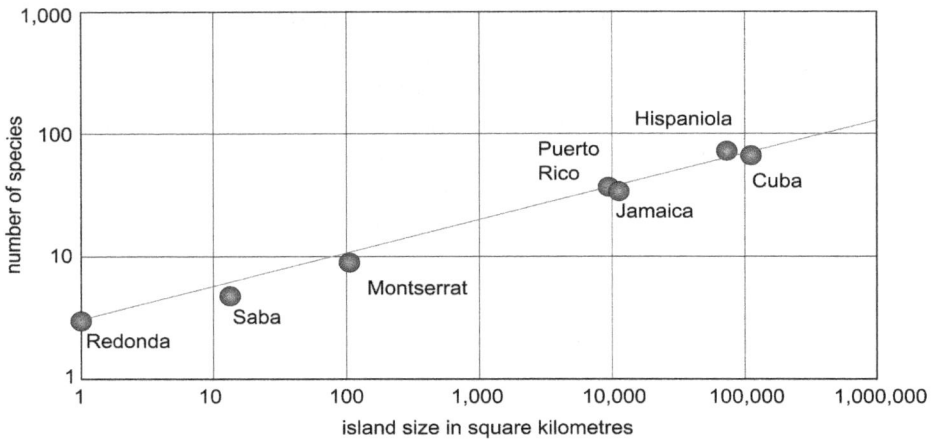

Figure 2.10   Relationship between the number of species present on a set of islands in the Caribbean and the areal size of those islands, as measured in log values. *Source:* **public domain**

distance from a continent influences the island's level of immigration and speciation. There-fore, all other things being equal, larger islands closer to continents would be expected to have higher species richness and lower species turnover. Smaller, more remote islands would have lower species richness and higher rates of turnover or extinction.

An application of one element of this model is found in Figure 2.10: the relationship be-tween the number of species present on a set of islands in the Caribbean and the areal size of those islands, as measured in log values (MacArthur and Wilson 1967). In a similar graph and using eight islands, Adsersen (1995) showed a strong direct relationship between the percent of endemic species on an island and the shortest distance to the nearest continent.

How Islands Acquire Species

We have already seen that islands such as Surtsey and those around Krakatau have proven to be exciting laboratories to better understand how places acquire species and, indeed, how evolution takes place. But how exactly do species arrive on islands? Sherwin Carlquist (1974), one of the most important figures in island biogeography, has summarized the many ways that animals and plants happen to find their way to islands in a chapter appropriately titled "Getting There is Half the Problem" (Carlquist 1965). He notes that for islands that were once part of continents and have become enisled due to the flooding of a land bridge or plate tectonic movement, the presence of most species is likely a result of their former link to the larger continent. Stated differently, many of these species were already present on the land be-fore they became islands. However, for those more distant oceanic islands, other mechanisms had to come into play. The first and simplest was seawater flotation. Although this would not allow mammals to travel long distances, some plant species are capable of travelling long dis-tances on the ocean currents and are still able to propagate after arriving on an island. Rafting is the second way in which islands might acquire species. As the term implies, this refers to the hitchhiking of plants and animals on debris washed out to sea. In an article in *Nature*, Ellen

Censky (1998) and colleagues make the case that green iguanas likely colonized Anguilla in the Caribbean by accidentally catching rides on logs or other debris during hurricanes.

Air flotation is the third mechanism for dispersal to islands. Although most plant species are designed to disperse seeds fairly close to the parent plant, the wind can be a powerful agent of dispersal over long distances. One of the earliest naturalists to reach the sterilized remnant of the island Rakata one year after the 1883 explosion of the Krakatau volcano reported seeing only one species on the island, a spider that had presumably arrived by ballooning across the sea on its own threads. More recently, using air traps on the newer island of Anak Krakatau, Thornton and others (1988) captured an average of twenty insects per square metre per day. They estimated that, at this rate, between five and fifty million insects per day could be arriving on the island from surrounding islands that were at least four kilometres away. One of the more likely forms of dispersal for slightly larger plant or insect species would be to hitch a ride in the stomachs, feathers, or feet of birds. For example, on Surtsey off the coast of Iceland, botanists captured ninety-seven snow buntings and were able to identify eighty-seven different kinds of seeds on or in the birds, not only from Iceland but from as far away as Great Britain and continental Europe (Lasky 2012). Not only do birds bring species to islands but, as we saw in an earlier section of this chapter, they can also provide guano, a rich fertilizer within which seeds are more likely to germinate and grow.

Of course the likelihood of success of any of these means of dispersal on any given day or even any given year is quite low, especially over long distances. The additional chance of a species surviving and propagating once it arrives on an island is even lower. However, when you consider that islands may have had thousands or even hundreds of thousands of years to acquire species, then the probability of rare events occurring that would allow for the establishment of a rich ecosystem becomes much greater. You would also expect this process of colonization to have occurred faster in those regions where there is a greater density of islands, such as with archipelagos, through a process known as "island hopping," than in regions where islands do not have any nearby neighbours. Of course, many species of plants and animals have arrived by accident or intent with the first human appearance on islands. Later, invasive species such as the rat, goat, and pig arrived with the first European travellers, and resulted in the decimation of many original species that had lost their ability to defend themselves against these new predators.

This discussion regarding how islands acquire species represents a good opportunity to introduce the contributions of the nineteenth-century naturalist and British scientist, Alfred Russel Wallace. Wallace may be the most significant historical figure in the field of island studies (see Claridge 2009; Quammen 1996; Raby 2002). In addition to his work that paralleled

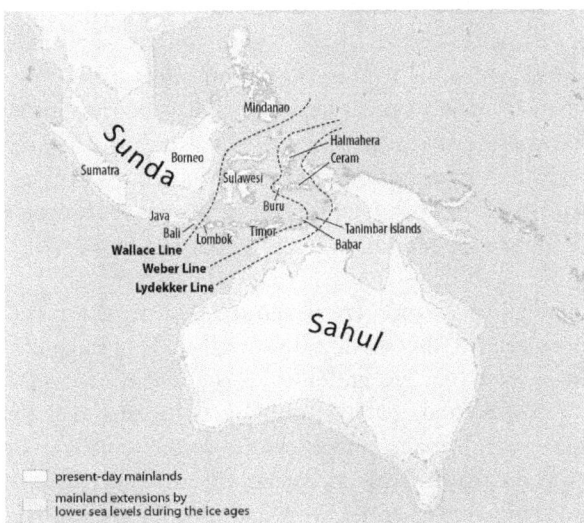

Figure 2.11 Wallace's Line. *Source:* **Maximilian Dörrbecker**

Charles Darwin's on the evolution of species in response to their surrounding environments and natural selection, Wallace is considered to be the founder of modern biogeography (and especially island biogeography) and was the first scientist to classify islands based on the role of isolation and climate change. He is perhaps best known for discovering the geographic boundaries between two distinct zoological regions in an archipelago. What is now referred to as Wallace's Line in the Malay archipelago in southeastern Asia, as seen in Figure 2.11, is a boundary separating the Asian faunal region (i.e., Sundra) that is part of the Malay Peninsula from the Australasian or Sahul faunal region.

The most striking differences occur across the Lombok Strait, a body of water only twenty kilometres across at its narrowest spot separating the island of Bali (as part of Asia) and the island of Lombok (as part of Australasia). Despite this relatively short span, Wallace found striking differences in the species of animals and plants on either side of the body of water. Initially, Wallace and other scientists believed that the great depth of the strait, at 240 metres, served as an effective barrier to prevent the two-way diffusion of most species. In retrospect, this was only part of the story. Sea-level changes and plate tectonics have combined to explain the significance of this barrier. The depth of this strait was such that, even during periods of glaciation when land bridges would have connected other islands in the archipelago, the Lombok Strait would still have existed, thereby discouraging the dispersal of species beyond these points. We also now know that many of the islands east of Bali, including New Guinea, are part of the Australian plate, while most of the islands west and north of Bali are part of the Eurasian plate. These two plates were once much farther apart and have been moving inexorably closer to each other over millions of years.

Endemism

One of the terms that has already been mentioned several times is endemism. But what exactly is endemism and why is it relevant to islands? The short answer is that endemism refers to a species within the same taxonomic group that is unique to a certain geographic space. Because islands have distinct, readily observable boundaries, endemic regions are often but not always associated with islands. As you can imagine, the scale or size of the region can greatly influence whether a species might be called endemic. Although interesting methodologically, we will not delve into the nuances of the problem of scale in this discussion. Also referred to as speciation, species can gain this endemic status either because they have evolved in place, or in situ, to the point where they are taxonomically different from similar species elsewhere in the world (i.e., neoendemic) or the species only exists in this one location because it has become extinct everywhere else in the world (i.e., paleoendemic) (see Whittaker and Fernandez-Palacios 2007).

Although endemism exists in continental regions, islands are without question the most prominent homes for species found nowhere else in the world. Bramwell (2011) suggests that there are about 50,000 flowering plant species that are endemic to the world's islands, constituting 35 percent of all world plants. Of this total, 20,000 (40%) are threatened with extinction. About 15 percent of the mammals, amphibians, and birds are found only on islands and 35 percent of all threatened bird species are endemic to islands (Pimm et al. 2006). Some of the best evidence for the exaggerated presence of endemic species on islands relative to mainland regions is provided by Gerold Kier and others (Kier et al. 2009). They looked at

ENDEMISM REFERS TO A SPECIES WITHIN THE SAME TAXONOMIC GROUP THAT IS UNIQUE TO A CERTAIN GEOGRAPHIC SPACE. BECAUSE ISLANDS HAVE DISTINCT, READILY OBSERVABLE BOUNDARIES, ENDEMIC REGIONS ARE OFTEN BUT NOT ALWAYS ASSOCIATED WITH ISLANDS.

ninety world biogeographic regions, fourteen of which were island-based, and compared the level of endemism among vascular plants and vertebrates across those regions. They found that island regions, and especially those found along the mid-latitudes, had a much greater number and proportion of endemic species across all groups of organisms. In fact, endemism richness on islands exceeded that on mainland regions by a factor of 9.5 to 1 (plants) and 8.1 to 1 (vertebrates).

One of the main factors associated with endemism is isolation: the more remote the island, the greater the proportion of species present would be classified as endemic. It is also apparent that greater isolation is also accompanied by a smaller number of species. In other words, where there are fewer species, you would also expect to find more gaps in the species food chain. This latter feature is important when we talk about many of the oddities that have fascinated people about animal and plant life on islands.

Undoubtedly, some of the mystery and allure of islands that remains to this day is a result of the endemic animal specimens and stories brought back to Europe by explorers and merchants of the sixteenth to nineteenth centuries. For the general public, our imagination has been stimulated because we have been exposed to living specimens, fossil records, and oral histories of mammals that are radically different from the versions we might have seen in children's picture books or zoos. In the popular book *The Song of the Dodo: Island Biogeography in an Age of Extinction*, David Quammen (1996, 17) refers to island biogeography as being "full of cheap thrills." Because islands are so populated with unique outlandish species, he explains, they have captured and continue to capture the imagination of scientists and the lay public alike. The development of these oddities or unique features is called adaptive radiation by evolutionary biologists but, more informally on islands, it has been referred to as the "island effect" (Berry 2007; Berry and Gillespie 2018) or as the "island syndrome" (Carlquist 1974).

These adaptations can take many forms and can occur much more quickly on islands than on continents. They include Darwin's famous example about the evolution of beak shapes in finches on the Galápagos Islands as well as flightlessness in birds and insects. However, the two changes that have captured the most attention and imagination are insular gigantism and insular dwarfism. As the terms imply, gigantism and dwarfism are situations where the body size of a species grows, respectively, much larger or much smaller than you would expect to find in the parent species on a continent. Examples of insular dwarfism include the pygmy hippo of Madagascar, the now-extinct dwarf elephant of Malta, Cyprus, Crete, and Sicily, and the world's smallest "bee" hummingbird on the island of Cuba. Examples of insular gigantism include the "elephant bird" and the "hissing cockroach" of Madagascar, and New Zealand's family of wetas or crickets.

These oddities in size are best known as the Island Rule or Foster's Rule, named after J. Bristol Foster and published in the journal *Nature* (1964). It states that, depending on the resources available to them on an island, small island animals tend to get larger and large island animals tend to get smaller. The smaller animals get larger because of the absence of natural predators, and especially carnivores. They also extend their range and magnitude of food supplies by feeding on those items that are available because of the gaps in the food supply chain. It also allows them to produce larger, healthier litters. The large animals get smaller on islands that have limited food supplies. In these situations, the smaller versions of the species are more likely to reproduce because they do not need as much to survive. It is also easier for them to hide in places where predators might be present. This simple explanation was later expanded and refined by Ted Case (1978), at least in part by looking at species on the islands near Baja, California.

The terms gigantism and dwarfism are sometimes applied too loosely to island species. For example, the famous komodo dragon in the Indonesian Islands and the land tortoises in the Seychelles and Galápagos Islands are often depicted as examples of gigantism. In fact, if the populations of these species were more widespread on continents, they might not be considered unusually large.

Island Biodiversity: A Different Kind of Hotspot

Earlier in this chapter, the concept of a geological hotspot was raised as one of the ways in which many oceanic islands have been formed. There is another phrase involving the use of hotspots that holds particular importance for islands: the "biodiversity hotspot." This concept was first expressed by Norman Myers (1988) to describe tropical forested areas of the world that had both very high concentrations of endemic species and were also at the greatest threat of depletion of these species. Although Myers' analysis was not specifically focussed on islands, five of the top ten–ranked tropical forests that he identified happened to be islands or island regions, including Madagascar, peninsular Malaysia, Borneo, the Philippines, and New Caledonia. Later work by Myers and colleagues (2000) expanded this to identify biodiversity hotspots throughout the world. Based on the number of endemic species of plants and animals, the ratio of the number of endemic species to area (i.e., the greater the ratio the "hotter" they are), and the threat to habitat, they suggested that there were twenty-five world biodiversity hotspots, of which eight were classified as extremely important; in their words, the "hottest of the hot spots." Given what we now know about endemism on islands, it should come as no surprise to learn that four of the top five of these are islands, consisting of Madagascar, the island archipelago of the Philippines, the Sundaland island region of Southeast Asia, and the Caribbean. Other researchers have focused their attention specifically on islands and have made the case that, when you look at the entire world, islands stand out even more as places with high levels endemism and a greater likelihood of species extinction. For example, it was pointed out earlier in this chapter that a large proportion of the world's flowering plants, mammals, birds, and insects are found only on islands and many of these are threatened with extinction.

Although the distribution of this endemism, and even the connection to threatened habitats, was not necessarily startling news, what Myers and others advocated for was the recognition that these rich regions of biodiversity should be given higher priority in the allocation of limited global conservation resources. In response to critics, he and his colleagues pointed out they were not suggesting that existing funds be shifted away from non-hotspot areas but rather, any new resources be invested in those areas of the world that would allow for the preservation of the greatest number of endemic species. This idea of biodiversity hotspots has been applied to develop conservation practices and policies for "fragmented landscapes" (Tabarelli and Gascon 2005), small areas or habitats surrounded by otherwise uninhabitable landscapes that contain endemic or socially valued species of plants or animals. For example, if a region is going to be developed, how large, in what shape, and where should a remaining fragment(s) be located to reduce the risk of extinction of certain species within that region? The identification and analysis of these "metaphorical" islands, sometimes referred to as fragmented communities, has been applied to manage many different kinds of landscapes, from deforestation projects in the Amazonian jungle to urban greenbelts threatened by suburbanization (Bennett 1998; Kattan and Alvarez-Lopez 1996; Marsh et al. 2003).

The preservation of biodiverse regions and endemic species may be rewarding from a social perspective but it may also be important economically for islanders and for the future well-being of humankind. Increasingly, the pharmaceutical and medical industries are relying on the compounds derived and synthesized from natural products to create the next generation of drugs, health, therapeutic, and personal care products. The concentration of endemic species found on islands and in coral reefs and marine areas surrounding islands is already proving to be valuable to companies in these sectors (Beattie et al. 2011; Gascon et al. 2015). What remains to be seen is the degree to which these and future benefits might accrue to the islanders themselves.

CONCLUSIONS

This chapter has shown that the appearance, modification, and disappearance of islands and island biota have been shaped by scale as well as the passage of time. From small micro-areas on individual islands to large regions of the world, and from events spanning millions of years to almost instantaneous cataclysmic incidents, the physical forces associated with islands have been intriguing features of our natural world.

It is also clear that the study of islands has reached well beyond the physical boundaries of the islands themselves to inform and advance human knowledge in many other fields. The fields of volcanology and the study of plate movement have benefited significantly from observing the birth, evolution, and mobility of islands. Glaciologists and geomorphologists have benefited from the many ways that glaciers, wind, and waves have acted on islands.

Perhaps one of the greatest contributions from studying natural processes on islands has been the contribution it has made to the field of evolution. Because of what takes place on islands, we are now better able to understand the factors that have influenced evolution from two threads of research. Conceived and used as natural laboratories, analyzing the changes taking place on newly created islands has served to show how organic life reappears and becomes established as a result of dispersion from other locations. Taking the form of a small closed ecosystem, we have also seen how islands have created the conditions for the many pathways of evolution of species based on the unique circumstances faced on individual islands.

Islands have also been instrumental to our understanding of and ability to adapt to climate change and extreme climate events. As the outcomes of human-induced global warming increase across the world, and as part of the wider field of human-environment interactions, this is an area where islands will continue to provide us with important lessons for climate change adaptation on continents. As far back as the 1980s, geographers such as Tim Bayliss-Smith (1988) and others were making the case that islands should be viewed by researchers and decision makers as living laboratories for the study of humans and the natural world, suggesting that these relatively closed contexts represent an exceptional opportunity, under controlled conditions, to study the interaction of the range of natural and social processes influencing these human-environment relationships. Taking this one step further, island research has informed zoology as well as conservation management and policy, by allowing us to see the impacts of the introduction of non-indigenous species on island species and by assessing how fragments of "habitats" on continents might be configured to preserve threatened species in a kind of metaphorical island. Because of the high level of endemism that exists on islands and their surrounding marine ecosystems, the next major contribution of island regions to the world may be in the field of microbiology through the discovery and production of natural products and by-products.

Key Readings:

Adsersen, Henning. 1995. "Research on Islands: Classic, Recent, and Prospective Approaches." In *Islands: Biological Diversity and Ecosystem*, edited by Peter Vitousek, Lloyd L. Loope, and Henning Adsersen, 7–21. New York: Springer.

Fernández-Palacios, Jose Maria. 2009. "Island biogeography, theory of." In *Encyclopedia of Islands.*, edited by Rosemary Gillespie and D.A. Clague, 486–490. Berkeley: University of California Press.

MacArthur, Robert and Edward Wilson. 1967. *The Theory of Island Biogeography*. Princeton, New Jersey: Princeton University Press.

Nunn, Patrick. 1999. "Geomorphology." In *The Pacific Islands: Environment and Society*, edited by Moshe Rapoport, 43–55. Hong Kong: The Bess Press.

Whittaker, Robert and Jose Maria Fernández-Palacios. 2007. *Island Biogeography: Ecology, Evolution, and Conservation*, 2nd Edition. Oxford: Oxford University Press.

Chapter Three

Images of Islands from Literature and the Popular Media

Islands are places of dreams and nightmares, utopias and dystopias. They are the glitzy hideaway of the rich and famous, but also the last refuge of the infamous. They are places of banishment, where people are silenced, but also of self-discovery, where people find a voice. They are the dumping grounds out there in the shadows of Biscayne Bay, Florida, but also the luminous flamingo-pink floating islands on the horizon that everyone can see, if they really want to. (Peckham 2002, 88)

INTRODUCTION

Our understanding and perceptions of islands inevitably influence our behaviours and actions. For example, if we perceive islands to be idyllic paradises where we can temporarily escape from our everyday obligations, as tourists we may be predisposed to seek out those places that represent that ideal. If we perceive islands as remote and isolated from the major world markets, we may be less likely to view them as places to establish a business. Or if we perceive islands to be dangerous and islanders to be socially insular and inferior, we may marginalize or malign the people who live there. These impressions or stereotypes that we hold regarding islands have been created, transmitted, and manipulated primarily through popular media in the form of literature, poetry, music, film, television programming, advertising and, more recently, social media. Of course, many of these impressions are created and disseminated by individuals who have only a fleeting association with islands, and whose knowledge and experiences are very different from those who have been born and raised on islands. As Steven Fischer (2012, 198) has noted, island literature is rarely by islanders about their islands—it is "about non-islanders' troubled encounters with unacceptable or misunderstood island demands and cultures."

The purpose of this chapter is to provide an overview of the ways in which various forms of media have portrayed islands, the impact that these portrayals have had on islands, and the ways that these might differ from the impressions of islanders themselves. It is impossible in one short chapter to encompass all of the examples of island-based imagery in the public arena. This is made even more difficult by the fact that recorded English, whether in print, video, or audio, barely scratches the surface of the images of islands. For example, although some oral storytelling such as the Icelandic Sagas or the accounts of the Lapita peoples of Oceania has survived the passage of time, the stories associated with the lives and cultures of many Indigenous island peoples, passed on orally for many generations, have been lost forever. Despite these gaps, by focusing on what might be the most recognizable and significant contributions regarding islands from various formats, including literature, music, art, and cinema, we may be able to discern some general patterns.

ISLANDS IN LITERATURE AND POETRY

Historical Literature and European Exploration

It might be difficult to make the case that there is a genre of literature that could be described as island literature. Despite this, islands have figured prominently in fiction and poetry for at least the past thousand years. Sam Taylor (2009) and Le Juez and Springer (2015) suggest that islands are more than just scenic backdrops; rather they are literary devices of contemplation, transformation, and social upheaval.

Some of the earliest fictional books in English are about islands, including Homer's voyages of Odysseus as written in the epic poem *The Odyssey*. Here, the Greek hero Odysseus and his crew encounter islands populated by mythical beasts and floating islands. The islands represent the unknown and the ability of the hero Odysseus to conquer it. The story of the fictional island of Atlantis, as told by the Greek philosopher Plato, portrays islands as idealized civilizations. This idea was expanded and given utopian characteristics by Sir Francis Bacon as the island of Bensalem in the book *New Atlantis* (1627) and then by Sir Thomas More in *Utopia*. All three of these books convey the authors' impressions of fictitious and utopian island societies.

Another early example of a description of island life comes from a series of books referred to as the Icelandic or Family Sagas and the earlier Greenlanders' Saga (Smiley 2001). Unlike the idealized heroism of Odysseus or the utopian societies in *New Atlantis* or *Utopia*, these sagas were written about the everyday lives of the early Norse explorers and settlers. Written in the thirteenth and fourteenth centuries, they were based on stories that had been passed down orally and in written fragments. These prose narratives chronicle the daily events of Vikings/Norse families who supposedly lived in and around Iceland and Greenland in the tenth and eleventh centuries. By chronicling explorers such as Erik the Red and his son Leif (the Lucky) Erikson, they are one of the more compelling pieces of proof that Vikings found and temporarily settled in an area of North America that they referred to as Vinland, likely somewhere along the American New England coastline, or in Canada in Nova Scotia, New Brunswick, or perhaps even Prince Edward Island (Barnes 1995; Magnusson

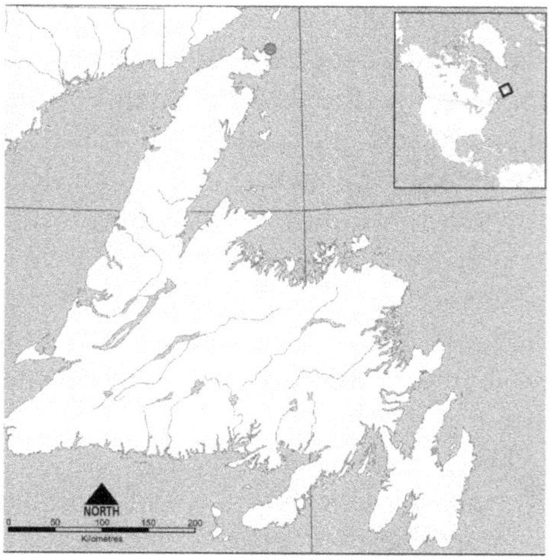

Figure 3.1 Location of L'Anse aux Meadows on map. *Source:* **public domain**

Figure 3.2   L'Anse aux Meadows. *Source:* **Dylan Kereluk**

and Pálsson 1965; Sigurdsson 2004). An archaeological site at L'Anse aux Meadows, in the northeastern part of the island of Newfoundland, is one of the few archaeological sites of the short-term colonization of this part of North America (See Figures 3.1 and 3.2).

One of the most significant contributions to the creation of mainlanders' stereotypes of islands were the so-called island adventure novels written in the eighteenth and early nineteenth centuries. The most prominent examples include *Robinson Crusoe* by Daniel Defoe (1719), *Gulliver's Travels* by Jonathon Swift (1726), *Swiss Family Robinson* by Johann David Wyss (1812), and *The Mysterious Island* by Jules Verne (1874). Kevin Carpenter (1984), as quoted by Elizabeth DeLoughrey (2007), has estimated that over 500 desert island stories were published in England during the period from 1788 to 1910. In fact, an entire academic school of literary commentary called the Robinsonades has emerged to interpret this body of fiction in light of prevailing social values. It is no coincidence that these novels emerged at the time of the European "Age of Discovery" that involved exploration, colonialism, and empire building (Rainbird 1999). Explorers, closely followed by merchants and missionaries, brought back stories and evidence of islands and island life. These included encounters with fantastical creatures unique to Europe, not altogether surprising given the endemism associated with island flora and fauna. Either by design or by accident these images and distortions of images found their way into adventure books.

Although each of these adventure stories had their own origin, morality, and plotline, they tended to share similar characteristics. First, although the stories appealed to readers of all ages, their target audience was young adults. Second, they describe wondrous, almost unbelievable places. In retrospect, we would categorize these clearly as works of fiction, and even fantasy. However, at the time, it was not unusual for the unknown to be filled with exaggerations and distorted imagery. For example, as seen in Figure 3.3, historical maps and globes occasionally contained drawings of sea monsters or other fantastical creatures (often versions of whales, squid, or walruses) in seas at the edge of the "known world."

Sometimes the island environments in these stories were seen to be dangerous, as in *The Mysterious Island*, while in other cases the islands were described as being more benign. In

Figure 3.3 Sebastian Münster's Map of the World (1553) with dragons. *Source*: **public domain**

Defoe's *Robinson Crusoe*, which was based on the true story of Alexander Selkirk who was stranded on the Juan Fernandez Islands off the coast of Chile for four years, both the island and individual are transformed, partly through the actions of the castaway and partly through the way the protagonist starts to perceive the surrounding environment. Although Crusoe first refers to his new home as the Island of Despair, later on he states that, "Now I look'd back upon my desolate solitary Island, as the most pleasant Place in the World, and all the Happiness my Heart could wish for, was to be but there again. I stretch'd out my Hands to it with eager Wishes. O happy Desart, said I, I shall never see thee more" (Defoe 1883, 164).

Islands as places of transformation, contemplation, or self-reinvention (Le Juez and Springer 2015) is a common theme in these adventure novels. On one hand, it reflects the sentiment that human beings are able to control their natural surroundings and remake them into an image that more closely reflects some European community ideal. For example, despite being populated by odd and dangerous creatures, Jules Verne suggests that *The Mysterious Island* can be tamed and civilized by human virtues. At the same time, experiences gained by a non-islander living on an island can represent a form of escape or refuge from the social norms and expectations associated with his (and it usually was a male) former life in the urbanized mainland. In this way, the apparent ability of the island to transform the individual is also an important common thread in these and other more recent works of fiction (Patke 2004, 178). Patke goes on to say that islands have been used as convenient contexts by poets and novelists to better understand the "psychic economy of human experience," in other words, what it means to live in a place, to leave it, to visit it, and to return to it, and even to be turned out by it.

This tension between island places as being simultaneously places of danger, mystery, fantasy, escape, and transformation continues to be used today by the travel and tourism marketing industry to entice vacation travellers to purchase their very own temporary island experience, more commonly known as a vacation. Even though the following quote from *Beyond the Floating Islands* is in reference to nineteenth-century literature, it could just as easily be used as part of a twenty-first-century tourism advertising campaign: "Islands are often represented as sites of some kind of magical transaction or exchange, places where individuals encounter different cultures and find that they can no longer relate in the same way to the places they have left" (Michelucci 2002, 8).

Many commentators believe that this type of story featuring adventures on desert islands was used to inculcate British youth to accept the virtues associated with British colonialism and empire-building. It has been suggested that this ability to tame and improve an environment supposedly untouched by humankind is used as an allegory to support a colonial ideology and Enlightenment (Scott 2014). As Diana Loxley (1990, 3) says in her introduction to an analysis of these stories, islands represent "a simplification of existing colonial problems and thus an ideological process of wish-fulfillment." As we will see in Chapter Four, it was not just nature that was being tamed on these islands but also the Indigenous peoples encountered by Western explorers. DeLoughrey (2007, 20) goes so far as to say that the combination of the desert-island and nautical-adventure forms of literary fiction "were vital to imagining this transoceanic empire."

It should be noted that this genre of island adventure stories did not end with the nineteenth century. Children's books of the twentieth century, including William Steig's *Abel's Island* (1976), Harry Mazer's *The Island Keeper* (1981), and Theodore Taylor's *The Cay* (1969) continue to employ the plot of individuals who are inadvertently confined to an island and find themselves and their values being transformed by their experiences (Gunstra 1985).

Of course, this genre of island adventure literature is focused almost completely on a colonial imperialist interpretation of islands and ignores or marginalizes locally based island authors. As Teaiwa (2010, 730) states, "In contrast to the canonized European and American texts forged out of imperialist adventures in the islands of the Pacific (from Samuel Wallis, Louis de Bougainville, and James Cook to Herman Melville, Somerset Maugham, Jack London, and even James Michener), the work of Indigenous authors often fails to escape the category of exotica." The excuse for marginalizing Indigenous literature is that the local cultures prior to European contact were oral. Despite this, the poet, novelist, and painter Albert Wendt has suggested that the rich history of storytelling, incantations, and poetry has found its way into current Pacific literature to the point that we now have an "oceanic imaginary" (Teaiwa and Marsh 2010; Wendt 1982).

The word utopia may be derived both from the Greek word *ou-topos* meaning "no place or nowhere," or the related word *eu-topos* meaning "a good place." Both of these interpretations may explain why islands, as separate and apart or distant from the mainland, were considered by novelists and philosophers as ideal locations for utopian societies. Two noteworthy examples are Sir Thomas More's *Utopia* and Aldous Huxley's *Island*. More's ideal society of Utopia was set on an island somewhere in the New World where city-states were distributed along the coastline of bays and harbours (Savory 2011; Scott 2014). In fact, this may be one of the first recorded (fictional) examples of a man-made island. The region that would become Utopia

THIS TENSION BETWEEN ISLANDS AS BEING SIMULTANEOUSLY PLACES OF DANGER, MYSTERY, FANTASY, ESCAPE, AND TRANSFORMATION CONTINUES TO BE USED TODAY BY THE TRAVEL AND TOURISM MARKETING INDUSTRY TO ENTICE VACATION TRAVELLERS TO PURCHASE THEIR VERY OWN TEMPORARY ISLAND EXPERIENCE, MORE COMMONLY KNOWN AS A VACATION.

was originally a peninsula. However, King Utopus had a twenty-four-kilometre-long channel dug that made the peninsula into an island. Utopia's change from being part of a mainland to being an island is supposed to reflect "the social desire for self-containment, autonomy, and unchangeable stability" where the dream of a utopia is born out of a less than utopian continent (Stephanides and Bassnett 2008). The implication is that islands are ideal because they are contained and separated from the impurities associated with mainland societies.

In Huxley's *Island*, published in 1932, an Englishman (William Farnaby) deliberately wrecks his boat on the fictional Polynesian island of Pala in an attempt to convince the island's queen to sell the island's oil rights to his oil baron employer. The island-as-utopia and mainland-as-dystopia dichotomy found in More's *Utopia* is also represented in the differences between Huxley's *Island* and his earlier, more well-known *Brave New World*. For example, while drugs were used in *Brave New World* to pacify the population, in *Island* they were used for enlightenment and the exploration of self-knowledge. A similar distinction is made by Huxley to the phenomenon of group living. In *Brave New World* group living is forced in order to eliminate individuality, while, in the novel *Island,* group living takes place so that children can be shielded from their parents' neuroses.

As we shall see in the next chapter, the early accounts by European explorers of many Pacific islands also painted a picture of these islands as paradises or utopias, an image in stark contrast to the dirty and polluted landscapes of much of urban, industrializing Europe of that time (Connell 2003b). This image of Pacific island utopias, whether through the pristine shorelines or the perceived promiscuous behaviour of the island's young women, has been one of the most pervasive and consistent messages regarding islands, and especially tropical islands, in literature and advertising since that time.

Islands have also figured prominently in several of the works by poet and playwright William Shakespeare. Two plays in particular deserve mention. *The Tempest* (circa 1611) takes place on a remote, mythical, enchanted island located somewhere in the Mediterranean Sea. As is the case with many other storylines involving islands, the play begins with a shipwreck and many of the ensuing scenes are about the interaction between the island residents (including the previously exiled Prospero) and the newcomers. The other Shakespearean play that employs an island as an important setting is *Hamlet*. It takes place in the castle Elsinore (also known as Kronborg) on the northeastern tip of the island of Zealand, Denmark, in the Øresund Strait. This location, at the narrow divide between the large inland Baltic Sea and the oceanic North Sea, may have been critical to the play, where the physical setting matches the spiritual dimension (Shell 2014, 153–154).

Islands and Twentieth-Century Literature

As more books, poems, and plays are written and find their way into the hands of a more literate population, islands have become a popular subject and setting for fiction. In the twentieth century, we started to see examples of islanders, such as the Nobel Prize laureate from St. Lucia, Derek Walcott, writing about islands on their own terms.

The short story "The Man Who Loved Islands" (1928) by British novelist, poet, and playwright D. H. Lawrence (1885–1930) can be seen as a classic example of the relationship that a non-islander has with the physical and social characteristics of islands (Michelucci 2002). Although the protagonist in the story, Cathcart, was apparently born on an island, he does not seem to understand the island or islanders. In this respect the story mirrors Lawrence's personal background. He had no personal connection to islands and this story is one of the few examples where he used an island as a setting. However, the themes that emerge in the story are reflective of prevailing attitudes at the time it was written and that still resonate today. For example, in the opening paragraph, Cathcart says that the island he was born on had too many people. "He wanted an island all of his own: not necessarily to be alone on it, but to make it a world of his own … A minute world of pure perfection, made by man, himself" (Lawrence 1928, n.p.). As was the case with Crusoe and the protagonists created by the early political philosophers, Cathcart believed that he could control nature and social relationships in this bounded space and, by becoming a master of the island, he could create a utopia.

Unlike the stories of being marooned on a tropical paradise, Cathcart came to the island(s) intentionally hoping to establish a thriving society and community. As the story unfolds, we see that Cathcart is unsuccessful in his repeated attempts to create a utopia on increasingly smaller islands. At first he blames his failures on those he has brought to the island to work for him. For example, he says that "The people were not contented. They were not islanders" (Lawrence 1928, n.p.). He also attributes his failures to the mystery and spirituality that surrounds the island, referring to old far-gone men, and men of Gaul. Eventually Cathcart goes bankrupt and he believes that his island utopia has been transformed into a dystopia. As with the famous sixteenth-century quote in the poem by John Donne (1988), "No man is an island entire of itself," one of the themes in this short story is that living in isolation prevents the social interaction that is essential for the wellbeing of the individual (Stephanides and Bassnett 2008).

Another twentieth-century short story that speaks to social life on an island is "The Boat," found in Canadian author and academic Alistair MacLeod's (2000) book *Island*. Born in the Canadian prairie province of Saskatchewan, MacLeod moved with his family back to their ancestral home on Cape Breton Island, Nova Scotia, when he was ten years old. Although he wasn't born on an island, it could be argued that his family roots that go back more than one hundred years and MacLeod's involvement in island life during his formative teenage years allowed him to develop a fuller understanding of island life that is reflected in his writing. "The Boat" is written from the perspective of a professor recalling his island childhood experiences. One of the themes that becomes apparent from this story is that life and livelihoods in small island communities revolved around the sea and, as the title suggests, the fishing boats that connected families to the sea.

> My earliest recollection of my mother is of being alone with her in the mornings while my father was away in the boat. She seemed to be always repairing clothes that were "torn in the boat," preparing food "to be eaten in the boat" or looking for "the boat" through our kitchen window which faced upon the sea. When my father returned about noon, she would ask, "Well, how did things go in the boat today?" It was the first question I remember asking: "Well, how did things go in the boat today?" "Well, how did things go in the boat today?"(A. MacLeod 2000, 3)

The story also shows the real life intergenerational changes that have taken place in many of these small island fishing communities as the younger generation figuratively and literally drifts away from the traditions that linked their ancestors to the island and their fishing-based way of life. "My mother had each of her daughters for fifteen years, then lost them for two and finally forever. None married a fisherman" (16). The tension in this transformation of a community way of life was also played out in the differences in attitude between the father and mother. When the young boy in the story was told by his father to go back to school rather than continue helping him in the boat, the narrator notes, "The next morning I returned to school. As I left, my mother followed me to the porch and said, 'I never thought a son of mine would choose useless books over the parents that gave him life'" (18). The tension in the relationship between islanders and outsiders, a prominent social theme regarding small islands to be explored later in this book, is also played out in this story. "The restaurant was run by a big American concern from Boston and catered to the tourists that flooded the area during July and August. My mother despised the whole operation. She said the restaurant was not run by 'our people,' and 'our people did not eat there, and that it was run by outsiders for outsiders'" (10).

> AS WAS THE CASE WITH THE EARLIER ADVENTURE NOVELS, MANY EXAMPLES OF TWENTIETH-CENTURY FICTION HAVE USED THE CONCEPT OF THE ISLAND AS A LABORATORY TO RECORD THE TRANSFORMATION OF THE HUMAN PSYCHE.

As was the case with the earlier adventure novels, many examples of twentieth-century fiction have used the concept of the island as a laboratory to record the transformation of the human psyche. William Golding's (1954) *Lord of the Flies* is perhaps the most cited example of this genre of island literature. In this novel, a plane crash leaves a group of British school children alone and isolated on a Pacific island. Because they are in this bounded space with no outside interference, we see how social relations deteriorate from a civilized beginning to a form of tribal primitivism. In so doing, this is another example of an island initially perceived of as a utopia becoming a dystopia for those marooned on it. Another critically acclaimed novel of the twentieth century that deals with the changing psychology of an individual on a small island is John Fowles' *The Magus* (1966). In this book the setting is the small fictional island of Phraxos off the coast of Greece, based on the author's earlier experiences on the island of Spetses, Greece. A young Englishman (Nicholas Urfe) who goes there to teach English is caught up in a psychological game of wits (i.e., the "godgame") with a resident of the island, a former Nazi collaborator. As we have now seen with many other island-based novels, the protagonist experiences a sense of self-discovery during his period on the island. This theme of island as a psychological laboratory is also evident in Dennis Lehane's *Shutter Island* (2003), the setting of a fictional hospital for the criminally insane. Readers might be more familiar with the 2010 film adaption of this novel starring Leonardo DiCaprio. Le Juez and Springer (2015, 1-2) suggest that works like *Lord of the Flies* and Shakespeare's *The Tempest* are really literary devices where islands are portrayed as "false havens where conventional laws and moral codes are put to the test."

We can't leave the literature of the twentieth century without commenting on the series of children's novels that are synonymous with my adopted home of Prince Edward Island, Canada. Lucy Maud Montgomery's (1908) *Anne of Green Gables* is set at the beginning of the twentieth century in a rural farming community on the north shore of Prince Edward

Island, Canada. It follows the story of Anne Shirley, a precocious eleven-year-old orphan who is adopted into the home of an elderly brother and sister. This series of books has sold more than fifty million copies worldwide and has been translated into more than thirty-five different languages. As of the early 1990s, *Anne of Green Gables* had been read by more people than any other Canadian-authored book (Baldwin 1993) and inspired a considerable tourist industry on PEI, especially among visitors from Japan (Squire 1996). There are several aspects to the literary *Anne,* as well as Montgomery's other novels set on PEI, that parallel the social and natural context of other islands. In Edward MacDonald's (2011, 75) words, Montgomery's novels serve as refuges for readers, "a terrain of human relationships where happiness and happy endings are possible while the other is a physical landscape where beauty flourishes and is shielded." In fact, Baldwin (1993) argues that one of the reasons that Anne may be popular in Japan is that both islands (PEI in the early 1900s and Japan after the Sercond World War) consisted of social contexts where you have to live peaceably with your neighbours by sacrificing your personal desires for the good of the community. Along with a powerful sense of community that is so often reflected in island literature and poetry, Japanese readers saw in Anne someone who had a close human connection with and idealization of nature.

Islands as Represented in Poetry

How important are islands to the expression of poetry? As was the case with island literature, Pete Hay (2003a) is of the opinion that island arts in general, including poetry, are relatively insignificant in the global cultural industry as a whole. Poetically, he refers to island arts as "eddies in the greater swirl of world culture" (Hay 2003a, 553–4). At the same time, there are many examples of poetry and literary writing that emerge from islands, either from island-based poets or from those who visit islands and find them distinctive and remarkable. As one example, the British Broadcasting Corporation (BBC) News Magazine reported that Iceland has more writers, more books published (one for every ten people), and more books read per person than anywhere else in the world (R. Goldsmith 2013). What is it about islands that invite this creative streak? Is it that the social and natural circumstances faced by islands and islanders, from exposure to natural catastrophes, oppression as a result of colonial rule, and the diverse physical environments all lend themselves to expression in the form of poetry? Philip Conkling (2007) has a unique perspective on this connection between islands and poetry. He says that poets are used to expressing themselves in fewer words (he refers to it as "a sparse form") so they have a natural affinity to small islands that, by their physical geography and boundedness, serve to concentrate space.

It may be that there is a spiritual quality to small islands that evokes a need for those who are artistically inclined to express themselves. A self-reflective passage in an article written by Conkling (2007, 195) may be a good example of how these kinds of experiences may contribute to poetic creativity. He recounts emerging from his tent on a cool, fog-shrouded morning on an island off the State of Maine:

> I was staring at two magnificent bald eagles: one a female, slightly larger than its mate that trailed just behind her. They flew by, wingtip to wingtip, out of the fog—as startled to see me as I them, and then careened sharply and were gone. I heard, and in the damp air imagined I felt, the rush of heavy air from their wings on my face. Although I'm not sure I knew it then, in that moment my life changed forever.

As suggested from that quote, and like the island adventure novels, many people believe islands have a transformative effect on their souls or psyches. The following short passage from Prince Edward Island author David Weale (1991, 81), in reference to his ten-year-old son during a drive along the shoreline, reinforces this belief:

He sat quietly in the seat just looking out the window at the passing landscape and seascape. I turned to look at him several times, but he didn't even notice. He was absorbed in the looking. Then it occurred to me what he was doing. He was taking in the landscape. He was, if you will, ingesting the Island. And that is exactly what happens when you live here for long—you take the Island inside, deep inside. You become an Islander, which is to say, a creature of the Island.

Is it any wonder then that so many people try to convey these emotions in whatever means possible, including words, music, and visual art?

Perhaps one of the most well-known phrases in English language poetry referring to islands is not really about islands at all. Instead, it uses the island as a metaphor to describe the human condition. This is, of course, the famous "No Man is An Island" poem by English poet and cleric John Donne (1572–1631), where the mutually beneficial relationship between an individual and society is compared to the relationship between an island and the nearby continent.

No man is an island entire of itself; every man is a piece of the continent, a part of the main; if a clod be washed away by the sea, Europe is the less, as well as if a promontory were, as well as any manner of thy friends or of thine own were; any man's death diminishes me, because I am involved in mankind. And therefore never send to know for whom the bell tolls; it tolls for thee. (Donne 1988, 5)

As with all poetry, it is difficult to suggest that poetry about islands takes any one form. Some island poetry is confrontational and abrasive, and may have been motivated because the island and/or the author were under the threat of obliteration by adjacent cultures (Hay 2003). These lines from Tasmanian poet Mary Kille (2011), from her poem "Under the Cape," address the loss of Tasmania's Aboriginal culture when European settlers arrived:

A hundred years of sweat
and unremitting toil,
when first the Cape was cleared with ropes and chains
and bullock teams, and men with whips,
as monumental stringybarks were felled and rolled
over the cliffs and crashed and splintered
on the grey volcanic rocks,
and so into the sea below.

The European eyes saw livelihoods
and profit, and a way of life
rewarding their prodigious industry.

And over many years the farmers set alight
the fringes of the Cape,
burning the blackwood and bursaria,
the dogwoods, daisybushes,
clematis and fireweed,

and the fragrant gums.

Was this a cleansing,
an erasing of the presence,
over thirty thousand years,
of those who'd lived beneath the Cape?
A way of holding back the void?

So little's left of Tommeginer time.

Other island poetry celebrates island life, and may be a reflection of the strong social bonds that are formed and the beauty of the natural landscapes. For example, the poem "Why I Stay" from PEI poet Jane Ledwell (2005, 15), encapsulates the tension between the two:

Why I Stay

because dunes move in stubborn rusting away from rock,
their isthmus kiss closing the mouth of a brackish bay

because at the island's slender throat, strait and gulf can almost
see each other, and desire will someday disintegrate to touch

because the pond's shadow darners weave evening out of trees
and mend the dark path to where northern lights burn rare and possible

because the herons build the marsh from twigs and brine
and sandpipers all together fear the sea's slenderest footfall

because my lungs are calloused with the work of this air,
my hands bruised with the work of this water

because I have already crushed too much stone to red dust
under my feet and I wear all this sand on my tongue

This short untitled poem by Jane Ledwell's father, poet Frank Ledwell (2002, 167), from his book *The North Shore of Home* also expresses the deep, sometimes pathological, connection some islanders have to their island homes:

For most
of us here
being islanders
is a ter-
minal condition.
But those who
go away
aren't cured.
They simply die
of the same
ailment
on alien soil.

Some island poetry uses the island as a metaphor, by using the isolation of island living to suggest that one might live mindfully, as exemplified in these lines from *The Grey Islands* by Newfoundland's John Steffler (1985, 9): "The island floating ahead of me like a moon, tugging me forward. Whatever it has in store … A way to corner myself is what I want. Some blunt place I can't go beyond. Where excuses stop." Still other island poetry embodies a sense of sadness or remorse that may echo the challenges currently facing many islanders and small islands. PEI poet Laurie Brinklow addresses loss in these lines from her poem "The Language of Seashells" (2012, 63): "Where the absences make you ache and you're forever reminded of them as you walk the bush or shore." Finally, the form of poetry known as haiku has been used to express the loss of those who no longer live on islands, as in Mike Hauser's (2016, n.p.), "There is an island that keeps calling out to me leave it all behind."

There are now several websites that have gathered together many examples of poetry on islands or by islanders. These also allow poets to disseminate their own work to a global audience. In the field of island poetry the most prominent of these websites is poetrysoup.com. An excellent source of contemporary poetry, artwork, music, and essays on island topics is also found on the website of the online magazine *The Island Review* at theislandreview.com. Established in 2013, the home page describes the magazine's mission as "a haven in the vast and stormy online ocean. We bring together great writing and visual arts from islands all over the world, and provide a second home for islanders and island lovers everywhere." By accepting contributions of stories, poetry, visual art, and music from a grassroots audience, the magazine attempts to provide a voice for islanders and those who have a passion for islands to connect with one another. A good example of this connection is from a 2013 posting that includes an interview with Scottish singer and songwriter Roddy Woomble, lead vocalist for the band Idlewild. He and his wife moved to the island of Mull in the Inner Hebrides to write and record music. In the interview, he expressed an opinion shared by many islanders to the inevitable question about a musician's ability to carry out his/her profession in a peripheral location. His response was, "This whole idea of being 'remote' is a strange one—remote from what?" (*The Island Review 2013*). Although a newcomer to island life, this statement suggests that Woomble has already grasped what many islanders have always felt: that although the world may view their islands as remote, inaccessible, and an impediment to functioning in the contemporary world, more often than not islanders do not share this opinion and even fail to understand the rationale for the question. To them, the island is the centre of their lives and provides everything they need.

THE MUSIC OF ISLANDS

The example used above suggests that poetry and music lyrics related to islands are intertwined. The concept of "island music" can be viewed in several ways. From one perspective, music and the creative arts in general seem to be a very large part of everyday life in many island communities. Whether this takes the form of the written word (e.g., storytelling), singing, dancing, or the visual arts, when island families and communities come together, the creative arts often play a prominent role. It is argued that there is a very strong relationship among music, identity, and place on islands, whether this is the Balinese musical culture on Lombok Island, Indonesia, Okinawan (Japanese) folk music, or Celtic music and dancing on

Cape Breton Island, Canada (Hudson 2006). The following quote, taken from an interview (Linnane 2000) with band member Paddy Moloney (in Irish *Pádraig Ó Maoldomhnaigh*) from The Chieftains eloquently expresses the importance of Celtic music to an (Irish) island community: "The music belongs in the various districts and the people. If you go to [County] Kerry, well, you automatically go to the middle of the town you can smell it, it's there."

From another perspective, some of the most recognizable and influential types of music worldwide have originated on islands or regions of islands and have then dispersed around the world. In this section, I am going to touch on three examples: Celtic music from Ireland and other Gaelic-speaking areas of the British Isles, reggae from Jamaica and the rest of the Caribbean, and Polynesian music from island regions of the Pacific, including so-called "Hawaiian" music. One of the commonalities of these three genres of music may be that they are conveying stories of community and place. Therefore, it should come as no surprise to see that they have emerged from islands that have strong storytelling traditions.

Although much of Celtic music is now being produced outside of the British Isles, it is still most closely identified with the island of Ireland (or *Éire* in Irish Gaelic) and Scotland, located in the northern part of the larger island of Great Britain. June Sawyers (2001, 8) argues that the roots of Irish music are in the twelfth century when the Normans arrived on the island. Even during the centuries of English invasion and subjugation, she says that, "The people in the countryside continued to keep it [Celtic music] alive during the centuries with their love songs (the most common), vision poems (called *aisling*), laments, drinking songs, and work songs." In this way, and especially among the rural population that was still largely illiterate, music was one of the few ways to transmit community stories and serve as a symbol of resistance. In some cases, such as in Tasmania in the 1840s, music was banned from pubs (Hay 2002). Baldacchino agrees that, while serving to preserve the history and culture of island peoples, songs and singing have been very important in serving as a form of resistance to colonial regimes (Baldacchino 2011).

Some of the most evocative Celtic music and poetry is associated with the vast emigration from Ireland to the Americas and Australia that took place as a result of the Great (potato) Famine in the 1840s. In the following anonymous poem, as quoted by Sawyers (2001, 228–229), a son is asking his father why they moved from Ireland to America:

> Oh, father dear, I often hear you speak of Erin's Isle,
> Her lofty scenes and valleys green, her mountains rude and wild,
> They say it is a lovely land wherein a prince might dwell,
> Oh, why did you abandon it? The reason to me tell.
>
> Oh, son! I loved my native land with energy and pride,
> Till a blight came o'er my crops—my sheep, my cattle died;
> My rent and taxes were too high, I could not them redeem,
> And that's the cruel reason that I left old Skibbereen.
>
> Oh, well do I remember the bleak December day,
> The landlord and the sheriff came to drive us all away;
> They set my roof on fire with their demon yellow spleen,
> And that's another reason that I left old Skibbereen.

The father did not want to emigrate from Ireland but he felt compelled to do so by circumstances beyond his control. This also suggests that the next generation, born and/or raised in the New World, is already losing that connection to their ancestral island home. Ironically, by expressing these sentiments in verse and lyrics, the music and words may help overcome this intergenerational loss of island memories.

With its origins in Jamaica, some have argued that reggae music was one of the most important cultural music symbols in the late 1960s and early 1970s, with Dagnini (2010) calling it a "cultural bombshell" and Keith Nurse (2007) describing it as one of the major genres of music. It has become one of the most, if not *the* most, familiar diasporic forms of music and has proven to be one of the most important elements in the Jamaican tourist industry (Connell and Gibson 2004). For example, it has been estimated that the Caribbean music industry has generated between USD 170 and 210 million in foreign exchange earnings and, in Jamaica alone, employs 15,000 people (K. Nurse 2007). Both the Jamaican government and the travel industry use reggae as an important part of their tourism marketing, as seen in the January/February 2010 issue of *Islands Magazine*. An article called "The Beat of Jamaica" brought a pop expert to the streets of Jamaica to showcase the origins and influences of reggae.

Part of the explanation for the growth of this genre of music outside of the Caribbean may be very similar to the relationships between the Irish diaspora and global reach of Celtic music: there was a large immigrant population living away from Jamaica that used music as a way to reconnect with their former island culture. In the case of reggae, this first- and second-generation Caribbean population was concentrated in the working-class neighbourhoods of London and southern England. This largely poor, disaffected, and younger group adopted this counterculture music and its iconic leader Bob Marley, and spread it to a larger audience, in turn influencing the growth of the skinhead and punk movements. An ongoing expression of the popularity of reggae and related music is reflected in music and cultural festivals in places that have a large Caribbean diaspora presence. For example, the annual Caribana Festival in Toronto, Labour Day carnival in New York and in Notting Hill, London, collectively draw 6.5 million attendees and have expenditures, respectively, of CAD 200 million, USD 300 million, and GBP 93 million (K. Nurse 2007). Connell (2011a) says that we should not be too surprised at the widespread popularity of reggae music, as well as Celtic and Polynesian music and dance. After all, singing about the homeland undoubtedly becomes more important for those islanders separated from their island birthplaces.

Figure 3.4   Elvis Presley poster Paradise - Hawaiian Style.
Source: **Chris Light**

These dual themes of the significance of locally produced music as part of the everyday lives of islanders and as a worldwide genre of music affiliated with a group of islands also applies to the music and people of the Pacific Islands. For example, on the Cook Islands, Alexeyeff (2004, 149) says that "island music is everywhere. The radio station, Radio Cook Islands, is turned up loud in most households, workplaces, and cars." As is the case on other islands, music in the Cook Islands is an integral part of functions such as weddings and community fundraising events. As is the case with other islands that have large diaspora populations, the many emigrants from the Cook Islands who now reside in New Zealand and Australia remain connected to their homeland through island music.

The argument has also been made that the music and dance as practiced by Polynesians on Tahiti, Tonga, Hawai'i, and New Zealand, and recorded in detail during the voyages of British Captain James Cook, influenced the European romantic composers of the early nineteenth century, particularly in

Figure 3.5 Dancers of the "Kodak Hula Show." *Source:* **Dennis Sylvester Hurd, public domain**

the adoption of exotic and mystical themes (Kaeppler 1978). As well, there has been a reciprocal impact on Polynesian music and dance from Europe and elsewhere. For example, the introduction of Protestant hymn tunes as well as the guitar and ukulele into many areas of Oceania came about because of European influence. The ukulele in particular seems to be one of the most powerful musical symbols of popular tourist Hawaiian music. This New World identification of Hawaiian music in turn influenced the development of American country-western music in the early twentieth century (B.B. Smith 1983).

Some have even suggested that music, as much as pineapples, sugar, and military history, has served as one of the most powerful tools in packaging and selling places such as Hawai'i as a tourism product (J. Schroeder and Borgerson 1999). For example, album covers and posters with titles like "Paradise-Hawaiian Style" and "The Lure of Paradise" portray Hawai'i as a utopia populated primarily by scantily clad young women (Figure 3.4) .

The "Hawaii Calls" radio show (1935–1975), hosted by Webley Edwards and broadcast weekly from Waikiki Beach, Honolulu, featured a regular selection of Hawaiian singers and at one point was broadcast on over 750 stations. Many people who were raised in the latter half of the twentieth century viewed Hawaiian pop culture through the lens of two iconic musical images: the Hawaiian-born singer and piano/ukulele player Don Ho (1930–2007) singing his most well-known song "Tiny Bubbles," and the "Kodak Hula Show" (also known as the Pleasant Hawaiian Hula Show and as seen in Figure 3.5), an outdoor presentation of Hawaiian music and hula dancing that ran for sixty years and was seen in person by an estimated twenty million people.

ISLANDS IN CINEMA, TELEVISION, AND ADVERTISING

Almost all of the most compelling works of island fiction have also been made into feature-length films, sometimes several times over, allowing them to be experienced by a much larger audience. For example, the Hollywood films *Jurassic Park* and *Lost World* are derived from Michael Crichton's novels of the same name. We also have to keep in mind that television series, documentaries, and even advertising all start in some form as written works, and at the very least, as screenplays. One of the prevailing themes of these books/screenplays/films is the transformation of the relationship between the "outsiders" and the island as it evolves from a place of paradise/utopia to a place of danger/dystopia. William Golding's *Lord of the Flies* has already been mentioned in this regard, having been made into a feature length film in 1963. The film *Cast Away* starring Tom Hanks is also a good example of this theme. The premise of this film is that the main character (Chuck Noland) survives on a desert island after his plane crashes in the South Pacific. As a result of his experiences on the island, he appears to become more self-reflective and is better able to communicate his feelings, both to inanimate objects while he is on the island (i.e., "Wilson" the volleyball) as well as his co-workers and former fiancée after his rescue. The trailer for the film foreshadows his personal transformation after being washed ashore by stating that "The end of a man's journey will become the beginning of his life."

> ONE OF THE PREVAILING THEMES OF THESE BOOKS/SCREENPLAYS/FILMS IS THE TRANSFORMATION OF THE RELATIONSHIP BETWEEN THE "OUTSIDERS" AND THE ISLAND AS IT EVOLVES FROM A PLACE OF PARADISE/UTOPIA TO A PLACE OF DANGER/DYSTOPIA.

The other popular way in which islands have been used in screenplays and books is as a place of forced exile, i.e., the "prison island." Rather than being accidentally confined by a quirk of fate or by nature (e.g., a plane crash or a shipwreck), in these cases islands have served as the settings for intentional confinement. As the site of a maximum security prison from 1934 to 1963, the small island of Alcatraz in San Francisco Bay has proven to be a popular setting for this theme. Now a museum, the island prison was the setting for several Hollywood films, including *The Birdman of Alcatraz* (1962) directed by John Frankenheimer and starring Burt Lancaster, *Escape from Alcatraz* (1979) directed by Don Siegel and starring Clint Eastwood, and *The Rock* (1996) directed by Michael Bay and starring Sean Connery and Nicholas Cage. Documentaries have also focused on Alcatraz, including the 2014 National Geographic documentary *Alcatraz: No Way Out"* and the 2015 BBC documentary *Prison Escape from Alcatraz.* Although the island was very close to urban San Francisco, the prison was reputed to be impossible to escape from at least partly because of the cold temperatures and strong currents associated with the surrounding water. This image of Alcatraz Island reinforces this strange duality about islands and connectedness; many of them are isolated and disconnected from the continents but they are also not too far from our consciousness.

The other book and film that made an island infamous as a prison was the notorious "Devil's Island" off the coast of French Guiana in South America. Based on the true-life experience of Rene Belbenoit, who was an inmate for fifteen years before escaping, it was later fictionalized in the book *Papillon* (1970) by Henri Charrière and then made into a film starring Steve McQueen and Dustin Hoffman.

It is not just works of fiction that have portrayed islands in cinema. There are many real life incidents, written in the form of non-fiction, that have been retold in the form of film. For example, *Mutiny on the Bounty* tells the story of the eighteenth-century mutiny against the infamous Captain William Bligh by his crew and the eventual settlement of the mutineers, first on Tahiti where they took a number of Tahitian wives/mistresses and then on Pitcairn Island. The other notable non-fiction book made into movie form is *Kon Tiki,* a story by Norwegian explorer and ethnologist Thor Heyerdahl, who travelled on a balsa wood raft from Peru to Raroia on the Tuamoru Islands in an attempt to prove that it was technically possible for Indigenous peoples of South America to have travelled to and populated islands in the south Pacific. One of the criticisms of the approach taken by Heyerdahl, in both the book and the film adaptations, is that it implies the discovery of islands in the Pacific by South American Indigenous peoples was more accidental than intentional. As we shall see in the next chapter, this accepts the narrow Westernized view of oceanic islands as being small dots in the middle of an expansive body of water. In reality, Polynesians were excellent seafarers who moved between islands throughout the Pacific for hundreds of years prior to European contact.

Many twenty-first-century youth have an image of Indigenous Oceanic culture as defined by Disney's 2016 animated film *Moana,* a story of a young Polynesian princess who appeases the ocean spirits by returning a heart to the goddess Te Fiti. Unlike the criticism surrounding many of Disney's earlier animation stories, this film was praised for infusing Indigenous stories, opening up new ways of thinking of the stereotypical tropical island paradise, using actors of Polynesian descent for almost all of the characters, informing a new generation about the complexity of Polynesian wayfinding, and consulting with an Oceanic Story Trust that included Pacific Islanders (Hamasher and Guedes 2017; Tamaira and Fonoti 2018). Others have objected that the film is yet another example of Western corporate cultural appropriation, a romanticization of the primitive, and that it reduces the culturally diverse and vast Pacific region into one homogenous entity (Diaz 2016; Ngata 2014; Yoshinaga 2019). Herman (2016, 35) is aptly quoted as saying, "having brown advisors doesn't make it a brown story. It is still very much a white person's story."

Some of the earliest examples of the currently popular genre of reality television shows employed islands as an integral part of the production. The longest-running English-language example is the Columbia Broadcasting System's *Survivor,* having started in 2000 and with thirty-nine seasons as of September 2019). Its precursors (e.g., *Expedition Robinson* in Sweden) and its imitators and successors (e.g., *Shipwrecked,* a Survivor-like show for teens; *Temptation Island,* with a dating and relationship dimension; the British Broadcasting Corporation's *Castaway 2000,* set in the cold-weather environment of the Outer Hebrides, Scotland, and the most recent *Love Island,* set on Mediterranean island of Mallorca) used the same formula. This island-based reality show model has been reproduced for television audiences in a similar format across many other non-English settings, including *Expedition Robinson* in many Scandinavian and northern European countries, *L'Isola dei Famosi* (The Island of the Famous) in Italy, and *Koh-Lanta* (named after an island off the coast of Thailand, the site of the first episode) in France. The boundedness of the island context, where the groups have to compete or collaborate to win, with no help from the outside world, is critical to the image portrayed on these shows. The twinned and sometimes contradictory notion of islands as places of danger and paradise is also an important part of many of these reality shows.

Two of the most popular island-based English-language television series are the situation-comedy *Gilligan's Island* (1964–1967) and the more recent dramatic series *Lost* (2004–2010). Although as a television series it lasted only three seasons, *Gilligan's Island* is arguably one of the most recognizable representations of outsiders' experience of island life from twentieth-century cinema and has become one of the most syndicated English-language television shows. For those not familiar with the premise of the show, a group of seven Americans is marooned on a desert island. Each episode of this situation comedy sees the group encounter new opportunities and obstacles to getting off the island. Morowitz (2010) has suggested that the show was about much more than "goofball humor." As with Defoe's *Robinson Crusoe* and the novels of Jules Verne, she says that this TV series was also about the desire to escape from Western civilization, the inability to really leave that world behind, and the quest to remake their (island) world in the image of the colonial society that they had left behind.

THE OTHER POPULAR WAY IN WHICH ISLANDS HAVE BEEN USED IN SCREENPLAYS AND BOOKS IS AS A PLACE OF FORCED EXILE, I.E., THE "PRISON ISLAND." RATHER THAN BEING ACCIDENTALLY CONFINED BY A QUIRK OF FATE OR BY NATURE (E.G., A PLANE CRASH OR A SHIPWRECK), IN THESE CASES ISLANDS HAVE SERVED AS THE SETTINGS FOR INTENTIONAL CONFINEMENT.

As was the case with many of the other island-themed works of fiction, the more recent television series *Lost* is as much about the turmoil and psychological transformation of the characters individually as it as about the group, with the bounded island serving as a backdrop. As such, and in terms of the ways it uses the island, it is not much different from the *Survivor* reality shows. As those who have seen the conclusion to this series can attest to, not only does the island serve as a psychological laboratory where the human condition is explored, but the island is the experiment. In Heidi Scott's (2014, 656) words, "islands are testing grounds of the individual mind and body. This holds on evolutionary, ecological, psychological, and theological levels." With all of these examples, we are reminded once again of how many fictional portrayals of islands intentionally or implicitly use the theme of an island as a closed system or social laboratory.

One of the early examples of a television series that combined the themes of "island as fantasy" and "island as transformation" was the American Broadcasting Corporation's *Fantasy Island* (1977–1984) where host Ricardo Montalbán greeted special guests weekly to his mysterious island, inviting them to live out their dreams. In most of the episodes the guests were forced to engage in self-reflection, leading to personal transformation, before realizing their dreams. In so doing, each of the episodes became mini-versions of the transformations that were being undertaken over a longer time period by Robinson Crusoe, Chuck Noland in *Cast Away,* and the young boys in *Lord of the Flies.*

Speaking largely from the perspective of twentieth-century cinema, Brislin (2003) suggests that Pacific Islanders are stereotypically portrayed in very few ways. They may be pleasant but ignorant natives; savage cannibals eventually overcome by superior Western firepower; sexy, uninhibited women willing to participate in sexual liaisons with Western men; or self-inflated and comical men who are easily fooled by Western men of superior intellect. Hawai'i in particular has been the focus of a large number of Hollywood-inspired television series that have employed these stereotypes, including two iterations of *Hawaii Five-0* (1968–1980, 2010–) and *Magnum P.I.* (1980–1988, 2018–). In Brislin's (2003) opinion, the islands and islanders in these shows are really just colourful backdrops, while the real focus is on the island visitor. The shows' worldview is one where the Western man (almost exclusively man, not woman) can play out his exotic and erotic fantasies on a stage where normal rules are suspended. While these observations were intended to apply to Pacific Islanders, it could be argued that these derogatory stereotypes have been applied in varying degrees to many other island peoples. Whether it is the "Newfies" of Newfoundland, Canada; the "kanaka," a historical term meaning "person" or "man" used by Europeans to describe Pacific Islanders (Keown 2007) and Pacific Island labourers; or the "sheep-shagger" used to describe Manx people (i.e., from the Isle of Man), there are numerous examples where islanders are viewed as slower and more dim-witted than their mainlander counterparts.

Many islands, and especially those island companies and government departments associated with the tourist sector, continue to perpetuate the stereotypes and myths of islands by branding themselves as tropical paradises in order to attract tourists and tourism-related investment to their shores. This has especially been the case for the Caribbean islands, where the public and private sectors have worked together to invest in public infrastructure and private hotel and resort facilities and have then flogged their islands as vacation destinations (B. King 1997).

Island branding has also extended to the real estate sector. One example of this form of island advertising comes from the website Private Islands Online and the *Private Islands* magazine, a print and online magazine featuring private islands for sale around the world. It also serves as a real estate brokerage firm (see privateislandsonline.com). The website describes 667 islands for sale and the online magazine uses the subheading "A Lifestyle for the Independent-Adventurous Personality." Another example is the Home and Garden Television (HGTV) network's show *Island Hunters*. Each episode has host Chris Krolow, who is also the publisher and CEO of *Private Islands* magazine, helping people "find their exotic dream property—their own private island."

Islands and Documentaries

For many reasons, islands have been a popular setting for documentaries, many of which focus on the uniqueness, fragility, and vulnerability of island ecosystems and societies. The Galápagos Islands in particular seem to be a favourite site for these film-based enquiries. For example, the three-part series produced by the British Broadcasting Corporation (BBC) describes the natural history of these islands through "Born of Fire," "Islands that Changed the World" (in reference to the Charles Darwin's observations), and "Forces of Change." In the documentary *Galápagos* you can "Follow the filmmakers from the Smithsonian Institute on a visual journey through the lush Pacific Ocean paradise that is home to some of the most precious flora and fauna on the planet." Or, from *Galápagos: Beyond Darwin*, you can "Explore the fascinating world of the Galápagos archipelago that Charles Darwin couldn't explore in his 1835 visit."

The impacts of climate change, and particularly sea-level rise, have provided a new subject area for island-based documentary films. Alexa Weik Von Mossner (2015) compares three documentaries that focus on the subject of climate change and small islands: Briar March's *There Once Was an Island* (2009), Paul Lindsay's *Before the Flood: Tuvalu* (2004), and Jon

Shenk's *The Island President* (2012). *The Island President* in particular seems to have been more successful in capturing the attention of a global audience. It follows the actions of Mohamed Nasheed, at that time the President of the Maldives (an archipelago in the Indian Ocean), as he travels to the United Nations' Copenhagen Climate Summit in 2009 and other international meetings pleading with the leaders of other nations to curb their consumption of fossil fuels before they risk the very survival of his country. In one speech, he states, "For us, this is more than just another meeting ... This is a matter of life and death." It may be that part of the reason that these kinds of documentaries have generated so much public attention is that it begins to unravel the prevailing myth about the pristine purity of the tropical island paradise while still conveying the theme of island vulnerability.

ISLANDS AND THE VISUAL ARTS

In the earlier section on island music, the point was made that music is a large part of the everyday lives of many islanders. The same could be said for the visual arts. The making of art seems to be more significant on small islands. Brinklow (2013, 40) says that this is true not only for those born on islands but also for those mainland-born artists who move to islands to practice their craft. In her words, "the creative spirit finds resonance in the state of being islanded. Artists find they make art they otherwise would or could not by living on an island."

Some of the most enduring collections of island visual art are represented in the drawings and paintings of early European explorers. In particular, the natural historians, botanists, and draughtsmen assembled large collections of drawings and paintings, not only of the flora and fauna but also of the landscapes and the people. Bernard W. Smith (1985) shows us many of these pieces of art associated with the Pacific islands, from the mid-1700s to the mid-1800s.

One of the most celebrated European artists whose work is associated with islands was the French post-impressionist Paul Gauguin (1848–1903). Gauguin first visited Tahiti in 1891 and lived there for most of the rest of his life. As is the case with many other island visitors, he came with a stereotype of Tahiti as a utopian paradise. He is quoted as saying, "I am going soon to Tahiti, a small island in Oceania, where the material necessities of life can be had without money....The Tahitian has only to lift his hands to gather his food, and in addition he never works" (Willumsen, as cited in Morowitz 2010, 4). As seen in Figure 4.6, Gauguin's paintings during this period are dominated by exotic, young, nude or partially nude Tahitian women. This subject matter likely contributed to reinforcing the stereotype in Europe of Pacific island women, and especially Tahitian women, as being sensuous and promiscuous. Once Gauguin moved to his island paradise, he quickly discovered that the main city of Papeete on the island of Tahiti is far from a utopia, noting in his writing that the city had already been corrupted by Western greed and consumption. Therefore, in order to find his utopia he moves to increasingly smaller and more remote islands within the larger French Polynesian archipelago. There is an interesting parallel here between Gauguin and D.H. Lawrence's fictional protagonist Cathcart in "The Man Who Loved Islands." In both cases, the central characters keep relocating to increasingly smaller, more remote islands, assuming that this farther distance from the impurities associated with mainland civilization will eventually lead them to their utopias.

Given the colonial impact on islands, many of the English-language impressions of the people of French Polynesia and other islands of Oceania have been influenced by Western artists. The reality is that this region consists of more than 20,000 islands and islanders who have lived there for thousands of years and who speak more than 1,800 different languages (Kjellgren 2007). Although much of the pre-European Oceanic art has been lost, sculptures and ceramic objects from the Lapita people originating in Southeast Asia still exist. One of the differences between "Western" art and Indigenous art of Oceania is that the latter is evaluated not just on the aesthetic enjoyment of the viewer but also by the relationship of the object to the artist and the context within which it is employed. Fischer (2005) calls Polynesian art "political art" in that it was intended to reflect the existing culture's belief systems and uphold the sacredness of ruling authority. Objects were intended to be used, whether that was for ceremonial purposes or within the household. As an example, much attention has focussed on the

Figure 3.6 Paul Gauguin's *Te aa no areois* (*The Seed of the Areoi*), 1892. *Source:* **public domain**

Mo'ai "heads" of Rapa Nui/Easter Island (Figure 3.7). Built primarily from the twelfth to the fifteenth century, the more than 900 statues measuring two to ten metres high were likely constructed to pay homage to gods or high-ranking ancestors (Fischer 2005; Kjellgren 2001).

Another form of visual art that incorporated islands is by the well-known environmental artists Christo and Jeanne-Claude. For a period of two weeks in May of 1983, the much publicized artists created and displayed a larger than life piece of art entitled *Surrounded Islands*. This was a very public display, consisting of 6.5 million square feet (or 603,870 square metres) of pink polypropylene fabric surrounding twelve islands in Biscayne Bay off the coast of Miami, Florida in the United States. The artists explained that the intent of the artwork was to reflect Miami and the way in which the people of the area lived between land and water. In a commentary, Peckham (2002) quotes the artists as saying that this work of art was supposed to reflect that islands are always enveloped by a sea of political and cultural contexts.

CONCLUSIONS

This discussion of art, literature, music, and other media suggests that islands are the homes of incredibly talented and creative peoples. It could be said that this is bound up in the heritage and collective history of the peoples of islands, many of whom were displaced from distant continents and sought out the arts as a way of remembering and passing on these memories to the next generation. We need to remind ourselves not to impose a Westernized, largely non-islander perspective on the role and value of storytelling, whether those stories are in the form of the written word, music, or film. For example, storytelling among Indigenous peoples of the Pacific islands is about much more than entertainment; it is about history, genealogy, customs, family, and land (Brislin 2003). Others have argued that there has been an explosion of artistic expression on islands in the past few decades, at least in part to assert a distinctive identity in response to a nationalist, or even internationalist expression of globalisation. Brinklow (2015) calls this "storying," or the pulling together of shared experiences firmly based in place, where the boundedness of the island provides a frame for the story.

Key Readings:

Conkling, Philip. 2007. "On Islanders and Islandness." *Geographical Review* 97 (2): 191–201.

DeLoughrey, Elizabeth. 2007. *Routes and Roots: Navigating Caribbean and Pacific Island Literatures*. Honolulu: University of Hawai'i Press.

Loxley, Diana. 1990. *Problematic Shores: The Literature of Islands*. London: MacMillan Press.

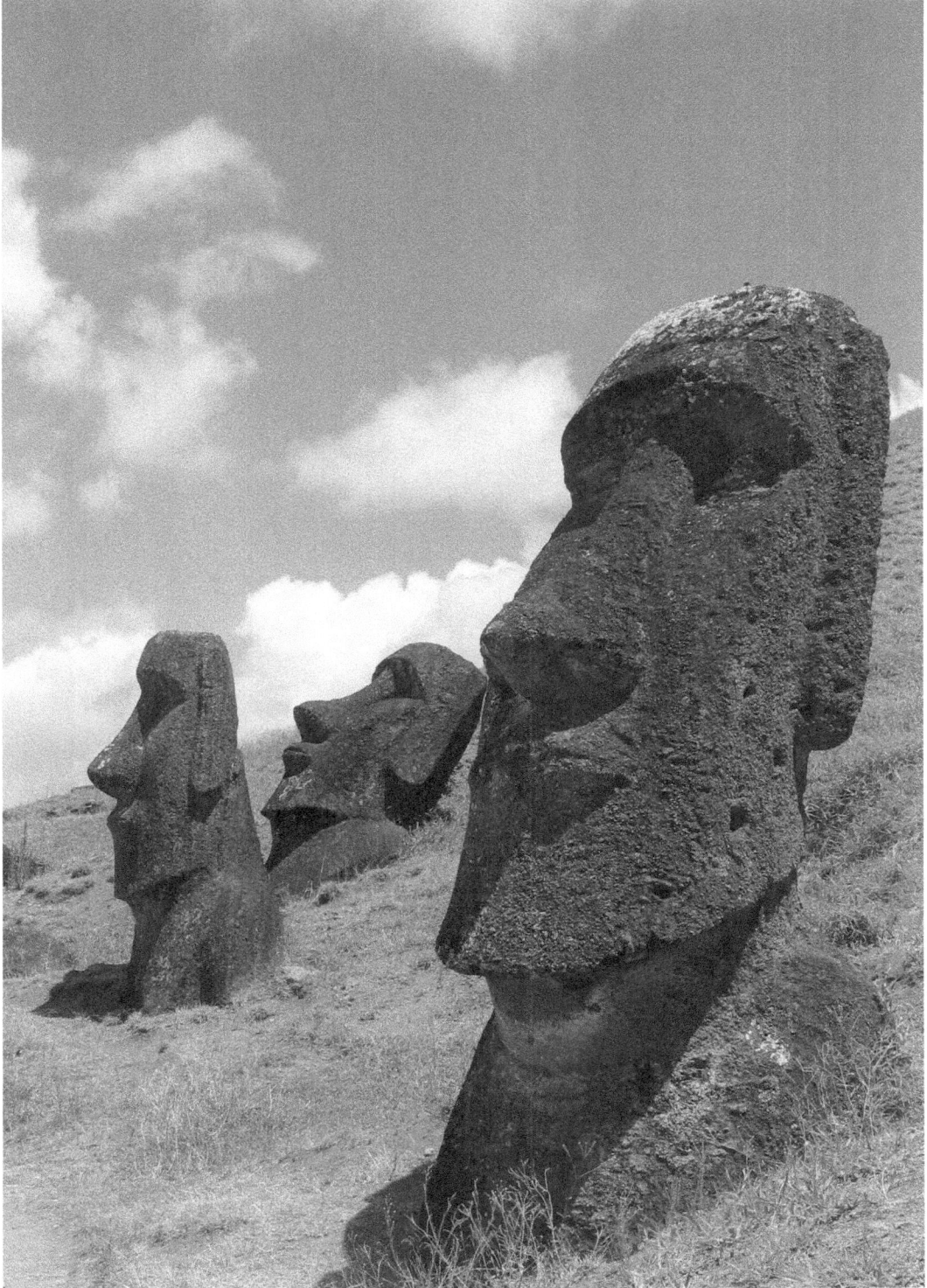

Figure 3.7 Mo'ai 'heads' of Rapa Nui/Easter Island. *Source*: **public domain**

Traditional Maori wood carved canoes at Waitangi, New Zealand. *Source*: **Patricia Hofmeester**

Chapter Four

The Settling of Islands and Indigenous-Outsider Interactions

Oceania was thus a wretched place, characterized by danger, poor living conditions, sickness, tropical torpor, degeneration, and sometimes death for white people. We tend to forget that these views, unfashionable as they are now, were predominant for well over a hundred years. (Howe 2000, 19)

INTRODUCTION AND OBJECTIVES

Settlement of the world's islands has taken place for the past 50,000 years at least in Southeast Asia and Australia, 15,000 years in the Caribbean, and 12,000 years in the eastern Mediterranean (Erlandson and Fitzpatrick 2006; Phoca-Cosmetatou 2011). There is even some evidence of earlier hominids reaching the island of Flores in the Indonesian archipelago 800,000 years ago (Morwood et al. 1998). Despite excellent oral traditions and storytelling cultures, the written evidence pertaining to the lives and livelihoods of most of this settlement has been lost. However, we can piece together an approximation of what may have taken place from the ethnography (through folk stories, traditions, artefacts, and myths passed from one generation of island peoples to the next) and also from the linguistic and archaeological record. The purpose of this chapter is to provide a selective summary of the prehistory and the history of some island settlements. Since some of the most traumatic and significant events in that history consisted of encounters between islanders and the first "outsiders," this chapter also describes some of the enduring impacts that these encounters have had on those islands and islanders.

Before embarking on this journey, it is useful to stop for a moment so that we might better understand the differences between prehistory and history. The word history comes from the Greek "historia," meaning inquiry, but which came to be interpreted as the existence of a written or recorded account of events. So the written word, as well as hieroglyphs, cuneiform, Greek, Latin, etc., all allowed events and stories to be recorded and passed on from generation to generation in much the same form and meaning as when they were first recorded. As such, any time before these events were recorded would be known as prehistory. Of course, this transition from prehistory to history would have taken place at different points in time for different societies around the world. As well, as a result of the contributions of researchers from many fields and the greater value and attention paid to verbal forms of communication, the line between prehistory and history has become much more blurred. For example, oral traditions that are retold in the form of stories can be a powerful and long-lived way of maintaining the cultural record of a society. The analysis of fossils and artefacts, through carbon-dating and other archaeological means, has also allowed us to better understand and interpret some aspects of ancient cultures even without the existence of the written word. Therefore, not only do we have to be cautious about the accuracy of the written word in depicting a historical event, but we must also not automatically dismiss prehistoric records and explanations just because there was no long-lived written record associated with these societies.

In this chapter and elsewhere in this book, the words "exploration," "settlement," and "discovery" have occasionally been used. In a discussion on the contact and subsequent relationship between Indigenous islanders and other groups that come after them, it is important to take a moment to question the meaning of these words. This is especially the case if the group that comes to control the terms and outcomes of the relationship is also writing the history of those encounters. So, for example, does mapping a region constitute discovery? And how do you "discover" a place that is already populated? And finally, what constitutes settlement?

Of course, these are very important questions that cannot be fully answered here. At best, we can provide part of the answer. For example, printed maps are only one way of recording and disseminating the site and situation of places. Mental or cognitive maps, with all the knowledge of space and places that are held and retold by individuals, may be just as effective in discovering, rediscovering, and understanding places. As well, the twig maps made by Polynesian navigators that included the accurate locations of many Pacific islands were an important part of the Polynesian seafaring tradition, even if they did not have the longevity and distribution of European paper maps. Also, when New World history books talk about Christopher Columbus "discovering" America, how can this be reconciled with the presence of the First Nations peoples who were already present on the Caribbean islands? And why in the public consciousness would the Vikings not be credited with their own "discovery" of North America that preceded Columbus's by almost 500 years?

Daniel Boorstin, the popular historian, may have provided us with elements of an answer when he wrote in *The Discoverers* (1983, 215), "What they [Vikings] did in America did not change their own or anybody else's view of the world … There was practically no feedback from the Vinland voyages. What is most remarkable is not that the Vikings actually reached America, but that they reached America and even settled there for a while, without discovering America." So discovery seems to be equated with both awareness and continuity of exploitation.

THE FIRST ISLANDERS

Islanders in the North Atlantic

From a European perspective, it has been said that world politics and economies have been centred around three major eras, each one based on a geographic region of water bodies and, by extension, the islands within these waters. Up until the European conquest of the Americas, Western life and trade had been centred around the Mediterranean and was known as the Thalassic period. Within the next hundred years or so, this had shifted once again, from the

Caribbean (i.e., Atlantic) to the Pacific and was known as an Oceanic era (Mintz 1991; Seeley 1883). One of the greatest changes between the first and second period was a shift in understanding. The ocean was no longer a barrier, it was now considered a pathway. However, even before this transition, the Atlantic Ocean and the islands of the North Atlantic were already being encountered, explored, and exploited by Northern Europeans, and interaction between these Norsemen and Indigenous peoples was taking place.

The first islands encountered and conquered by the Norse or Vikings from present-day Norway or Denmark were likely the Orkneys and Shetlands to the north of the Scottish mainland, and the Hebrides to the west of this region. Starting with raiding parties in the ninth century AD, the Vikings would have displaced an existing Pictish population, using settlements on the islands as bases to raid elsewhere in Scotland, Ireland, and the coast of Norway. The Picts they encountered had arrived on the Orkneys and Shetlands from elsewhere in Britain at least 5,000 years earlier. Although the current stereotype is that the Viking economy was based on raiding and plundering of nearby coastal communities, the kingdoms (or earldoms) that were created relied on a more diverse set of activities including mixed farming (grains and livestock), trapping, fishing, and trading (Barrett et al. 2000; Crawford 1987). As to what happened to the Pictish peoples, as has been the case throughout the history of islander-outsider encounters, they suffered a fate not unlike that of many other Indigenous peoples in the Caribbean and Pacific and all but disappeared.

> IT IS IMPORTANT TO TAKE A MOMENT TO QUESTION THE MEANING OF THE WORDS "EXPLORATION," "SETTLEMENT," AND "DISCOVERY." DOES MAPPING A REGION CONSTITUTE DISCOVERY? AND HOW DO YOU "DISCOVER" A PLACE THAT IS ALREADY POPULATED? AND FINALLY, WHAT CONSTITUTES SETTLEMENT?

In the islands north of the mainland of Scotland, our understanding of settlement patterns comes from treasures or hoards that were associated with burials and military excursions, and from oral histories that were later recorded in written form as the *Orkneyinga Saga*. As the Vikings moved progressively west, their oral history reflects this in the *Icelanders* and *Greenlanders Sagas*. These consist of a retelling of the lives of the Vikings who lived in Iceland and Greenland during the period 800–1,000 AD, and suggest that Iceland was settled by Norse farmers and traders in the period 860–930 AD. Later that century, overpopulation and a lack of food led to further exploration west to Greenland where a community of up to 4,000 people survived until the 1400s. In approximately 1,000 AD Leif Erikson (the son of Erik the Red) retraced the route taken by the trader Bjarni, past Baffin Island to a place they called Vinland, a "land of grassy meadows, with rivers full of salmon, and enough other resources to encourage over-wintering." The Sagas suggest, and some physical evidence corroborates (e.g., the L'Anse aux Meadows site at the northern tip of Newfoundland), that various Viking ships landed and overwintered at sites along the coast of what is now the Atlantic provinces of Canada and the New England states. Although there is no archaeological evidence to confirm the precise location of Vinland, it may have been in Nova Scotia, New Brunswick or, less likely, Prince Edward Island. However, since these visits were short-lived and did not lead to a permanent settlement of Europeans in North America, the words "discovery'" and "settlement" are less often associated with the Norse than they are with other Europeans 500 years later.

ISLANDERS IN OCEANIA

When discussing the settlement patterns and encounters in the Pacific Ocean, it is important to think about the geographical labels we commonly use. Following the rise of European economic, military, and political control in the Pacific, the region was cartographically divided into three areas. Beginning in the mid-nineteenth century, these zones were increasingly referred to as Polynesia, Micronesia, and Melanesia. As Figure 4.1 shows, the area of Polynesia consists of over 1,000 islands and formed a rough triangle defined at one corner by New Zealand, at another corner by Hawai'i, and at the third corner by Rapa Nui/Easter Island. The region of Micronesia was to the north and west of Polynesia and was bordered by the Philippines to the west and New Guinea to the south. Melanesia consisted of an arc of islands directly to the northeast of Australia in an area defined by New Guinea to the west and Fiji and New Caledonia to the southeast.

As was raised in the introduction, when translated into English, these labels roughly translate into "place of many islands" (Polynesia), "place of small islands" (Micronesia), and "islands of the black-skinned people" (Melanesia). Although they may still be useful for us to visualize broad geographic regions, they have less value in developing an understanding of distinct cultural regions. In his book *The Quest for Origins*, Kerry Howe (2003, 25) indicates that "Polynesians were regarded as relatively superior; Melanesians as smaller, darker, inferior; and Micronesians as not very important at all owing to the supposed minuscule size of their island homes." At least three criticisms can be levelled at this three-part regionalization. First,

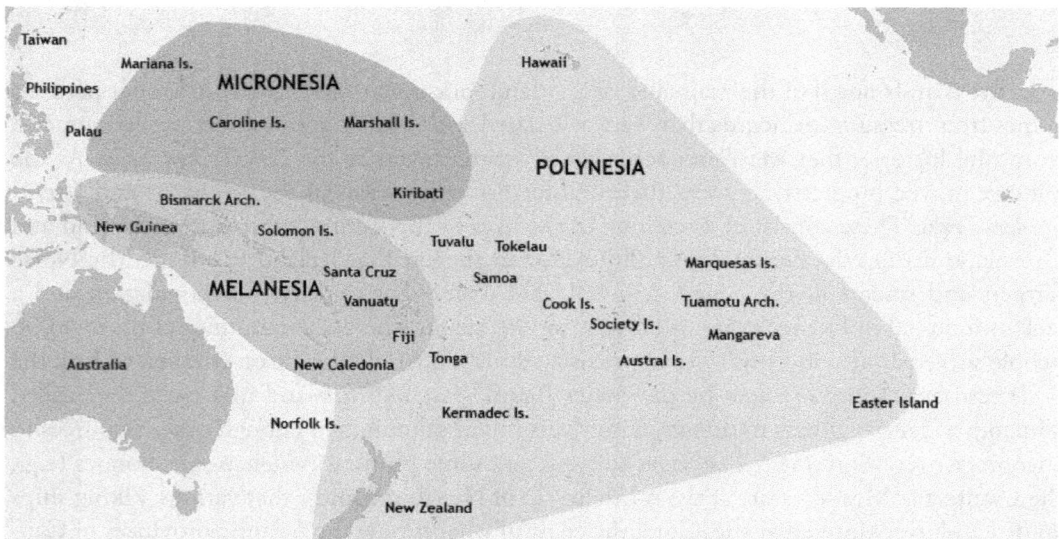

Fig 4.1   Three major cultural areas in the Pacific Ocean. *Source:* **public domain**

there is no evidence that ethnicity or race follows this categorization. Indeed, there are many cultural and racial characteristics that are shared across the regions. Second, although some Melanesians welcome an identifier that distinguishes them from others in Oceania, others are highly critical of a label that distinguishes them on the basis of the colour of their skin and denigrates them historically in comparison to Polynesians (Lawson 2013). Finally, the labels implicitly reflect an outsider's cultural perspective that the islands of the Pacific are tiny dots of land in a vast ocean, difficult to find, remote, and marginalized from the centres of civilization. In fact, the very notion of calling it a Pacific region has come to mean a vast area of sea whose boundaries are defined by the edges of the continents and where the islands contained within the region are relatively insignificant. Hauʻofa (1994, 150) suggests that this kind of regionalization reflects an economistic and geographical determinism, treating islands as if they are, "much too small, too poorly endowed with resources, and too isolated from the centers of economic growth for their inhabitants ever to be able to rise above their present condition of dependence on the largesse of wealthy nations." This dependent, marginalized world view is not something shared by many islanders themselves. As we shall see later in this chapter, for much of the period prior to encounters with outsiders, the peoples of the Pacific Ocean had a very rich understanding of their region and had the skills and knowledge to navigate throughout it in a purposeful manner. Their universe stretched as far as they could navigate and to the

> THE PEOPLES OF THE PACIFIC OCEAN HAD A VERY RICH UNDERSTANDING OF THEIR REGION AND HAD THE SKILLS AND KNOWLEDGE TO NAVIGATE THROUGHOUT IT IN A PURPOSEFUL MANNER. THEIR UNIVERSE STRETCHED AS FAR AS THEY COULD NAVIGATE AND TO THE DEPTHS OF THE SEA UPON WHICH THEY DEPENDED FOR SURVIVAL.

depths of the sea upon which they depended for survival. As such, many islanders have now rejected this three-part regionalization in favour of the more inclusive Oceania, reflecting the original path of settlement and the belief that the ocean connects all islands and all the peoples of the region. In some cases, this has been adapted to suggest that there is a Near and Remote Oceania to reflect the historical path of migration and settlement by the first Indigenous peoples. Near Oceania would include Papua New Guinea and the Bismarc Archipelago, with Far Oceania extending east to include the islands of Fiji, Tonga, Samoa, and Tahiti.

It is generally agreed that the first settlement of the islands of Oceania from mainland Asia took place approximately 750,000 to 800,000 years ago. Settlement by these predecessors to the Papuan and Lapita people initially was limited to the islands closest to the Malay Peninsula where they developed their seafaring skills, in what Irwin (1989, 168) calls a "voyaging nursery." According to Nunn (2009) and others, about 40,000 years ago they would have reached the Australia-New Guinea landmass. Evidence taken from oral histories on Taiwan and islands of the Indonesian archipelago suggest a presence of the Lapita or Austronesian people (based on their language characteristics) by about 7,000 years ago with the islands of Kiribati and Samoa being settled by between 6,000 and 5,500 AD. Islands farther to the east including Vanuatu, New Caledonia, Fiji, and Tonga were generally smaller and farther apart and therefore took longer to settle. However, by the period of 1,600 to 1,000 years ago the Hawaiian archipelago to the northeast and New Zealand and Rapa Nui/Easter Island to the southeast had been discovered and settled. This settlement of most of western Oceania in such a short period of time stands as one of the most impressive colonization movements in human history, especially given the size of the craft and the initial uncertainty about what they were going to encounter (Nunn 2009). Early Polynesians also moved west and likely reached Madagascar in the western Indian Ocean by about 2,000 years ago (Irwin 1989; Levison, Ward, and Webb 1972; Ward, Webb, and Levison 1973).

So when European explorers such as Captain Cook visited many of these Pacific islands in the 1700s, the Lapita Polynesians and their descendants had been living and communicating across Oceania for between 25 and 150 generations.

Indigenous Populations in Oceania

It is difficult to be confident about the numbers of Indigenous peoples who lived on the islands of the Pacific prior to European contact. Dening (2007) suggests that at the time islanders first encountered Europeans on the island of Tahiti it was estimated that the Tahitian population was about 100,000. And Johann Forster, a naturalist who was on the second of Cook's voyages to Tahiti, estimated the population at 121,500 (Kirch and Rallu 2007). Missionaries who visited the islands thirty years later estimated the population to be only 16,000 and by 1829, this number had been reduced to 8,658 (Kirch and Rallu 2007). To give you a comparison, the population of Tahiti is currently about 180,000. It has also been estimated that the population of Hawai'i upon Cook's first encounter (1778) was at least 400,000 (Kirch 2007).

How quickly did populations decline after European contact? This is a very difficult question to answer and requires specialized study on individual islands. For example, the population decline among the Maori of Aotearoa/New Zealand in the second half of the nineteenth century has been estimated at about 50 percent (Pool and Kkutai 2011). However, populations in smaller islands suffered greater declines. For example, the Chamorros people of the Mariana Islands saw their numbers decline by 90 percent in only one generation as a result of conflicts with the Spanish, forced relocation, and the spread of infectious diseases (Rogers 1995). On the Marquesas Islands in what is now French Polynesia, it was estimated that there were 45,000 Indigenous people living there at the start of the 1900s, a reduction of 95 percent from the pre-contact population (Kirch and Rallu 2007; Rallu and Ahlburg 1999). Although much of this decline can be attributed to infectious diseases such as tuberculosis and smallpox introduced by the Europeans, a rapid decline in the fertility rate because of the transmission of venereal disease among women was also to blame for the population decline. There are records from ships' captains such as Cook ordering his men to not engage in sexual activity with the native women. Nonetheless, sexual activity inevitably took place, leading to what Crosby (1986) called depopulation as a form of ecological imperialism, thereby allowing Europeans to move more aggressively into the Pacific with minimal resistance.

There has also been some discussion that these populations may have fluctuated significantly depending on the environmental conditions of the time. One school of thought is that populations were much greater in the period 1300–750 AD, a time associated with warmer temperatures and higher sea levels. Not only were these higher temperatures conducive to more hospitable living conditions, but the conditions also led to a stronger and more complex long-distance trading system and less intertribal conflict (Nunn et al. 2007). Following the Little Ice Age around the Year 700 AD, the region entered into a prolonged period of colder temperatures and lower sea levels that continued until only 200 years ago (Nunn and Britton 2001). This period resulted in population declines as carrying capacities decreased. It also may have led to a breakdown in some of the established societal structures and greater levels of intertribal warfare (Nunn et al. 2007). Therefore, when Europeans were first "discovering" many of the Oceanic islands, they were seeing these societies in a very different and more tenuous condition than they had been 500 years earlier. They also encountered island cultures that had lost the ability and/or the need to engage in long-distance navigation (e.g., on Rapa Nui and Palau), leading them to the mistaken assumption that these islands must have been populated accidentally (Erlandson and Fitzpatrick 2006).

Given the way that Europeans viewed the status of their society relative to other world cultures, and the hardships suffered by the European expeditions, it is not surprising then that the accepted wisdom for the next 450 years was that most of the Pacific had been populated by accident. Over a long period of time, there were likely many examples where islands were discovered by accident, especially over shorter distances. However, there is now overwhelming evidence that many discoveries and subsequent marine voyages were quite purposeful.

So how did Polynesians go about populating the Pacific so thoroughly? First, you have to understand that the Polynesian seafarers had a very different conceptualization of the sea than most Westerners. In fact, Elizabeth Deloughrey (2007) has used the "roots and routes" metaphor to describe how Indigenous identities were both grounded on their islands (roots) and also used the sea to connect the various island cultures (routes). So instead of considering the sea as a barrier, it was perceived as a passageway. Levison, Ward and Webb (1973, 64) contrast the European and Polynesian perception of the sea as follows:

> The European, at sea in a small vessel, tends to envisage his situation as one in which his craft moves towards, passes by, and then away from fixed islands. The islands are secure and he is in motion. But Gladwin describes how the Puluwat navigator, once on course, inverts the concept and in his navigational system, considers the canoe to be stationary and the islands to move towards and past him. Such a vision seems to reflect a high level of security and confidence in the self-contained little world of craft, crew, and navigational lore.
>
> We accept that the risk and dangers of the sea which seem to weigh heavily on the minds of continental men are not given such emphasis by island navigators today. And we may surmise that a western Pacific Islander in the past might well sail east or south or north in search of new land, confident in the belief that, as usual, islands would rise over the horizon to meet him.

In addition to this worldview, in his book *We, the Navigators*, David Lewis (1994) points out the many strategies that Polynesians used to maintain long-distance communication and transportation lines across the Pacific, including the kinds of vessels that were constructed, their ability to steer by the stars and keep course by the sun, the reflection and refraction on ocean swells and the wind, "dead reckoning," knowledge of the habits of various seabirds, and even the presence of phosphorescence in the sea and the reflection of atoll lagoons off of low-lying clouds. Oliver Allen (1980, 98) said, "When nearing islands beyond the horizon, they could actually smell land, feel the echoes in the water from the swells bouncing off atolls,

Fig 4.2 Stamp from the Marshall Islands, featuring a stick chart. Source: **dustin77a/Shutterstock**

and see the greenish reflection of forests on the underside of clouds." Because they are often referred to as canoes in English, it is sometimes difficult to picture the size and capacity of these seafaring vessels (called va'a). They were fast, safe, and large, either double-hulled or single-hulled, and often fitted with outriggers. With a platform between the two hulls, they may have been able to carry fifty to eighty people, cargo of up to 30,000 kilograms, and could cover distances of 160 kilometres in twenty-four hours (Dening 2007, 294). As seen in figure 4.2, some of these navigational devices were represented on this stamp from the Marshall Islands. Each of the nodes on the twig map on this stamp represent the location of different Pacific islands.

Dening (2007, 293) includes a map by Emily Brissenden that shows the voyages that Pacific Islanders may have undertaken prior to European contact. The different birds drawn on the map correspond to the species that travellers would have looked for to indicate the presence of specific islands, while the dotted areas around the groupings of islands would be the approximate targets of the trips given their ability to read the sea and the sky. This evidence suggests a much more coherent and predictable region of islands than the specks of land in the vast seas often conveyed by the early European explorers.

It was not uncommon for Western ships in the Pacific to have included a seafaring islander or navigator as a member of their crew. For example, British Captain Cook had on board a former high chief from Raiatea, a noble named Tupaia. Raiatea is the second largest island in present-day French Polynesia and is considered one of the major ancient migration staging points to other islands in the central Pacific Ocean. Tupaia was apparently able to draw a map of the Pacific that extended for 4,200 kilometres and included every major group of islands except for Hawai'i and Aotearoa/New Zealand. Perhaps as a reflection of Westerners' reliance on navigational technology and attitudes regarding Indigenous knowledge, and even when

Western missionaries and merchants developed stronger abilities in native languages, rarely did anyone ask these Indigenous islanders about their methods of navigation (D. Lewis 1994).

Could Polynesia Have Been Settled from the Americas?

The research expedition of Norwegian Thor Heyerdahl (1914–2002) serves as a good and recent example of the reluctance of Westerners to accept the theory that Polynesians were quite capable of populating the Pacific islands from the west to the east in a systematic and deliberate manner. An ethnographer, zoologist, and geographer by training, Heyerdahl saw similarities between South American and Polynesian customs and artefacts that suggested to him that the pre-Incan peoples may have settled at least some of the Pacific islands. More specifically, he believed the presence of the South American sweet potato on Fatu Hiva, one of the Marquesas Islands in French Polynesia, was an indicator that the island had been settled from the east. Therefore, he attempted to prove that it would have been technically possible for South Americans to have crossed the Pacific from east to west. He and a five-man crew built a raft of balsa logs using drawings made by Spanish conquistadors and, as closely as possible, the tools and equipment that would have been available at that time. Leaving from Peru in 1947, they travelled 6,900 kilometres in 101 days, eventually landing at Raroia in the Tuamotu Islands. Based on the book Heyerdahl (1950) wrote describing the trip, the movie *Kon-Tiki* (i.e., the name of the craft; named after an Incan sun god) won an Academy Award for best documentary feature film in 1951. Although the book and subsequent films captured the imagination of Western audiences, almost all researchers agree it is extremely unlikely that the Pacific Islands were first settled from South America. Evidence from archaeological records, word analysis, and the origins of introduced plants and animals all point to initial settlement from the Southeast Asian archipelago (D. Lewis 1994).

Heyerdahl's relationship with island settlement patterns did not end with this Pacific expedition. Later, he repeated the premise of the earlier expedition, this time attempting to prove that Africans could have technically travelled across the Atlantic Ocean and settled on the islands of the Caribbean. In this case he built a raft made from reeds and sailed west from Morocco. After an unsuccessful voyage in 1969, he tried again the following year, eventually sailing 6,000 kilometres in fifty-seven days and landing on the island of Barbados. Even though research has proven that the skills and knowledge base of Indigenous peoples was superior to intellectual Western society at the time, in principle it was still difficult for many in the West to accept the idea that the Polynesians (and Caribs) they encountered may have been at least their equals in ocean navigation.

Islanders in the Caribbean

As was the case in the Pacific, just prior to European contact, islands in the Caribbean were settled by many different Indigenous groups. Although there may have been three major linguistic groups, all of them originated on the South American mainland. A small group of Ciboney (or Siboney) people settled on parts of Cuba and Hispaniola. The Arawak (or Taino) people were the main group on the Bahamas, the Greater Antilles, and Trinidad, and the Carib people were the major group on the Virgin Islands, the Lesser Antilles, and the northwest tip of Trinidad. Early Spanish explorers were told by the Taino that they had been driven out of most of the rest of the Caribbean by the Caribs (Rogoziński 1994; S. Wilson 1999). We know that there was a thriving population of native peoples throughout the Americas prior to 1492, estimated to be as high as 112 million and as low as 8 million (Denevan 1992). As this wide range suggests, there is not much consensus on the actual numbers of pre-European-contact Indigenous peoples on these islands. The Spanish estimated that at the end of the fifteenth century there were between three and four million people living on Hispaniola, and Cuba had a population of between 225,000 and 300,000 (Rosenblat 1992). Girvan (2014) estimates that the population of Indigenous peoples in the Americas decreased by 90 percent within the first two centuries of European conquest. Regardless of the exact number, populations of Indigenous peoples were reduced drastically in a relatively short period of time following initial contact with Europeans. For example, on Hispaniola the population of natives may have only been about 60,000 in 1508 and not much more than 500 by 1548. There were many causes for these population declines, including diseases introduced by the Europeans, and the choking off of native foodstuffs that led to starvation, war, and slavery.

ISLANDER AND EUROPEAN ENCOUNTERS: EXPLORERS, MISSIONARIES, AND MERCHANTS

Islander Encounters with European Explorers

When were the first encounters between Indigenous peoples and Europeans in the Caribbean and Pacific? In the Caribbean, this began with Columbus in the late fifteenth century and continued with the Spaniards and the Portuguese. Columbus described the Caribbean islands as green and fertile and the natives as timid, guileless, content, generous, and having the potential to be affectionate servants of Spain and Christians (Howe 2000, 11). Outside of the United States, Columbus is viewed as a villain, including being referred to as the father of the slave trade for his treatment of Indigenous peoples (Davidson 1992). The movement of Europeans into the Pacific islands did not start for more than one hundred years after the first encounters in the Caribbean. The Portuguese were the first Europeans to colonize the Philippines and other nearby islands. In fact, Ferdinand Magellan's first encounter with Pacific Islanders resulted in the first of many cultural misunderstandings. When his ship reached the Marianas Islands in 1521, the Chamorro people came aboard and began, according to the Europeans, stealing their possessions. However, unlike the European model of private property ownership, the Chamorros practiced a system of communal property sharing. As such, once they shared their food with the Portuguese, the Chamorros believed that the newcomers' material possessions were then to be shared with them. Magellan promptly named the Marianas "the Ladrones" (from the Spanish meaning bandit or mercenary), or the "The Island of Thieves" (Chappell 1999; Day 1987).

The Dutch were also early European entrants into this region, landing and mapping Tasmania, New Zealand, Tonga, Fiji, and the Bismarck Archipelago by the mid-1600s. It is about this time that the English and French became much more active in exploring, mapping, and claiming islands in the Pacific. It should also be noted that for the first 250 years of European presence in the Pacific, very little attention was paid to most of the islands. Not only were they considered dangerous, but they were distractions from the main goals of the time: discovering a Northwest Passage to reach the "spice islands" of the Moluccas more efficiently, and/or discovering the supposedly wealthy southern continent of *Terra Australis Incognita* (Howe 2000, 2003). In this respect, attitudes were similar to European exploration of the Caribbean region and indeed all of the Americas.

North American grade school history lessons tell us that the Caribbean and the Americas were not intentionally sought out by the Europeans. At first they represented a barrier to a quicker passage to the riches of the Orient. Later, they were used for a quick return on investment and a quick return home (Gillis 2003). Therefore, in both oceans, the Europeans did not initially seek new lands but rather new routes to old lands (Parry 1974 as cited in Gillis 2003). The final decades of the eighteenth century experienced the greatest intensity of European contact with islanders of the Pacific Ocean. The most well-travelled explorer during this period was the British Captain James Cook who encountered, meticulously mapped, and inventoried many islands and their peoples during three expeditions from 1768 to 1780.

> EPELI HAUʻOFA (1994) BELIEVED THAT THE CONDESCENDING AND BELITTLING ATTITUDE OF EUROPEAN EXPLORERS, MISSIONARIES, AND TRADERS HAS HAD A LASTING AND NEGATIVE EFFECT ON HOW PACIFIC ISLANDERS VIEW THEIR OWN HISTORY AND TRADITIONS.

In many ways, the attitude of Europeans towards Indigenous islanders during these ages of European "discovery" and "enlightenment" was much the same as attitudes regarding Indigenous peoples encountered on mainlands. Similar to that which was described above, when Columbus arrived in what was to become San Salvador and the Bahamas in the fifteenth century, he described the people he met as being "friendly and well-dispositioned," having "handsome bodies and very fine faces," with eyes that were "large and very pretty," and he praised their "docility" (Barratt 2003). Given the importance of religious conversion associated with many of these early trips, he speculated, "I think they can easily be made Christians, for they seem to have no religion" (Girvan 2014, 50). More than 300 years later, when the British ship HMS *Beagle* landed at Tierra del Fuego, the island on the southern tip of South America, Darwin described the peoples he encountered as "miserable degraded savages," and "I could not have believed how wide was the difference between savage and civilized man: it is greater than between a wild and domesticated animal, inasmuch as in man there is a greater power of improvement" (Grigg 2010, 13). Indeed, Darwin compared the Indigenous peoples to devils he had seen in plays. Epeli Hauʻofa (1994) believed that the condescending and belittling attitude of European explorers, missionaries, and traders has had a lasting and negative effect on how Pacific Islanders view their own history and traditions. He notes that a number of Pacific societies still think of their pre-European contact as one of darkness and savagery while the "Christian era" has brought with it civilization and light. At the same time, some Indigenous islander groups that have rightly repudiated the ills brought on by colonialism in favour of a society based on a fictional pre-contact utopia fail to realize that inter-island warfare was not uncommon and life was tenuous and violent for many of the people who populated the islands of the Pacific (Younger 2009).

Colonialism and Islands

For a long period of time, many of the European colonial powers were reluctant to assume responsibility for island territories. They were often perceived as being too distant, too costly to maintain, having few resources to extract, and having local markets that were too small to benefit colonial companies (Bouchard and Crumplin 2010). In the Caribbean throughout the sixteenth century, the Catholic church worked collaboratively with governments to establish territorial domain over islands. Papal decrees or "bulls" gave sweeping power to explorers to claim new territories in the name of a given European imperial power, primarily Spain or Portugal (Mantz 2003). This was done less for economic reasons and more to spread Christianity to the Indigenous peoples they encountered. Of course, as noted earlier in this chapter, most of the Indigenous population died as a result of the spread of diseases and enslavement. As other European powers became involved, and the attention of Spain and Portugal turned to mainland Latin America, the plantation economy became a central part of colonial relationships on many Caribbean islands.

This changed in the latter half of the nineteenth century through the First World War. The reasons for European colonialism of world islands were often similar to the reasons for colonial intrusions on mainland territories. In some cases, there may not have been an identifiable direct benefit. However, a country would sometimes establish a political presence on an island to prevent a competing European power from doing likewise. In other cases, resources were identified that would benefit the government or businesses affiliated with that colonial power. For example, being able to provide the Chinese market with sea otter pelts, sandalwood for incense, sea slugs, pearls, and turtle shells may have justified developing some kind of political control over island territories in the Pacific Islands which were also becoming increasingly valuable as strategic provisioning stations for trading vessels transiting the Pacific Ocean.

Reilly Ridgell (1995) presents two examples of ways in which Pacific islands were acquired by colonial powers. The first is Samoa. Germany, Britain, and the United States all wanted the island, primarily because of the quality of the harbour at Pago Pago and its potential use as a coal fuel stop for steamships. Each power had aligned itself with a different chief in order to try to gain the upper hand and had sent their own warships to enforce their claims on the territory. However, when a typhoon swept across the island in 1889 sinking the German and US ships, it was left to the British. Subsequently, the British traded its claim for German claims in Melanesia and the Germans and Americans split the island between them. Following the First World War New Zealand won the German portion of Samoa. We now have a situation where there is still an American Samoa and an independent nation of Western Samoa. This trading of island territories was not uncommon throughout the colonial period and similar imperial transactions occurred throughout the colonial history of St. Lucia, Prince Edward Island, and many other island territories. The second example pertains to Hawai'i. The Hawaiian chief Kamehameha and his family had built a strong kingdom that lasted from 1795 to 1893. During this time the sugar plantation owners had established a profitable business selling sugar back to the United States. When the McKinley Tariff of 1890 was passed in the United States making foreign sugar more expensive relative to American sugar, the foreign business leaders overthrew the current queen (Lili'uokalani) and then called on the United States for protection and annexation. Once it was annexed as a territory in 1898, the plantation owners could once again ship sugar profitably to the mainland United States.

WHEN THE MCKINLEY TARIFF OF 1890 WAS PASSED IN THE UNITED STATES MAKING FOREIGN SUGAR MORE EXPENSIVE RELATIVE TO AMERICAN SUGAR, THE FOREIGN BUSINESS LEADERS OVERTHREW THE CURRENT QUEEN (LILI'UOKALANI) AND THEN CALLED ON THE UNITED STATES FOR PROTECTION AND ANNEXATION.

In addition to gathering flora and fauna to take back to the Old World, Columbus and other European explorers of the Caribbean took back many Indigenous peoples. From his journal, Columbus writes, "If it pleases our Lord, I will take six of them to Your Highnesses when I depart, in order that they may learn our language" (Girvan 2014, 50). From the island of Hispaniola, thousands of Taino "Indians" were sent to Spain as slaves. As was the case elsewhere in Central America at that time, the Indigenous population on the Caribbean islands was quickly decimated by slavery and disease. It has been estimated that, within the first sixty years of Europeans landing on Hispaniola, the population of Taino peoples declined from about 250,000 to only several hundred. Ironically, elimination of the first peoples of the Caribbean was so severe that the region became "underpopulated," thus requiring slaves to be brought from Africa to work on the plantations. Gaspar (1991) describes conditions on the small island of Antigua as a case study in how many of these islands were being transformed by European presence. After driving off the Indigenous peoples in the 1630s, the English transformed this island into a monoculture sugar plantation that required an ongoing supply of slaves to operate. Within one hundred years, eighty-four percent of the population of Antigua consisted of slaves (Gaspar 1991, 132).

The other encounter that is noteworthy was that which took place between Europeans and the Indigenous Maori in New Zealand, otherwise known as Aotearoa or "Land of the Long White Cloud." It is generally agreed that Polynesians discovered and settled the islands of New Zealand by about 1100 AD. This was later than elsewhere in Oceania because it was one of the few island groupings located outside of the tropics (Irwin 1989). Its relatively extreme southern location required the ability to navigate more complex and risky wind conditions. As a temperate island, it may also have been difficult for peoples accustomed to a tropical lifestyle to adapt to this more temperate environment (A. Anderson 1997; Nunn et al. 2007).

The first record of a European sighting of New Zealand was in 1642 by the Dutchman Abel Tasman, for whom Tasmania, the island off the south coast of Australia, is named. He did not land during this visit because of the violent encounter he had with the Maori natives as he approached the island (John Wilson 2005). The British Captain James Cook and the French Captain Jean François Marie de Surville, leading separate expeditions, both landed on New Zealand in 1769, more than 100 years after Tasman's original encounter. It was estimated that there were between 100,000 and 250,000 Maori living on the two main islands of New Zealand at the time leading up to these encounters. From the 1790s on, European whalers, traders, and missionaries increasingly stopped at New Zealand, mapping the coastline and interior and searching for resources and converts to Christianity. The traders also provided muskets and modern weaponry to the Maori as trade items. One of the consequences of this trade was an increased intensity to the long-standing and often violent conflicts between different Maori tribes.

A milestone in New Zealand history and relationships between Europe and the Maori was the 1840 signing of the Treaty of Waitangi between British Captain William Hobson and many of the tribal chiefs in order to preempt further French settlement and claims on the islands. In the English version of this treaty, it ceded sovereignty of New Zealand to Britain, and the right of preemption of all Maori-owned land (i.e., the right to buy the land before it is offered to anyone else). It also gave the Maori full rights of ownership over their land, forests, fisheries, etc. and provided the Maori with full rights as British subjects and Crown protection (Orange 2015). Unfortunately, the Maori version of the Treaty failed to convey the meaning of these provisions, resulting in ongoing differences in interpretation. Only thirty-nine chiefs signed the English version while almost all of them signed the Maori version. It is noteworthy that in the Maori version of the Treaty the word sovereignty is translated to *kawanatanga* or governance. It has been speculated that some Maori believed they had not given up ownership of the land in the Treaty (New Zealand Ministry for Culture and Heritage 2017). At the signing of the Treaty in 1840 there were still fewer than 2,000 Europeans living on NZ, concentrated mostly in Wellington, but this number grew to an estimated 22,000 within the next ten years. Shortly thereafter, the relationship between the European settlers and the Maori became increasingly confrontational, with some, but not all, tribes fighting the British. One of the famous battles, and a continuing symbol of pride for the Maori, was the Battle of Gate Pa in 1864. Despite having superior military equipment and numbers of soldiers, the British military force suffered heavy casualties before the Maori withdrew. It has been argued that their ability to withstand European settlement and military force on this and other occasions occurred in part because of their larger numbers, because they already had developed a warlike culture as a function of intertribal conflicts, and because they had a strong spiritual attachment to the land (Belich 1986; Dalton 1966). Despite this victory, with increasing numbers of British settlers and military support, the Maori were eventually defeated. By the 1858 census, only 56,049 Maori were counted.

Missions and Colonialism

As has already been alluded to, converting island Indigenous peoples to Christianity, especially on government-sponsored missions, was a key objective of colonial trips. This objective started with the first encounters and increased in importance as Europeans established a presence on the islands of the Caribbean and the Pacific. When the Caribbean was being mapped and colonized, exploration was linked very closely to both the needs of the monarchies and the needs of the Vatican. For example, a papal bull or decree written in 1455 authorized the Portuguese Crown to "subdue, enslave or conquer any Pagan or Muslim peoples whom the Portuguese may encounter on their voyages of discovery between Cape Bojador [off West Africa] and the nebulously defined region of 'the Indies'" (Vallandingham 2013, 260). By the sixteenth century the Crown had taken over much of the role of the Vatican, claiming the lands in the name of the monarch and sovereign. Throughout this period, missionaries representing many denominations were being sent all over the world in order to create a universal Christian empire.

> DURING THE EIGHTEENTH CENTURY, THE PREVAILING EUROPEAN PERCEPTION OF ISLANDERS WAS ONE OF THE "NOBLE SAVAGE" WHO WERE MORALLY PURE, LIVED IN HARMONY WITH NATURE, AND EXPERIENCED LITTLE INTERNAL STRIFE (CONNELL 2003B).

In the Caribbean and the Pacific, these missions were closely connected to trade objectives, with denominations linked to the regional land claims being made by the Crown. For example, Catholic priests were in the Philippines together with the Spanish traders. The Russian Orthodox church was in the Alaskan islands with the Russian fur traders and the Lutherans were in New Guinea. Part of the increased motivation for this conversion had to do with the changing cultural attitudes towards island peoples. During the eighteenth century, the prevailing European perception of islanders was one of the "noble savage" who were morally pure, lived in harmony with nature, and experienced little internal strife (Connell 2003b). There was admiration for their leaders, their physical beauty, and their marine skills. This attitude toward islanders was closely related to perceptions of Pacific islands as a kind of utopian Eden, with Tahiti being the archetype of this perspective. It is not unrelated that this paradisiacal picture was in sharp contrast to the polluted, industrial landscapes of urban England and France at the time, where living and working conditions were dangerous and unhealthy for much of the population. Part of this positive attitude towards the islands and islanders was reflected in the Anglicized names given to the islands, including the Friendly Islands (now known as Tonga) and the Pleasant Islands (now known as Nauru). Of course, the reality was that many island societies at this time were not as pure as portrayed, with intertribal warfare, a fear of spirits and leaders' decisions, and tenuous living conditions being part of everyday life. It is also the case that as the realities of early encounters on islands such as the Canaries off the coast of what is now Morocco became more well-known, they lost their mythical status as paradises. The names Danger Island (in the Chagos Archipelago of the Indian Ocean) and Savage Island (now known as Niue) attest to this evolving perception. However, in the minds of Europeans, with each new, more remote island "discovered," hope for a new utopia is created, first in the Caribbean and then in the Pacific (Gillis 2003).

Bernard W. Smith (1985) distinguishes the European attitude towards Pacific peoples as one of either soft or hard primitivism, both in time and space. "Soft primitivism" was more

> AT THE START OF THE NINETEENTH CENTURY, ATTITUDES
> TOWARDS AND INTERACTIONS WITH INDIGENOUS PEOPLES
> BEGAN TO CHANGE . . . INDIGENOUS PEOPLES WERE
> INCREASINGLY BEING REFERRED TO AS IGNOBLE, VIOLENT,
> UNCULTURED, AND UNPREDICTABLE. IN OTHER WORDS, THE
> NATURAL ENVIRONMENT MIGHT STILL BE IDYLLIC, BUT THE
> PEOPLES, AS A RESULT OF THEIR EXPOSURE TO EUROPEANS,
> WERE NOW SEEN TO BE IMPURE OR TAINTED.

closely associated with inhabitants of the Society Islands in what is now French Polynesia, while hard primitivism was associated with Indigenous Fuegians (from Tierra del Fuego off the southern tip of South America), Maori, and the Australian Aborigines. The hard primitive characteristics of austerity and fortitude were more closely associated with Christian values during the last decade of the eighteenth century. With the decline in the belief of a supreme being (i.e., deism), any belief in pagan worship was viewed by Europeans as repugnant, and contributed to Pacific Islanders overall being perceived as depraved and ignoble (B.W. Smith 1985).

At the start of the nineteenth century, attitudes towards and interactions with Indigenous peoples began to change, leading to an increase in the number of missionaries. One of the reasons for this was the changing nature of the European settler population and the perceived impact they were having on the islanders. Many of the European traders, sailors, and settlers of this period were from the lowest socioeconomic groups and were therefore illiterate and in ill-health (e.g., scurvy). Darwin described the English settlers in the Bay of Islands, the former capital of New Zealand, as being "the very refuse of society." At the same time, resistance to a European presence by the Indigenous peoples was increasing on some of the islands (e.g., Polynesian guerilla warfare against the French on Tahitian islands). With the use of weapons obtained in trade from the Europeans, the level of intertribal violence was becoming greater and more visible. Rather than being described by the European press as noble and pure, Indigenous peoples were increasingly being referred to as ignoble, violent, uncultured, and unpredictable. In other words, the natural environment might still be idyllic, but the peoples, as a result of their exposure to Europeans, were now seen to be impure or tainted. From the perspective of European outsiders, this period has been given many labels, including a "paradise lost," a "shattered dream," and a "fall from Grace." It has been argued that the event that best symbolized this transition was the killing of British Captain Cook by the local Indigenous peoples in Hawai'i in 1779 (Connell 2003b). Of course, as many researchers have now pointed out, if the original label of paradise was a fabrication, then there really was nothing to be "lost" (Howe 2000). However, if islanders have now "fallen from Grace," it meant that missionaries were therefore even more necessary to save islanders from themselves and from their interactions with European civilization.

Attitudes were not uniformly the same towards all islanders across the entire Pacific. For example, Polynesians were generally described in more positive terms as being more docile and intelligent. Their governance system, with powerful leaders exercising centralized control, was something that was more familiar and understandable to Europeans. Coincidentally, they were also lighter-skinned and therefore closer in appearance to the Caucasian Europeans. At the other extreme, Melanesians from New Guinea were generally described as savages. Because their governance system was not as hierarchical as in Polynesia, societies were viewed as being more fragmented and out of control. As the name of the region implies, the Indigenous peoples in this region tended to be more dark-skinned, leading one to believe that racism entered into this difference in attitudes. This quote from the Encyclopedia of Missions (1913) helps describe some of these attitudes:

> The Natives of the Hawaii Islands belong to the Malay race, modified by the Polynesian type. Physically, they are among the finest races in the Pacific, and they show considerable intellectual capacity. Previous to the introduction of Christianity they were not much more superior in moral character to any of the other savages in the Pacific. Polygamy, infanticide, and polyandry all prevailed. The idolatry of the Kanakas, as the Natives are called, was barbarous and bloodthirsty, for human sacrifices were frequently offered during the sickness of a chief, at the dedication of a temple, or the inception of war. On the other hand, the natives are even-tempered, light-hearted and a pleasure-loving race.

In New Zealand, the Anglican missions had two objectives. As was the case with missionaries elsewhere, one of these was to convert the Indigenous people to Christianity. The second, and perhaps less transparent objective, was to keep law and order among the European settlers in the absence of any official and consistent presence by the English judiciary. Despite these overarching objectives, the relationships with the Maori and the impacts upon Maori life often depended on the personalities and approaches of the specific missionary leaders. For example, Samuel Marsden (1765–1838) was credited with establishing a stable and flexible relationship with the Maori. Although he was not very successful in converting many Maori to Christianity, he was considered to be more pragmatic. He facilitated trade between the Maori and the English settlers and traders, and provided muskets to the Maori in return for food for the hungry settlers. He also introduced the written word and helped develop a written Maori language so Indigenous stories could be conveyed to a wider audience. His approach was that teaching language skills was a precondition for conversion. In contrast, his successor to New Zealand, Henry Williams (1792–1867), was considered to be much more rigid in his dealings with the Maori. For example, he stopped the trade of muskets to the Maori and, in return, they stopped providing food to support the English settlements. From the perspective of the Indigenous islanders, the vast majority of the encounters with missionaries and, by extension, with Europeans in general, led to profoundly negative outcomes. First and foremost, almost all encounters brought with them significant numbers of deaths as a result of the introduction of diseases for which the islanders had no immunity. Examples have already been provided of populations that were decimated within a generation of encountering Europeans for the first time, leading to breakdowns in the social systems and jeopardizing the very survival of many island communities. Increased numbers of deaths also occurred as a result of the distribution of European weapons to inter-tribal combatants. For those Indigenous societies that did survive, their native cultural practices were condemned and outlawed in favour of the adoption of European cultural practices. Europeans viewed islander conceptions of time and work as idleness and laziness, encouraging them instead to adopt a Protestant work ethic. This quote by author Mark Twain (2018, n.p.) from one of his lectures, regarding the Sandwich Islands

(Hawai'i) and the impacts of the missionaries circa 1872, provides a third-party humorous interpretation of some of these impact:.

> Nearby is an interesting ruin—the meager remains of an ancient temple—a place where human sacrifices were offered up in those old bygone days … long, long before the missionaries braved a thousand privations to come and make [the natives] permanently miserable by telling them how beautiful and how blissful a place heaven is, and how nearly impossible it is to get there; and showed the poor native how dreary a place perdition is and what unnecessarily liberal facilities there are for going to it; showed him how, in his ignorance, he had gone and fooled away all his kinsfolk to no purpose; showed him what rapture it is to work all day long for fifty cents to buy food for next day with, as compared with fishing for a pastime and lolling in the shade through eternal summer, and eating of the bounty that nobody labored to provide but Nature. How sad it is to think of the multitudes who have gone to their graves in this beautiful island and never knew there was a hell.

Missionaries also marginalized local spiritual beliefs and practices, including islanders' own idols/gods and animistic beliefs, in favour of Christian religious practices. In Hau'ofa's (1993, 149) "Our Sea of Islands," he argues that much of the blame for the way islanders view themselves now can be traced back to those earliest encounters with European explorers and missionaries. "The wholesale condemnation by Christian missionaries of Oceanic cultures as savage, lascivious, and barbaric has had a lasting effect on people's views of their histories and traditions," making them believe that anything pre-European contact was associated with savagery and anything afterwards was associated with the light and civilization of Christianity."

Many of the secular cultural practices were also criticized or prohibited, including nudity, tattoos, the use of natural drugs such as betel-nut chewing, and feasting. This latter issue has continued to the present day, leading to some of the highest levels of obesity and health-related outcomes in the world. Traditional social and governance structures that emphasized communal decision-making and assistance were changed to one that favoured a nuclear, family-based structure. Finally, as Hau'ofa alluded to in an earlier section of this chapter, the loss of traditions and cultural practices has led to a collective malaise among islanders in Oceania that has pervaded all aspects of their societies. The tension between Indigenous and Christian belief systems should not come as a surprise. As Hau'ofa and others have pointed out, Europeans routinely belittled the cultures and traditions of Pacific Islanders.

Some research has suggested that there were positive outcomes for islanders as a result of their encounters with missionaries (Stevens-Arroyo 1993). For example, European missionary-linguists provided the means to record oral Indigenous stories in Indigenous written languages and set up the printing press (Keown 2007). In so doing, it has allowed these stories to be preserved for future generations and to be disseminated to non-Indigenous peoples. Also, even though early European visitors used slavery as a tool themselves, when this practice was abolished elsewhere in the world it also led to the curtailment of slavery among Indigenous tribes. These external influences may have also led to a reduction in some of the local practices such as infanticide, cannibalism, and intertribal warfare. Perhaps most significantly, the prominence of Christianity now constitutes a powerful social force bringing together many island communities. At the same time, it is hard to conceive that these achievements in any way make up for the cultural and physical genocide experienced by native islanders as a result of their encounters with outsiders.

CONCLUSIONS

One of the consistent themes in island studies has been the relationships between islanders and outsiders. It could be argued that the tensions associated with these encounters began with European colonialism. As much as we might talk about the resilience of islanders to external forces, most of these historical encounters did not end well for the Indigenous peoples living on islands. In most cases, island populations were decimated by disease, war, and occupation. Indigenous use of lands and fishing rights that might have been organized in a communal tenure system were swept away by colonial settlers and their governments. Missionaries led the way in replacing Indigenous spiritual beliefs with Christian teachings, and resources, such as guano, were exported with little compensation to islanders. In some island regions, such as in the Caribbean, there is little left of the culture of the first peoples to settle these islands. Instead, islander culture is an amalgam of the various groups that have occupied these islands over the past 500 years. In some regions of Oceania, the situation is a little different. Despite being shaped by foreign governments, settlers, merchants, and missionaries, Indigenous cultures continue to play a role in the lives of many of the peoples. We will learn more about this in the next chapter.

Key Readings

Allen, Oliver. 1980. *The Pacific Navigators*. Alexandria, VI: Time-Life.

Connell, John. 2003. "Island Dreaming: The Contemplation of Polynesian Paradise." *Journal of Historical Geography* 29 (4): 554–581.

Nunn, Patrick. 2009. *Vanished Islands and Hidden Continents of the Pacific*. Honolulu: University of Hawai'i Press.

West Point Lighthouse, Prince Edward Island. *Source:* **Verena Joy**

Chapter Five

Islands, Islandness, and Culture

Wherever an Islander may travel he remains an Islander. He speaks of his home as "The Island" without explanation or qualification, and he uses the term as if less to describe a body of land surrounded by water than a state of life. (Greenhill and Giffard 1967, Prologue)

INTRODUCTION

It is well beyond the scope of this book to provide a comprehensive discussion of culture and how this term has been used theoretically and in practice. However, it is still useful to provide some introductory comments on the concept in order to understand how it may pertain to islands and islandness. The seminal work by Raymond Williams (2006) is a good starting point. He stated that there are three ways of looking at culture: as an idealized set of values, as the body of work that describes human thought and experience, and as a description of a particular way of life. This last interpretation of culture is the one we will be focusing on here.

Culture has further been described as "the accepted and patterned ways of behaviour of a given people ... as a result of belonging to some particular group" (Peck 1998, n.p.) and as the shared features which encapsulate people together in a community (Shah 2003). It is also relational. In other words, you may not be aware of your culture until you encounter other communities that do not exhibit the same characteristics, practices, or customs (A. Cohen 1985).

There are many forces that shape a group's culture, including religion, language, gender, education, race, shared lived experiences, historical events, and social background. Many of the topics we have covered so far in this book can also be linked to culture, including how islands and islanders are represented in art, music, and literature, as well as how societies are shaped by encounters with external colonial powers. This chapter addresses culture as it has been applied to island studies. We'll start by examining the concept of islandness and how it might shape islander attitudes and behaviours, both at the level of the individual and for island society as a whole. Given the importance of place in island studies, we will also touch on how an islander's sense of belonging and sense of place shapes their perceptions and behaviours. We will then look at the relationship between the field of cultural anthropology and island studies, and particularly how each has influenced the development of the other. This chapter then examines the patterns of some of the major cultural features on islands, such as language and religion, to determine how similar (homogenous) or different (heterogeneous) islands really are. This chapter concludes with a discussion of the relationships between islanders and outsiders and the tensions that sometimes arise between them.

ISLANDNESS, CULTURE, AND PLACE

ISLANDNESS IS ASSOCIATED WITH A SENSE OF PLACE AND BELONGING THAT CONNECTS PEOPLE AND COMMUNITIES WITH THEIR NATURAL AND SOCIAL SURROUNDINGS IN TANGIBLE AND INTANGIBLE WAYS.

Every discipline has sets of words and concepts that seem to define it. Although the interdisciplinary field of island studies is still very young, it is already starting to develop its own jargon. "Islandness" is one of those words that sets island studies apart from geography, sociology, environmental studies, or urban studies. On the surface the concept of islandness appears simple. However, as you dig deeper, it becomes more confounding and ephemeral. Early on, many people equated islandness with an island's physical features, including small size and isolation (C. Anckar 2008). Increasingly the term is becoming associated with a state of mind, both for individual islanders and for the larger community. For an individual, islandness can be considered a construct of the mind (Platt 2004), a heightened sense of attachment to place, a closeness to nature (Conkling 2007; Royle and Brinklow 2018; Stratford 2008; Vannini and Taggart 2013) and an awareness of the boundaries between the land and the sea (Baldacchino 2005a; Hay 2006). The description by David Weale (1991, 6) is particularly evocative. He says that the island "becomes part of your being—a part as deep as marrow, and as natural and unself-conscious as breathing." Conkling (2007, 191) reflects this peculiar sense of belonging and place by stating that "[Islanders] share a sense of islandness that transcends the particulars of a local culture. Islandness is a metaphysical sensation that derives from the heightened experience that accompanies physical isolation ... [it] amplifies a sense of place that is closer to the natural world because you are in closer proximity to your neighbors ... Islandness thus helps maintain island communities in spite of daunting economic pressures to abandon them."

This last definition suggests that the characteristics associated with islandness provide an island society with a greater ability to respond to external crises, including those that emerge from economic and cultural globalization and extreme climate events. Socially, this also means a closeness and solidarity (Péron 2004). Islandness is increasingly being identified with research in the social sciences, and representations of islandness often appear in the humanities through creative modes of expression such as visual arts, poetry, fiction, and storytelling. We saw some of this in Chapter Three. A good example of the expression of islandness is provided in Figure 5.1. This set of islander "characteristics" is intended to show how islanders may view themselves.

Before we assume that this sense of islandness is limited only to those from the western world, consider the opening sentence of Albert Wendt's (1998, 202) "Towards a new Oceania," where he says, "I belong to Oceania—or, at least, I am rooted in a fertile portion of it—and it nourishes my spirit, helps to define me, and feeds my imagination." This intangible characteristic of islandness is a trait of both individual islanders and of communities as a whole. This should come as no surprise, since many of the other features that define and influence culture, such as race, religion, language, ethnicity, and shared experiences, also influence individual and group behaviours and attitudes. Islandness is also associated with a sense of place and belonging that connects people and communities with their natural and social surroundings in tangible and intangible ways. As Doreen Massey (2005, 125) suggests, "places are collections of those stories, articulations within the wider power geometries of space." On islands, Taglioni (2011, 47) similarly describes islandness as "the sum of representations and experiences of islanders, which thus structure their island territory." Of course, a sense of place exists in many other locales, including farm towns in the Prairies, coastal outports, and northern Indigenous communities. However, on a small island, the sense of place is affected by the presence and role of water and the boundary between water and land. Islanders are rarely out of sight of the shoreline and for many, the sea plays an important part in their social and economic livelihoods.

This "boundedness" can have a profound yet subliminal impact on the collective state-of-mind. In his classic book *Topophilia*, geographer Yi-Fu Tuan (1974, 118) sees islands as one of the four natural environments that come closest to humanity's imagination of the ideal world, symbolizing "innocence and bliss, quarantined by the sea from the ills of the world." There is also a strong sense of solidarity to one another, and to generations of kin and community who have shared those same island experiences, something that Wynne (2007) refers to as an intense feeling of collective identification. Perhaps that is what sets islanders apart from those who were raised outside shared social ties and geographic constraints.

One of the confounding aspects of islandness is that it is often still confused with insularity or isolation. You may recall a brief review of the differences between insularity, isolation, and islandness from an earlier chapter. Historically, the term insularity was usually used in refer-

A SOMEWHAT HUMOROUS LIST OF ISLANDER CHARACTERISTICS

- Independence—small boats and social circles demand it if a personality is to survive.
- Loyalty—ultimate mutual care and generosity, even between ostensible enemies.
- A strong sense of honor, easily betrayed.
- Polydextrous and multifaceted competence, or what islanders call handiness.
- A belligerent sense of competition, interlaced with vigilant cooperation.
- Traditional frugality with bursts of spectacular exception.
- Earthy common sense.
- Opinionated machismo in both the male and female mode.
- Live-and-let-live tolerance of eccentricity.
- Fragile discretion within a welter of gossip.
- Highly individualized blends of spirituality and superstition.
- A complex oral tradition, with long memories fueled by a mix of responsible record keeping and nostalgia.
- And finally, a canny literacy and intelligence.

(Putz 1984, 26, as quoted in Conkling 2007, 192)

ence to the physical distance of an island to the mainland or other islands. But it is increasingly being used pejoratively or in derogatory ways to describe what are perceived to be narrow attitudes or closed-mindedness of islanders. Although the term isolation has also acquired a strong social meaning, it is now often identified with the level of connectedness or accessibility in the movement of people, products, and ideas. Moreover, both insularity and isolation are relative, situating islands and islanders in relation to other people and places. Islandness has less to do with the outside world and more to do with how islanders characterize themselves and their own identity.

ISLANDERS AND EARLY SOCIAL ANTHROPOLOGY

One of the most frequent and early representations of islander identity by researchers, both in scholarly literature and, by extension, in the popular press, has emerged from the work of cultural and social anthropologists or ethnographers. In an introduction to a special section of the journal *Social Identities*, called "Managing Island Life," Jonathan Skinner (2002) reminds us of a few of these more powerful and enduring representations. They include the island natives of the western Torres Strait north of Australia as described by British biologist and ethnologist Alfred Haddon at the end of the nineteenth century. Haddon wanted to document the customs, myths, and legends of the people before they were lost through their interaction with European traders and Christian missionaries. As further examples, Alfred Radcliffe-Brown travelled to the Andaman Islands in the Indian Ocean in the early 1900s to study kinship patterns. New Zealand-born and Oxford-trained Raymond Firth lived among

Figure 5.1   American cultural anthropologist Margaret Mead, circa 1950. *Source:* **Edward Lynch, public domain**

the Polynesian peoples of the tiny atoll of Tikopia in the Solomon Islands in the 1920s and 1930s and wrote about them extensively throughout the rest of his career. The Trobriands of Papua New Guinea were observed and studied by Polish-born Bronislaw Malinowski, and the highland Merina people of Madagascar were the subject of study by Maurice Bloch in the 1970s. This list could easily be expanded to include many other Western-trained scholars of the nineteenth and twentieth centuries who have lived among and written about Indigenous island populations.

Perhaps one of the most famous and influential examples of this genre of islander characterization was by Margaret Mead (see Figure 5.1). Her ethnographic account of the sexual rituals of adolescent women on the island of Samoa (as seen in Figure 5.2) was widely communicated to the general public in her 1928 book *Coming of Age in Samoa.* Early in her professional career, Mead spent nine months living among a tribe on the small island of Ta'u (or Tau) in Samoa. She was espe-

SAMOAN TAUPO GIRL.

Figure 5.2 Adolescent Samoan girl, c. 1896. *Source:* **public domain**

cially interested in the sexual rituals of young women and observed that they had few of the inhibitions and anxieties over their sexuality that were common among Western adolescent girls of that time. In what we now refer to as the "nature versus nurture" debate, Mead (1928) used the experiences of this island group to challenge conventional wisdom by stating that culture rather than biology largely influenced sexual behaviour. Perhaps the reason her work stands out is that she included chapters in her book that extrapolated her findings to the context of childrearing in the western world. Her research was also used to support the position of the sexual liberation movement in the 1960s.

Although many of Mead's conclusions regarding the Ta'u islanders were later contradicted by the anthropologist Derek Freeman (1999), she and many of the other expeditionary social anthropologists of the past 100 years set the stage for the emergence of social anthropology as a field of study. This tradition of what Baldacchino (2007) refers to as descending on islands is part of a larger practice of islands and islanders being used in research as living laboratories, where one is presumably able to study a people or plant or animal in a "pure" state, free from the influences of the mainland, and thereby make observations that can be extrapolated to a broader society. In the eyes of the social and cultural anthropologists, explaining the social phenomena of "primitive" people who were presumably untainted by civilization could allow us to better understand Western societies and human behaviour in general. For Baldacchino (2006c), the island-based fieldwork by Mead, Malinowsky, Firth, and others led to the birth of ethnography and social anthropology as discrete fields of study. Fitzpatrick (2007) goes so far as to suggest that a reliance on these contexts by early ethnographers meant that islandness implicitly became a part of anthropology. It also reinforced the concept of small islands as social laboratories (Mead 1957).

In much the same way that early "Robinsonade" adventure novels shaped Western perceptions of islanders in the eighteenth and nineteenth centuries, it can be argued that research by social anthropologists and ethnographers on small island populations thrust island societies, with or without their permission, into the mainstream of intellectual and public discourse in the twentieth century. Even Margaret Mead's work on Samoa, taken out of context, fueled the notion of promiscuous islanders and sensual tropical paradises that we commonly associate with tourism branding and marketing (Baldacchino 2004).

HOMOGENEITY AND HETEROGENEITY ON ISLANDS

The topic of islandness raised earlier suggests that there may be some shared intangible characteristics among islanders which set them apart from those born and raised on the mainland. In this section, we will look a little more closely at the cultural similarities and dissimilarities of islanders.

In the introduction to this chapter, it was suggested that some of the principal indicators of culture include religion, language, race, ethnicity, and kinship. If we look at the distribution of some of these features on islands, can we come to any conclusions about how alike islanders are as a whole? In terms of language, there are many examples where a variety of distinct languages are spoken on islands in close proximity. Although there are only six official languages among the islands of the Caribbean (i.e., Dutch, English, French, Spanish, Haitian, and Papiamento—the latter being a Creole-based language spoken by about 330,000 people in the Netherland Antilles), there are more than seventy languages spoken in the region. Premdas (1996, 2) notes that despite the myth of the Caribbean as homogenous, "perhaps, no other region of the world is so richly varied." He uses the example of the southern Caribbean (e.g., Trinidad) and northern South America (e.g., Guyana and Suriname) where you see Asian Indians, Africans, Chinese, Syrians, Lebanese, Jews, Portuguese, Europeans, Amerindians, and various other mixes and combinations of people. He attributes this wide diversity in cultural characteristics to colonization and the plantation economy, resulting in an immigrant society "with weak social cohesion and community organization" (Premdas 1996, 8).

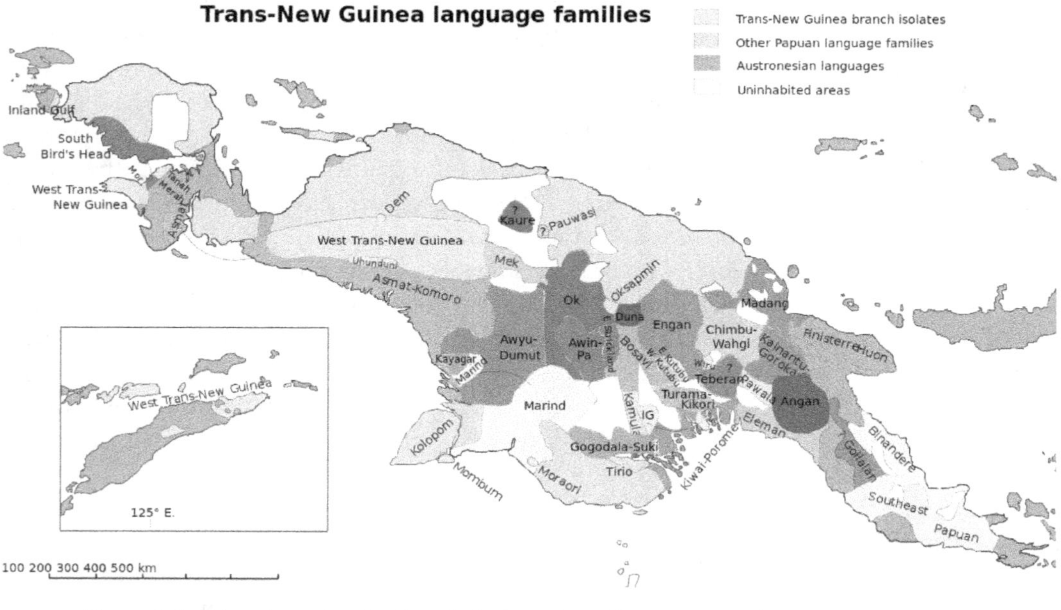

Figure 5.3 Map depicting the linguistic diversity of Papua New Guinea. *Source:* **Kwami**

Language heterogeneity exists on other island groupings. For example, as illustrated in Figure 5.3, it has been estimated that there are 780 languages spoken in the country of Papua New Guinea on the island of New Guinea. This linguistic diversity is also reported by the US State Department (2001, n.p.):

> The Indigenous population of PNG is one of the most heterogeneous in the world. PNG has several thousand separate communities, most with only a few hundred people … The isolation created by the mountainous terrain is so great that some groups, until recently, were unaware of the existence of neighboring groups only a few kilometers away.

The implication from this passage is that the physical geography of the island has contributed to a lack of mobility and interaction among the various tribes and therefore a greater diversity of language groups. Another example is on the archipelagic Oceanic nation of Vanuatu (see Figure 5.4). On the 65 inhabited islands of the Vanuatu archipelago spread over 12,274 square kilometres, 113 languages are spoken by a population of 221,000. In this case, immigration and settlement over the past 3,000 years may have led to these linguistic differences. Although the water in between these islands may have indeed served as a metaphorical highway connecting people for trade and support, especially during crises, it may have also contributed to these linguistic distinctions.

It is not always water that contributes to these cultural language distinctions. On the small British territorial islands of Jersey and Guernsey just off the coast of France, English has become the dominant language. However, for the past thousand years, versions of the Norman language (a mixture of French, English and Norse) and its various dialects were an important part of the islands' cultures. It has been said among locals that if someone is attuned to the language, they could tell where a native speaker was from to within one mile of their home (Sallabank 2011). At the same time, distance between islands may not appear to be an obstacle to the dissemination and homogenization of language in all island regions. For example, at the time of European contact in Oceania, the spoken languages on islands in the vast Polynesian region were so alike that islanders were still able to communicate with one another, despite being separated by thousands of kilometres of sea.

In some parts of Oceania, the everyday use of Indigenous languages is threatened by urbaniza-

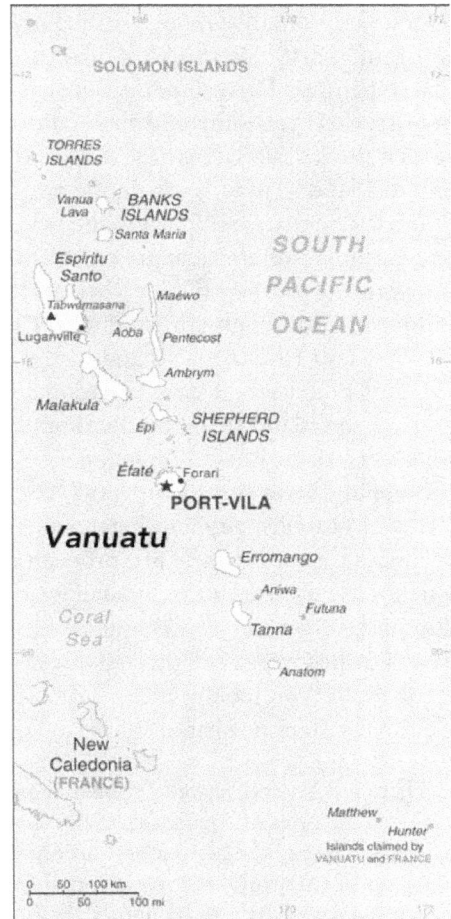

Figure 5.4 Map of Vanuatu (from Wikipedia Commons. *Source:* **public domain**

LANGUAGE ON SMALL ISLANDS ALSO SERVES AS ONE OF THE KEY FEATURES IN DEVELOPING AND MAINTAINING A STRONG SENSE OF ISLAND IDENTITY.

tion and the subsequent need to replace home languages with global languages like English and French, so that the diverse groups from across these vast archipelagos can communicate with one another in these urban centres. This has been compounded by the fact that local elites tend to favour these global colonial languages, especially since many of them have been educated at foreign English- and French-speaking universities. Concurrently, there are recent examples from New Caledonia, the Solomon Islands, and Vanuatu, where cultural centres and elders are recovering and teaching the traditional, Indigenous languages to a new generation of Indigenous islanders (UNESCO 2010).

Language on small islands also serves as one of the key features in developing and maintaining a strong sense of island identity. For example, although the Faroe Islands in the North Atlantic is a semi-autonomous territory of Denmark, the use of the Faroese language, together with other cultural symbols, has greatly contributed to creating a common identity (Hayfield and Schug 2019). In this particular case, a shared language is not the only feature symbolizing island identity. There appears to be a shared "sense of place" that influences all aspects of their politics, their economy and their culture (Hovgaard 2002; Wynne 2007).

As is the case with language, religious affiliation can be quite heterogeneous among islanders on one island or islands within the same region. Among the more than 6,000 inhabited islands of the Indonesian archipelago, the belief systems of Islam, Protestantism, Taoism, Shintoism, and Catholicism are strongly represented, including many coexisting among those living on the island of Borneo (which also contains the country of Brunei and parts of Malaysia and Indonesia). Although Christianity is the most prominent religion for most of those living on New Guinea (the largest island in the archipelago), there is a very distinct spatial pattern in religious practice that crosses national borders, with the northeastern and southwestern portions of the island being predominantly Catholic, and the southeast and northwest are areas where Protestantism is dominant.

Stepping back and looking at Oceania as a whole allows us to examine these concurrent forces of homogeneity and heterogeneity. In Fischer's (2002) *A History of the Pacific Islands*, he argues that at the time of European contact, the peoples of Micronesia and Melanesia were much more diverse culturally than were the people of Polynesia. However, especially as a function of the movement of peoples associated with colonialism, island populations in Polynesia have also become more diverse. Even with all of this diversity, Smith and Jones (2007, 18) argue that if you look at present-day Oceania, there are underlying features shaping a shared identity that are not based solely on language and religion. They state,

There is a sense of shared regional identity in the Pacific Islands which is both a product of the shared histories of Pacific societies and a response to the special character of the region. The region has one of the highest proportions of Indigenous peoples within national populations in any region of the world and has amongst the highest proportion of people living within traditional governance systems and amongst the highest proportion of land and sea remaining under traditional management of any region of the world.

Cook Islander farmer holds a watermelon. *Source:* **ChameleonsEye**

The last line of the quote may be important in understanding a shared islander culture, at least in this region of the world. Smith and Jones (2007) point out that more than 90 percent of arable land in the Pacific Islands is still held in a more traditional tenure system that includes a communal approach as opposed to a Western private-property approach. Oral traditions remain important and there is still a strong social bond with the land and the surrounding sea. Therefore, although there may be considerable differences among islanders of the Pacific in the languages they speak or in their belief systems, there appears to be an Oceanic identity that transcends these traditional cultural indicators.

Although many social groups can claim an affinity to the land as an aspect of sense of place, this also seems to be a common and enduring feature shared among islanders. For example, it is said that most Tahitians' sense of place is defined by their relationship to the land (Kahn 2000). As discussed above, there are many other examples of Pacific Islanders who have expressed similar sentiments (Feld 1996; Kahn 1996). Perhaps what sets islanders apart from other social groups that lead lives with similarly close attachments to the land is that the affinity to place does not stop at the shoreline but extends to the surrounding marine region. Perhaps even more important than on the land, a customary marine tenure system that manages resources such as community-based traditional fisheries has a long history and extends throughout South Pacific societies (Lam 1998). The interplay and interrelationship between sea and land has played such an important historical, economic, and cultural role in the lives of island communities that even with the boundary between land and water, islanders' sense of place encompasses both environments.

ALTHOUGH MANY SOCIAL GROUPS CAN CLAIM AN AFFINITY TO THE LAND AS AN ASPECT OF SENSE OF PLACE, THIS ALSO SEEMS TO BE A COMMON AND ENDURING FEATURE SHARED AMONG ISLANDERS.

ISLANDER-OUTSIDER RELATIONSHIPS IN THE TWENTY-FIRST CENTURY

It may be that the most distinguishing social features of islanders are their attitudes and actions towards those who visit the islands, either for a short period of time (e.g., tourists) or for much longer periods (e.g., immigrants). The relationships and tensions between long-time residents of communities and so-called outsiders exist in many places. However, there seems to be a heightened awareness and visibility to the interactions between islanders and mainlanders.

Even the labels given to non-islanders give you a clue to some aspects of a simmering us-versus-them relationship. For example, on Cape Breton Island, Prince Edward Island, and Newfoundland (and to a lesser extent the rest of mainland Atlantic Canada), those not born on the island are referred to as come from aways or CFAs. Even if you were born on the island, it is not unusual for you to be a CFA if your parents were not born on the island. On the island of Nantucket on the northeastern seaboard of the United States, where you find the popular vacation community of Martha's Vineyard, those who move to the island may be called wash-ashores. In the archipelagic Orkneys off the coast of northern Scotland, those who have moved to the island might be ferry-loupers, a term that has been in use for nearly two centuries. On the divided island of Cyprus, there has been a long-standing antagonism between Greek Cypriots and mainland Greeks, who are referred to as *kalamaradhes* or quill-wielders, a term that originated in the 1880s when school teachers arrived on the island from peninsular Greece to teach the islanders proper Greek (Dubin 2009). It was also applied to Greek army officers stationed on the island in the 1960s and was sometimes prefaced with the label *pousti*, literally translated to mean sodomite or "untrustworthy bastard" (Dubin 2009). In Hawai'i the word *haole* has been used by those of Hawaiian descent, many of whom themselves had immigrated to Hawai'i to work on the plantations, to describe anyone not of that cultural group, and especially to describe white Americans and Europeans (Rohrer 2010). Although the term preceded contact with Europeans, it came to mean more than just a difference in skin colour. It emerged after centuries of colonialism, and especially after the missionaries attempted to ban many of the traditional Hawaiian practices. It is now considered synonymous with a set of attitudes and behaviours that are at odds with Hawaiian values (Rohrer 2010).

There are many more examples of these labels for non-islanders; however, I will end with perhaps one of the most formal and codified terms used to distinguish islanders from outsiders. On the British Virgin Islands (BVI), a territory of the United Kingdom in the Caribbean Sea, the terms belonger and non-belonger are used roughly to distinguish between those born or not born on BVI. This is not used merely informally or in the vernacular. The labels are embedded in the territory's constitution and have repercussions for the right to work, the right to buy property, and the land taxation rate. This dual set of rules—one for islanders and one for outsiders—is not uncommon among small islands. For example, the Finnish Åland Islands

THE COMMON THREAD IN THESE TERMS IS THAT THEY ARE USED TO DISTINGUISH US FROM THEM. EVEN ON THOSE ISLANDS WHERE THE TERM IS NOT BLATANTLY DISCRIMINATORY, AS WITH THE WASH-ASHORES AND THE COME FROM AWAYS, THE UNDERLYING MESSAGE IS THAT YOU ARE NOT ONE OF US.

have strict rules limiting outsiders from buying land, presumably to protect Åland's cultural identity (R.C. Williams 2018).

The common thread in these terms is that they are used to distinguish us from them. Even on those islands where the term is not blatantly discriminatory, as with the wash-ashores and the come from aways, the underlying message is that you are not one of us. Rabin (2009) makes the argument that these social divisions on islands are more complicated than a simple us versus them dichotomy, and in fact can be used to mask some very significant differences that exist among islanders themselves, including political and religious affiliation, and community of birth or kin/family.

We also have to be careful that we do not assume simplistically that being an islander or an outsider is a binary choice. There are many versions of islanders, including those who have relocated to islands later in life (e.g., as retirees), seasonal tourists with summer or second homes on islands, and those born on an island who may have moved away but still think of themselves as islanders. On the Swedish island of Gotland, for example, true Gotlanders include those who were born on the island and may or may not continue to live on the island. Then there are the summer-Gotlanders who commute to the mainland from their island summer homes, and mainlanders who moved to the island later in life (Ronström 2012). In order to distinguish between those who may call themselves Gotlanders merely because it is their place of residence, a new "ur-Gotar" category of islander has emerged. These are usually rural farmers whose families have lived on the island for at least three generations, who speak an older dialect and lead a traditional life (Ronström 2012). As such they are perceived to be a more exclusive and authentic version of a Gotland islander. All of this shows that the term "islander" is more complex and contested than we might at first assume.

If you looked at these questions concerning the relationships between islanders and outsiders theoretically, it could be argued that many small islands are among the best examples of places that demonstrate the kinds and degrees of social capital discussed by Robert Putnam (2000) in his popular book, *Bowling Alone*. Islanders may have developed a highly sophisticated level of social bonding capital, an important asset that has been employed to withstand crises brought from the mainland. In the twentieth century, many of these crises are associated with various aspects of globalization. However, island communities also appear to have weak social bridging capital, a quality that is critically important in embracing diversity and difference, and building an integrative, tolerant society (Cave, Brown, and Baldacchino 2012). Cave et al. go on to note that those who may want to stay are often subtly excluded from fully participating in an "island way of life" (2012). This may become increasingly problematic for those islands that find themselves with a declining population, either as a result of lower birth rates and/or youth migration, and whose governments are attempting to reverse this trend by encouraging non-islanders to immigrate to their islands.

A Case Study of Social Relationships Between Islanders and Outsiders: Grand Manan Island, New Brunswick, Canada

The small island of Grand Manan, located near the mouth of the Bay of Fundy and a ninety-minute ferry ride from the rest of the Canadian mainland province of New Brunswick (see Figure 5.5), serves as a good example of a place where tensions sometimes arise between islanders and those from away, including their own mainland-based governments. The popula-

Figure 5.5 Salmon fish farms on Grand Manan Island, New Brunswick, Canada. *Source*: **Russ Hienl/ Shutterstock**

tion of the island is approximately 2,400 and this has remained fairly stable for the past 100 years. The anthropologist Joan Marshall (1999a, 1999b, 2008) has studied the island community in detail, and describes it as having a strong sense of community, as well as a spirit of independence and flexibility. In other words, it has an abundance of social bonding capital and many of the characteristics of islandness raised earlier in this chapter.

Over the past twenty-five years, the world has intruded into the social and economic life of Grand Manan Island, creating tensions between the traditional and the modern lifestyle. Economically, this has played out in the growth of corporate salmon aquaculture farms in the bays surrounding the island, in the increasing demand for tourist access, and in the evolution of harvesting of rockweed. As the name suggests, rockweed is a seaweed that grows on rocks in the intertidal zone and it has been harvested for many years by small-scale local fishing families. It has been sold as a health food, a nutritional supplement for pets/humans, and a stabilizer in cosmetics and food. In raw form, it acts as an organic fertilizer and is spread on fields and gardens. It is also an important part of the marine ecosystem, converting inorganic nutrients into organic nutrients for animals, fish, and birds, and providing shelter for juvenile fish, crustaceans, and sea urchins. Traditionally, the rockweed harvest has been an important part of Grand Manan's social economy and has been harvested using a form of community commons which allows families who rely on seasonal industries to incorporate one more source of income. In the 1990s, in the guise of economic development and with little local public consultation, the provincial New Brunswick government licensed an out-of-province company to harvest rockweed on the island, largely for export to the United States (Marshall 2008). Despite widespread local protest that centred on who ultimately controls decision-making along their coastline, and uncertainty regarding the impacts on the fishery and other marine crops, the local community was unable to have the corporate harvest stopped. Perhaps more pervasively, these external influences have challenged gender relations. With the loss of home-centred roles and community centres, women have been forced to adopt new spaces to maintain their social connections. For males, the shift from being fishermen to being aquaculture employees has challenged their own notions of identity and masculinity (Marshall 2001).

CONCLUSIONS

Examining the relationship between culture and place is nothing new and is hardly unique to islands. However, on many small islands the attachment to the land and the sea and the close and dense network of social relations among islanders represents an additional cultural dimension not observed in most mainland communities. This islandness may be why island jurisdictions often have some of the highest rates of volunteerism, democratic participation (e.g., through voting), and civic engagement, while at the same time exhibiting characteristics of a less tolerant nature to those from away who may wish to settle on their islands. Despite differences in religion, language, political leanings, or community, this sense of social solidarity may have contributed to the historical resilience of island societies but may also threaten their future in a globalized world.

Key Readings

Marshall, Joan. 2008. *Tides of Change on Grand Manan Island: Culture and Belonging in a Fishing Community*. Montreal: McGill-Queens University Press.

Banana plantation in St. Lucia. *Source*: **Brian Snelson**

Chapter Six

Geopolitics and Island Governance

The history of the Mediterranean islands is like an enlarged photograph of the history of the Mediterranean region as a whole. As the least-weighty fragments of land, history has tossed them hither and thither. (R. King 1993, 22)

INTRODUCTION

Islands have been a constant player in geopolitics. They have been in the way of, and a part of, the advances and retreats of imperial nations. They have been contested by larger mainland countries and transnational corporations (TNCs) for what they have, for where they are located, or for what they might have in the future. Islands have been used to quarantine asylum seekers, to test atomic devices, and to exile political opponents. At the same time, island governments have used their political and geopolitical position to negotiate advantages for islanders. Increasingly, island states and territories appear to be exercising a kind of political and economic flexibility that is unexpected given their size and which allows them to achieve goals that much larger mainland regions may find difficult. As such, islands can be resilient while still displaying vulnerable characteristics. They can be remote yet also be instrumental in global political change.

This chapter examines the politics of islands. It starts with a discussion of geopolitics and how modern geopolitics is reflected in the economic realities facing small islands. It examines the grand period of growing island independence and the emergence on the world stage of small islands as a cohesive group capable of influencing global change. It also suggests that islands are becoming increasingly comfortable with the niche role they play in negotiating relationships with other jurisdictions.

GEOPOLITICS AND ISLANDS

It may be easier to understand the political relationships between islands and the rest of the world if we start with the concept of geopolitics. One definition of geopolitics is: "The study

> **GEOPOLITICS TRIES TO UNDERSTAND NATIONAL FOREIGN POLICY AND THE RELATIONS BETWEEN STATES. IT INVOLVES THE APPLICATION OF STATE POWER OVER OTHER STATES, MOST OFTEN IN ECONOMIC OR MILITARY TERMS.**

of the relationship among politics and geography, demography, and economics, especially with respect to the foreign policy of a nation" (Free Online Dictionary 2019b). The study of geopolitics tries to understand national foreign policy and the relations between states. It involves the application of state power over other states, most often in economic or military terms. It also encompasses the role that the private sector may play in partnership with the state in order to achieve mutually desirable goals.

Geopolitics was originally used in the early twentieth century by the British geographer Sir Halford MacKinder to describe what he called the heartland theory. This theory can best be characterized by the following quote from MacKinder (1942, 106): "Who rules East Europe commands the Heartland. Who rules the Heartland commands the World-Island. Who rules the World-Island commands the World." MacKinder's heartland or pivot area consisted of Russia and Eastern Europe, an area that was difficult to attack given the state of military technology at the time. His metaphorical World-Island was represented by all the other continents of the world that were relatively accessible by water and therefore vulnerable to military attack by sea. When this geopolitical concept was combined with the idea of *lebensraum* or living space/habitat developed by German political geographers such as Friedrich Ratzel and Karl Haushofer, the outcome was a galvanizing ideology used by Nazi Germany to justify German expansion across Europe leading up to the Second World War. Applied this way, these two concepts suggested that a nation-state could be like a living creature that had to compete and expand into new territories in order to survive. Although it could be argued that this was just another form of imperial colonialism similar to the way European powers such as England, France, Spain, and the Netherlands had previously explored and claimed territories during the earlier Age of Exploration, there was an important difference. During that earlier imperial age, the conquered lands would never be more than colonies of the empire. In the case of Nazi Germany's geopolitical expansion, the conquered territories were to become part of the German nation-state.

Other than the use of the term world-island as a metaphor for spatial central-peripheral state relations, islands were not particularly important in this early geopolitical story. However, in the sense that island populations, economies, and relationships have been shaped throughout history by the actions of larger continental states, geopolitics has been an important feature of island futures.

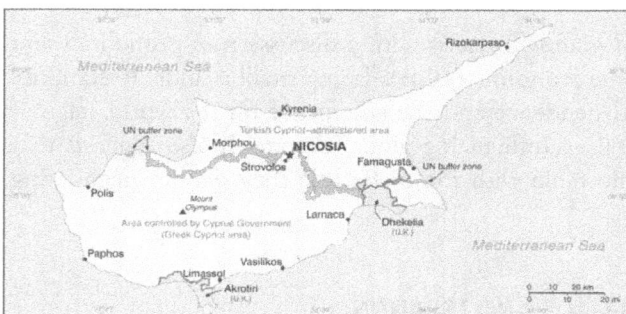

Figure 6.1 Map showing the political division of Cyprus (c. 2010). *Source*: **CIA World Factbook, public domain**

Stephen Royle (2001) has argued that throughout history many islands have been powerless to withstand the military and economic might of larger, continental states, and have therefore been subjected to repeated invasions and conquests. The examples of the island of Cyprus in the eastern Mediterranean Sea and of Saint Lucia in the Caribbean illustrate the degree to which larger powers have shaped the destinies of islands.

In the case of Cyprus, foreign powers have invaded the island at least thirteen times in the past 3,000 years (see Figure 6.1 and sidebar below). These waves of invasions started with the Mycenaeans and the Achaeans in the fourteenth to eleventh century BC and ended with the invasion by Turkey in 1974 (Frendo 1993). This has left the island politically divided along the "Green Line," influenced strongly in the northeast by Turkey and in the south by an independent nation-state that is heavily influenced by Greece.

A second example of island vulnerability in the face of external military intervention may be associated with the island of Saint Lucia. As Figure 6.2 shows, Saint Lucia is located in the Caribbean Sea's Windward Islands and is part of the archipelago known as the Lesser Antilles. It has a current population of about 175,000. The island was originally settled by the Carib people who resisted efforts by British and French to gain control over it during the first half of the seventeenth century. After finally succumbing to European colonization, over the next 400 hundred years this tiny island would change hands between the British and the French at least fourteen times before it gained its independence in 1979 (see sidebar on next page).

Island Geopolitics in the Twenty-First Century

Unequal power relations and geopolitics in the world today is not solely a function of military might. With the global reach and influence of transnational companies (TNCs) and other actors associated with globalization, small islands may find themselves unable to prevent external forces from controlling their economy and culture. In Royle's (2001, 60) words, "Powerlessness is not normally associated with military battles, it is the daily dependence on

INVASIONS OF CYPRUS

- 14th–11th C. BC – Mycenaean & Achaeans
- 9th C. BC – Phoenicians
- 7th C. BC – Assyrians
- 4th C. BC – Greeks
- 3rd C. BC – Egyptians
- 58 BC–395 AD – Romans
- 7th–10 C. AD – Byzantines & Muslims
- 1191 AD – Richard the Lionheart of England
- 13th–16th C. AD – Sold to French nobleman Guy de Lusignan & family
- 1498–1571 AD – Venetians
- 1571–1878 – Turks
- 1878–1960 – British
- 1960 – Gained independence from Britain
- 1974 – Invasion by Turkey and the island was partitioned

Source: Compiled by author

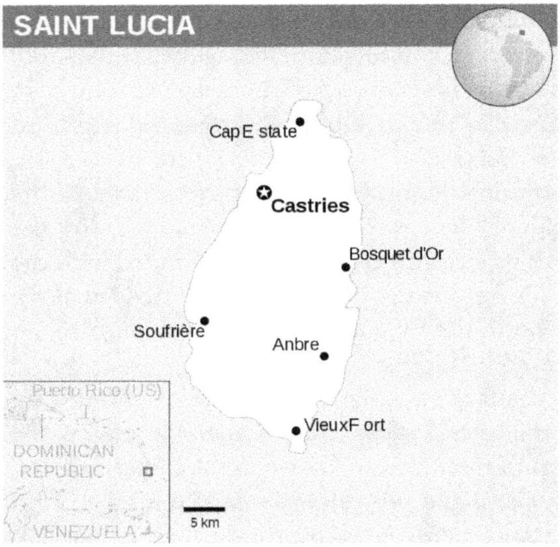

Figure 6.2   Location of the island of Saint Lucia.
Source: **OCHA**

CONTROL OF SAINT LUCIA

* **1635 Officially claimed by France**
* **1639 English settlement**
* **1643 French settlement**
* **1664 English fort (1,000 men)**
* **1674 French crown colony**
* **1723 Neutral territory (agreed by Britain and France)**
* **1743 French colony (Sainte Lucie)**
* **1748 Neutral territory (agreed by Britain and France)**
* **1756 French colony (Sainte Lucie)**
* **1762 British occupation**
* **1763 Restored to France**
* **1778 British occupation**
* **1783 Restored to France**
* **1796 British occupation**
* **1802 Restored to France**
* **1803 British occupation**
* **1814 British possession**
* **1979 Gained independence as a nation-state**

Source: Compiled by author

more powerful forces off the island to take decisions that affect the life and livelihoods of islanders." One might think of this as a modern version of geopolitics, where states or groups of states cooperate with transnational corporations and international regulatory bodies to control the development of smaller (island) states and territories.

Another example from the Caribbean, in this case using the three island groupings of Saint Lucia, Dominica, and Saint Vincent & the Grenadines, may help to illustrate how this modern geopolitics works. All of these islands are now independent nation-states, with Saint Lucia and Saint Vincent & the Grenadines gaining independence in 1979, and Dominica achieving it in 1978. However, in addition to specializing in tourism, their economies have traditionally been highly dependent on the export of agricultural products, and especially bananas. Throughout the 1990s, bananas accounted for more than half of all the export revenues earned by these islands (Mlachila, Cashin, and Haines 2010). In fact, the term "banana republic" was coined to refer to poor developing countries that relied on a single export commodity (Josling and Taylor 2003). The former Prime Minister of Saint Lucia was famously quoted as saying, "bananas are to us what cars are to Detroit" (Green 2001, 128). In the late 1990s, the World Trade Organization (WTO), after being lobbied by the United States and TNCs that owned large banana plantations elsewhere in Central and South America, supported a ban on the import of bananas to the United States from the islands of the West Indies, thereby lessening the competition that these companies would face for importing to the American market. The Minister of Commerce, Industry and Consumer Affairs on Saint Lucia best characterized the powerlessness of the island to this situation when he stated, "Globally, we're just a lonely pawn on a gigantic chessboard, surrounded by kings, queens, and rooks who are waiting their moment to pounce" (Royle 2001, 141–2).

Despite the fact that revenues from banana production in Saint Lucia declined by 11 percent per year during the 1990s, banana sales still accounted for 12 percent of the island's GDP, 40 percent of its export earnings, and 30 percent of direct and indirect employment, as sales have shifted to markets in Europe (Arias, Dankers, Liu, and Pilkauskas 2003). However, even these markets are now threatened by lower-cost banana imports from elsewhere in Latin America.

ISLANDS AND FOREIGN MILITARY BASES

> WHEN FOREIGN MILITARY BASES ARE SITUATED ON ISLANDS, THEY OFTEN CREATE A POLITICAL AND ECONOMIC DILEMMA FOR ISLANDERS AND ISLAND GOVERNMENTS.

Although direct military intervention may not be as common now as it was in the past, many islands still play an important strategic role for the military superpowers of the world. For example, at various times the United States has had military bases on islands such as Cuba (Guantanamo Bay), Bermuda, the Turks and Caicos, Japan (including Okinawa), the Philippines, Ascension Island, Diego Garcia, the Marshall Islands, Guam, and Greenland. When foreign military bases are situated on islands, they often create a political and economic dilemma for islanders and island governments. On one hand, they can bring economic benefits to local communities. This occurs in the form of employment for local residents and increased spending by purchasing goods and services from local businesses. Rent may also be paid to the island government as part of the agreement to allow the presence of a foreign military base (Biglaiser and DeRouen 2007; Heo and Ye 2017). The existence of the base may also deter other neighbours from encroaching on the island's territory. Finally, there may be other concessions that are associated with the presence of the base that benefit the economy of the island nation, including preferred trading arrangements and low or no interest loans or other forms of aid.

Conversely, there may also be disadvantages to this military presence. The agreements for these bases may lock an island into political alignment with a more powerful country, often for decades or generations. Therefore, it may reduce an island's flexibility to make arrangements with other countries that are not aligned with that superpower. The long-term foreign military presence is also a constant reminder to islanders that their territorial sovereignty is being compromised. Therefore, it is not uncommon for there to be locally based opposition to military bases, and this opposition spikes when foreign soldiers are accused of crimes against local citizenry (Vine 2019). For example, there have been frequent protests by local Okinawans in response to the continued presence of the American military base on the island of Okinawa, Japan. In Okinawa, there are two layers of resistance: not only to the American military presence but also to the central Japanese government and its perceived complicity with the Americans (McCormack 2018). The "mainland" islands that tend to hold the balance of economic and political power may also be considered outsiders to those on the smaller,

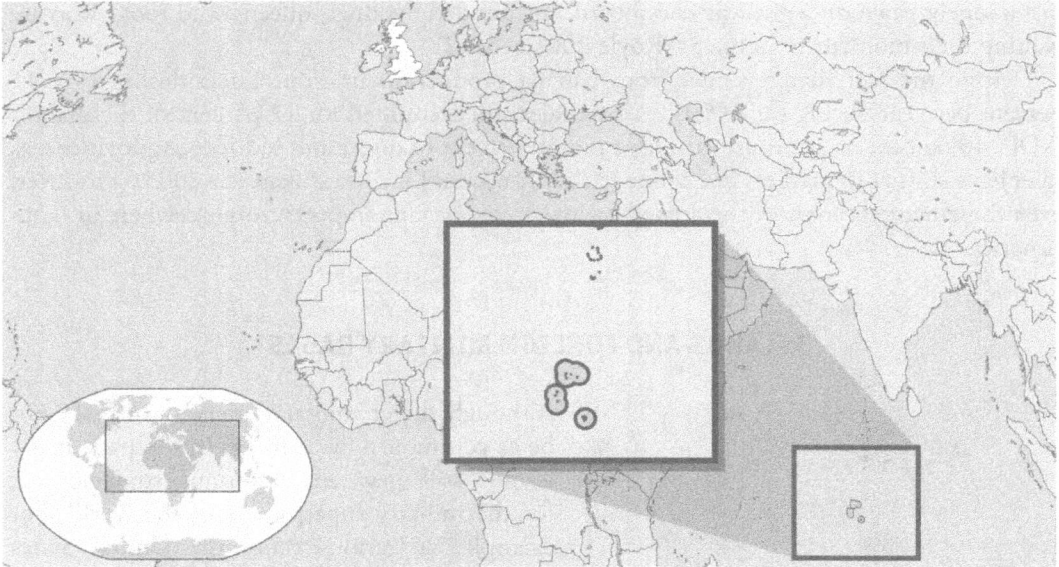

Figure 6.3   Location of the Chagos Archipelago. *Source:* **TUBS**

more remote islands of an archipelago, with values and agendas that do not match those on the smaller islands.

The island of Diego Garcia in the Indian Ocean is an important example, for it illustrates how islands are affected by the relationships between superpowers and how island populations are often unable to use international law to gain control over their own future. As shown in Figure 6.3, Diego Garcia is a small atoll in the much larger Chagos Archipelago, located 1,800 kilometres south of India, 4,700 kilometres west of Australia, and 3,500 kilometres east of the coast of Africa. For over 150 years, it was part of the British Indian Ocean Territory (BIOT) and was considered a Dependency of the British island colony of Mauritius. In 1965, the United Kingdom (UK) purchased Diego Garcia and the other islands of the Chagos Archipelago from the island of Mauritius, which was itself a territory of Britain at the time. The next year Britain entered into an agreement with the United States to lease the island for fifty years, with an option for an additional twenty-year extension. Although there were no permanent inhabitants on the island when it was encountered by the Portuguese, French, and British in the sixteenth to eighteenth centuries, a settlement of approximately 1,500 Chagossians (also known as *Îlois*, a French Creole term meaning islander) lived on the island shortly after it was turned over to the United States. These islanders were descendants of labourers who were brought to the island by the French and British to work in the coconut oil plantations.

Although it is unclear whether the US or the UK precipitated the move (Snoxell 2009), all of these inhabitants were forcibly relocated to Mauritius, the Seychelles, or other islands in the Chagos Archipelago, with the last person departing in 1973 (González-Salzberg and Hodson 2019; Snoxell 2008). Although a modest monetary compensation was given to some of these deportees, there has been ongoing litigation by descendants of the Chagossians to return to their island. Some have argued that the United Kingdom's declaration in 2010 (with the tacit approval of the United States) that the waters around Diego Garcia and the surrounding BIOT be protected as a marine reserve was really just a political manoeuvre to discourage the descendants of Diego Garcia from being allowed to return to the area because they could no longer make a living by fishing the surrounding waters (Sand 2011). Most recently, the International Court of Justice has told the UK to end its administration of the Chagos Islands and return them to Mauritius as soon as possible (Yiallourides 2019).

This section has demonstrated that geopolitics is not just an antiquated concept that describes the military strategies of superpowers in the twentieth century. It still applies to political relationships in the twenty-first century, as islands continue to be at the centre of power relationships among nations, global organizations, and companies.

DIVIDED ISLANDS

One of the other political features associated with islands are what has been referred to as "divided islands" (Baldacchino 2013). These are islands that are formally divided and administered by two or more nation-states. As the sidebar below shows, there are only ten examples of divided islands in the world.

Although the reasons for these divisions are often unique and rooted in the circumstances associated with the development of these places, many of the divisions are associated with a geographic clustering of people based on religion (e.g., Ireland: Catholics and Protestants) or ethnicity (e.g., Cyprus: Greek and Turkish influences). Many of them also became divided politically after some kind of external influence, including the territorial claims of two or more colonial powers, or the migration or invasion of a foreign power. Perhaps the more intriguing question should be, out of all the thousands of islands in the world, why are there only ten that are officially partitioned between two or more nations? In *The Political Economy of Divided Islands*, Baldacchino (2013, 3) speculates that this relative scarcity of islands bisected by national boundaries may occur because they are so often imagined as absolute spaces or imaginable wholes, "an island is easier to hold, to own, manage or manipulate, to embrace and to caress." It may also be because what makes up a community's sense of identity, or what could be referred to as its nationality, fits so closely with the finite jurisdictional boundaries or

DIVIDED ISLANDS

- Hispaniola (Haiti and Dominican Republic)
- New Guinea (Papua New Guinea and Indonesia)
- Ireland (United Kingdom and the Republic of Ireland)
- Timor (Indonesia and Timor Leste)
- Cyprus (Republic of Cyprus and the Turkish Republic of Northern Cyprus; this latter is only recognized officially by Turkey
- Borneo (Indonesia, Brunei, and Malaysia)
- Tierra del Fuego (Argentina and Chile)
- Saint-Martin/Sint Maarten (France and the Netherlands)
- Uzedom/Uzman (Germany and Poland
- Heixiazi/Bolshoi Ussuriyski (China and Russia)

Source: Compiled by author

the boundedness of islands. Baldacchino goes on to say that with the rise of the nation-state as a political concept, the number of examples of divided islands has continued to decrease.

Beyond these "pure" examples of islands that share international land borders, there are many more that have ambiguous, semi-autonomous political relationships with neighbours or former colonial powers. This next section explores these more complex political relationships.

ISLAND DECOLONIZATION AND SUPRANATIONAL ISLAND ORGANIZATIONS

With the massive wave of decolonization that took place following the Second World War, many more small islands gained their political independence. To give you some perspective, when the United Nations was formed in 1945 with fifty-one members, only six (11.8%) were islands. By 2019, the number of UN members had grown to one hundred and ninety-three and forty-five of these (or 23.3 %) were either a single island or, more commonly, groups of islands (Watts 2009).

Since the war, formal groups of small islands have become more prominent on the international political stage. For example, the Small Island Developing States (or SIDS) was accepted by the United Nations as a credible, coherent group of jurisdictions on the basis of similarities in their environments and development at the 1992 UN Conference on Environment and Development in Rio de Janeiro, Brazil. This was reinforced with the 1994 Declaration of Barbados and the Barbados Programme of Action (United Nations 1994). These were important steps in raising the profile of small islands because prior to this point their populations and economies meant that as individuals they were often viewed by international organizations (e.g., the IMF, Worldbank.org) and more powerful states as "mere pawns in a game played out be larger states" (Scheyvens and Momsen 2008, 504). With a shared voice on an international stage, SIDS started to take a more prominent role in areas such as economic and environmental development (Campling 2006). This growing South-South cooperation, even among island states in different oceans, meant that they were less likely to be ignored.

Figure 6.4 Map of Small Island Developing States (SIDS) as listed by UNESCO and UN-OHRLLS as of August 2014. *Source:* **Osiris**

Because there is no formal agreement on the criteria for SIDS membership, different sources will have different lists. The United Nations Conference on Trade and Development (or UNCTAD) counts twenty-nine island nation-states as facing a greater risk of marginalization due to their small size, their remoteness from large markets, and a high economic vulnerability to economic and natural shocks beyond domestic control. Another report published in 2011 by the United Nations Office of the High Representative for the Least Developed Countries, Landlocked Developing Countries, and Small Island Developing States counts thirty-eight UN member countries and a further twenty non-UN island territories as SIDS (UN-OHRLLS 2019) (see Figure 6.4). Their international prominence has occurred in no small part to their advocacy given the threats to their livelihood and existence as a result of the impacts of human-induced global warming. Leading up to the 2009 Climate Change Conference in Copenhagen, the then President of the Maldives, Mohamed Nasheed, gave several impassioned speeches, reminding the world that his nation of low-lying atolls in the Indian Ocean would no longer exist if the world did not take action to reduce greenhouse gas (GHG) emissions. The story of his plea for action was captured in a highly acclaimed 2011 documentary entitled *The Island President*.

> WITH A SHARED VOICE ON AN INTERNATIONAL STAGE, SIDS STARTED TO TAKE A MORE PROMINENT ROLE IN AREAS SUCH AS ECONOMIC AND ENVIRONMENTAL DEVELOPMENT (CAMPLING 2006). THIS GROWING SOUTH-SOUTH COOPERATION, EVEN AMONG ISLAND STATES IN DIFFERENT OCEANS, MEANT THAT THEY WERE LESS LIKELY TO BE IGNORED.

Two UN reports are especially important to mention with respect to SIDS. The first is the Barbados Programme of Action (1994) that looked at SIDS in terms of their sustainable development efforts. The Mauritius Strategy for Implementation (2005) was intended to take this one step further, articulating specific strategies for SIDS to achieve sustainable development. The more recent international meeting on the sustainable development of SIDS took place on Samoa in September 2014 and produced a "Pathway" that has been used by many of the states to cooperate on some of the current issues facing them, including climate change, sustainable energy, disaster risk reduction, marine science, and sustainable tourism (Esteves et al. 2019)

It is notable that recently the terminology has shifted away from the use of the term SIDS and particularly the negative connotation associated with the words "small" and "developing," to the phrase "large ocean states" (Chan 2018; Jumeau 2013) at least partly because the small land area of islands and archipelagos does not adequately reflect the vast ocean spaces surrounding and connecting these islands, especially since the implementation of Exclusive Economic Zones. As the Seychelles Ambassador for Climate change and SIDS Issues stated, "We are the ocean people … The oceans define who we are and the coastal and marine environment is an integral part of our island lifestyle" (Jumeau 2013, 2).

Subnational Island Jurisdictions

Although the number of independent island nation-states has increased dramatically, there are many more islands in the world that maintain a politically ambiguous relationship with a parent country or metropole. In a 2007 research project carried out through the island studies program at the University of Prince Edward Island, Kathy Stuart (2009) and her student colleagues estimated that there were 116 subnational island jurisdictions (or SNIJs). In fact, a consensus is emerging among island studies researchers that, and despite the urgings of the United Nations and its decolonization process, gaining full autonomy as a nation-state may not be in the best long-term development interests of islands in the early twenty-first century and these islanders are perfectly content to negotiate within the current semiautonomous arrangement.

So what are these "quasi-independent" islands? Unfortunately, there is not a simple answer to this question. They include islands that are fairly autonomous within a larger federation such as the states of Hawai'i in the United States and Tasmania in Australia. Pete Hay (2002) refers to these as "outrigger islands," a nautical term that seems to be an apt Oceanic metaphor. SNIJs may also include territories, dependencies, or autonomous regions that are remnants of a colonial past, such as Martinique, Guadeloupe, and French Polynesia (France), the British Virgin Islands, Cayman Islands, and Anguilla (UK), Greenland (Denmark), the Azores (Portugal), and the Canary Islands (Spain). Some have an ongoing territorial or colonial relationship, such as the American territories of Guam, American Samoa, Puerto Rico, and the US Virgin Islands. They also include oddities, such as the United Kingdom's distant and tiny Pitcairn Island, the home of the descendants of the British ship HMS *Bounty* mutineers, as well as the Isle of Man and the Channel Islands of Guernsey, Jersey, and Alderney that are much closer to mainland France than they are to Britain. In the Pacific, the Cook Islands and Niue are jurisdictions "in free association" with the unlikely neocolonial country of New Zealand. Åland, an island archipelago in the Baltic Sea, is an autonomous region of Finland, but the citizens of this archipelago identify much more with Sweden culturally and linguistically than they do with Finland.

The kinds of relationships and dependencies that SNIJs have with their mostly larger nation-state parents can vary widely. In many cases the subnational island depends on the metropole to provide diplomatic and military support in the event of an attack, as was the case when the United Kingdom came to the aid of the Falkland Islands/Malvinas in 1982 after the islands were occupied by Argentina. In other cases they provide financial support in the event of natural disasters. For example, when the small Caribbean island of Montserrat suffered the dual disasters of Hurricane Hugo in 1989 and the Soufrière Hills volcano explosion in 1995, the United Kingdom, albeit reluctantly, contributed large amounts of capital to repair the island infrastructure, and relocated nearly two-thirds of the island population, mostly to the UK (Philpott 1999; Taylor 2000). Large amounts of capital can also flow to overseas island territories on a regular basis to assist in providing public services such as health, education, water and sanitation, transportation infrastructure, and public administration. For example, in 2014 the French Development Agency spent over 1.5 billion Euros on French overseas territories and departments, primarily on loans and loan guarantees (Agence Française du Développement 2015). Because of preferential rates of taxation and duties, imports from the metropole to the island may be less expensive and island businesses may be able to export products to a large market without the same level of duties and fees that might be imposed on exports to other parts of the world.

Many SNIJs have used their ambiguous status to develop a thriving economy as offshore financial centres, exploiting tax, insurance, and investment loopholes, guaranteeing a high degree of confidentiality and a stable regulatory environment. For example, despite having a population of less than 60,000, in 2017 the Cayman Islands had 65,000 registered companies, including more than 280 banks, 700 insurers, and 10,500 mutual funds (Central Intelligence Agency 2019b). The fact that the citizens of many of these island jurisdictions hold passports to their parent nation-states means that they are able to move back and forth relatively easily. For example, as of the 2006 New Zealand Census, 58,000 Cook Islanders, almost 22,500 Niueans, and 1,400 Tokelauans were living in NZ. This constituted, respectively, 2.7 times, 10.4 times, and 4.9 times the total populations still living in the SNIJs themselves (Stats NZ 2006). Although this may lead to a form of brain drain with the most skilled and educated leaving home, it can also serve as a safety valve if there is a high rate of unemployment and lead to substantial amounts of earned income in the form of remittances flowing back to households, communities, and governments in the SNIJs. Leslie and Prinsen (2018) suggest that territories such as New Caledonia and French Polynesia are quite successful at negotiating assistance for development projects that may not be in the best interests of their metropole.

> MANY SNIJS HAVE USED THEIR AMBIGUOUS STATUS TO DEVELOP A THRIVING ECONOMY AS OFFSHORE FINANCIAL CENTRES, EXPLOITING TAX, INSURANCE, AND INVESTMENT LOOPHOLES, GUARANTEEING A HIGH DEGREE OF CONFIDENTIALITY AND A STABLE REGULATORY ENVIRONMENT.

A more complete discussion of the political and economic advantages to SNIJs maintaining a shared autonomous relationship to a larger and more powerful nation-state is found in Baldacchino (2006d), G. Fisher, Britto, and Thomas (2015), and Winters and Martins (2004).

One of the emerging themes on the economic capacity of islands is that governments and businesses on SNIJs may employ a highly sophisticated political and economic knowledge and flexibility to extract as much benefit as possible from the relationship. Despite the expectation by the United Nations Decolonization Committee that it would be a natural progression for former colonies to seek independence as an ultimate development goal, in referendum after referendum, many small island jurisdictions continue to vote overwhelmingly to retain the status quo with their "parent" country. In a 2013 referendum in the Falkland Islands, 99.8 percent of eligible voters (with only three dissenting votes) opted to stay within their current arrangement with the UK (Dodds 2013). In response to the question "Do you want New Caledonia to achieve full sovereignty [from France] and become independent," 56 percent of New Caledonians voted "No" (Connell 2019). Finally, in Montserrat, David Taylor (2000, 338) states that there is a tacit agreement among politicians and voters not to ask for independence: "Most politicians might like it in their heart of hearts ... but they know that people would not in general vote for it." Some island jurisdictions, such as French Polynesia, even object to being on the Decolonization Committee's list of places that are awaiting the ultimate development prize of political independence. Those who study the international relationships of small states are saying that in an era of globalization, where transnational corporations and international trade agreements are becoming more common, the role of the nation-state is evolving, "from a world of nation states to a world of constrained state sovereignty and increased inter-state linkages" (Watts 2009, 24). However, when we move beyond this anecdotal evidence, does this status really benefit these islands? McElroy and Pearce (2006) have compared groups of independent island states to "dependent" SNIJs across a range of economic and social indicators and concluded that the latter are statistically better off when measured in terms of things such as GDP per capita, life expectancy, and infant mortality rate, but not characteristics such as literacy and unemployment rates.

We should also be asking ourselves about the value of this continued political ambiguity for the metropoles. In other words, what do the former colonial empires get out of maintaining this relationship? In the case of island military bases, one of the reasons is obvious. It allows military superpowers to have a territorial and strategic presence far from their home continent (e.g., Guam or Diego Garcia for the US, Ascension Island and the Falklands for the UK, Réunion and New Caledonia for France). It also allows them to carry out social and economic policy that might be politically unpopular at home. As described below, nuclear testing by France and the United States in the Pacific might be the most extreme example of this benefit. More recently, Australia removed its territory of Christmas Island from its migration zone and

> WE SHOULD ALSO BE ASKING OURSELVES ABOUT THE VALUE OF THIS CONTINUED POLITICAL AMBIGUITY FOR THE METROPOLES. IN OTHER WORDS, WHAT DO THE FORMER COLONIAL EMPIRES GET OUT OF MAINTAINING THIS RELATIONSHIP?

then entered into an agreement to transship any asylum seekers picked up on the high seas and process them on the island before shipping them on to third countries. In a similar fashion, the United States used the military detention centre on Guantanamo Bay, Cuba, to detain alleged combatants from the war in Afghanistan.

SNIJs also allow governments the opportunity to give wealthy individuals and influential corporations an opportunity to escape the higher tax and regulatory environments of their home countries. These offshore financial centres, such as the Isle of Man and the British Virgin Islands, attract considerable wealth from the UK and elsewhere. In many cases, the metropoles could change the agreements they have with their SNIJs to close off the tax loopholes on these islands but they choose not to do so, instead engaging in periodic crackdowns on tax avoidance, tax evasion, and money laundering.

For the most part, former colonial empires such as France and the United States have taken a very different approach to SNIJs than have the United Kingdom and the Kingdom of the Netherlands. For the latter pair, near the end of the twentieth century they were trying to shed themselves of their governance responsibilities and the costs associated with governing all of these overseas territories, and instead were encouraging the SNIJs to seek independence. Of course, there are exceptions to this trend. When it is in the domestic political interests of a former colonial empire to intervene, they are quick to do so. Such was the case when Argentina invaded the Malvinas/Falklands in 1982 and Margaret Thatcher's Conservative British government engaged in a short-term war with the South American country to defend their territorial citizens and at the same time, some would argue, increase support for her party at home (H. Clarke et al. 1990). Even France has had an ambivalent history with its overseas territories. After Algeria gained independence in 1962, France shifted the site of nuclear testing to its possessions in the South Pacific and conducted hundreds of underground and aerial nuclear detonations on the atolls of Mururoa and Fangataufa, French Polynesia, between 1966 and 1996. This out-of-sight, out-of-mind atomic bomb testing by France (and to a lesser extent by the United States in the "Pacific Proving Grounds" of the Marshall Islands and Bikini Atoll) continued despite broad regional condemnation of the practice. Notwithstanding this use, in the mid-1970s, French President Charles de Gaulle referred to France's island territories as "les poussières de l'Empire" and "les confettis de l'Empire" (translated respectively as the dust and the confetti of the Empire) and future French President Valéry Giscard d'Estaing referred to them as "les danseuses qui coûtent cher," or "the costly dancing girls" (D. Fisher 2011). Of course, those attitudes have changed somewhat given the large deposits of nickel being mined on places like New Caledonia, the shift in economic power to the Asia Pacific region, and the potential marine resources associated with the 7.3 million square kilometres of Exclusive Economic Zone around these Pacific territories. Reflecting this more recent change of attitude, in 2010 the then-President of France Nicolas Sarkozy referred to his country as "*France des trois océans*," or the "France of three oceans" (D. Fisher 2012) and, in visits to Martinique and Guadeloupe in 2009 during labour unrest, he said that negotiations between the territories and France were not about independence but rather about the "right amount of autonomy"(Reuters 2009).

With few exceptions, France and the United States have retained their SNIJs and continue to support them financially. In some cases, they are also incorporated in different ways. For example, the French Overseas Departments and Territories elect representatives who sit as members of the French National Assembly and the Senate and the residents of these territories have the right to vote in European Parliament elections. So, although the total number of elected officials is still a very small share of the French Legislature and Senate, at least in theory they can influence French political decision-making.

These kinds of relationships do not only exist between mainland empires (or former empires) and small island outposts. In fact, the relationships among islands within an archipelago sometimes mimic the relationships between islands and their metropoles, with an asymmetrical political and economic relationship between the "main island," as defined by population, economic production, or areal extent, and the smaller peripheral islands. There are several examples, such as in the Shetlands and the Orkneys, where this main island is called Mainland, a reflection of their dominance in the archipelago.

A STRUCTURAL PERSPECTIVE ON ISLAND GOVERNANCE SYSTEMS

The kinds of political roles that small islands play in the world can be incredibly confusing and frustrating to understand. For example, what is the difference between a protectorate and an autonomous region, and how do you distinguish a French overseas department from an overseas region or an overseas territory? Edward Warrington and David Milne (2018) attempted to provide a different kind of classification in order to reduce this complexity; one that uses a political economy approach and recognizes that the political position of an island in the world, or its governance, can evolve over time. They based their categorization on four key elements of the island: 1) the geostrategic role and imperial connections, 2) the constitutional order and political system, 3) dominant policy concerns, and 4) an island's identity or world view, or how it sees itself in the world. Using this as a guide they suggest that there really are seven types of island governance.

The first is called a **Civilization** type of island. This is the rarest form of governance but the most influential globally. Geopolitically, these kinds of islands have dominated the rest of the world politically and economically and are capable of changing the world order. They have an enduring political stability and are able to adapt to political crises. Because they dominate the world economically, their most pressing political concern is to safeguard international trade and domestic production. They have a strong sense of cultural identity linked to empire and project a coherent sense of nationhood to the rest of the world. The best island examples of this type of governance are Great Britain and Japan during periods throughout the eighteenth to twentieth centuries.

Warrington and Milne's second form of island governance is called a **Fief or Fiefdom**. Geopolitically, these islands are extremely vulnerable and marginalized. These islands often experience repression and exploitation by both imperial powers and a local elite. The political system is characterized by weakness, illegitimacy, and a wide power imbalance between a small group of wealthy elites and a large marginalized population. The dominant policy concern is to stay in power. These islands tend to have an incoherent or fragmented national identity, especially as it pertains to addressing the needs of the large, lower-income population. One of the best recent examples that fit this category was the nation of Haiti (see Figure 6.5) on Hispaniola during the regimes of Francois "Papa Doc" Duvalier (1957–1971) and his successor son, Jean-Claude "Baby Doc" Duvalier (1971–1986).

The third major form of island governance is a **Fortress**. Examples might include Malta at the end of the eighteenth century, Singapore prior to the Second World War, and Hong Kong prior to its administrative transfer to the People's Republic of China. All of these islands served as garrisons or outposts, either militarily or for trade and communications. Therefore, geopolitically, their location is critical. They tend to embrace an imperial constitutional system, with a government that is competent, paternalistic, and interventionist. Their dominant policy concern is to maintain their "gatekeeper" role as a regional economic outpost and they identify themselves as an outpost version of the imperial power, mimicking the governance structures and symbols of their colonial state and suppressing the growth of any kind of Indigenous national identity.

The fourth kind of island is known as a **Refuge**. Geopolitically, these islands have been forced into this role because of their confrontations with another dominant political power. So, for example, Taiwan (see Figure 6.6) and its relationship with the People's Republic of China (i.e., mainland China), and Cuba's relationship with the United States during much of Fidel Castro's reign, have made them refuges. The internal political system in these kinds of islands is often one of authoritarian rule, where there is an enduring sense of a state of emergency that discourages constitutional and democratic reform. Internal policy concerns are overwhelmingly focussed on foreign affairs, both in terms of how actions elsewhere might

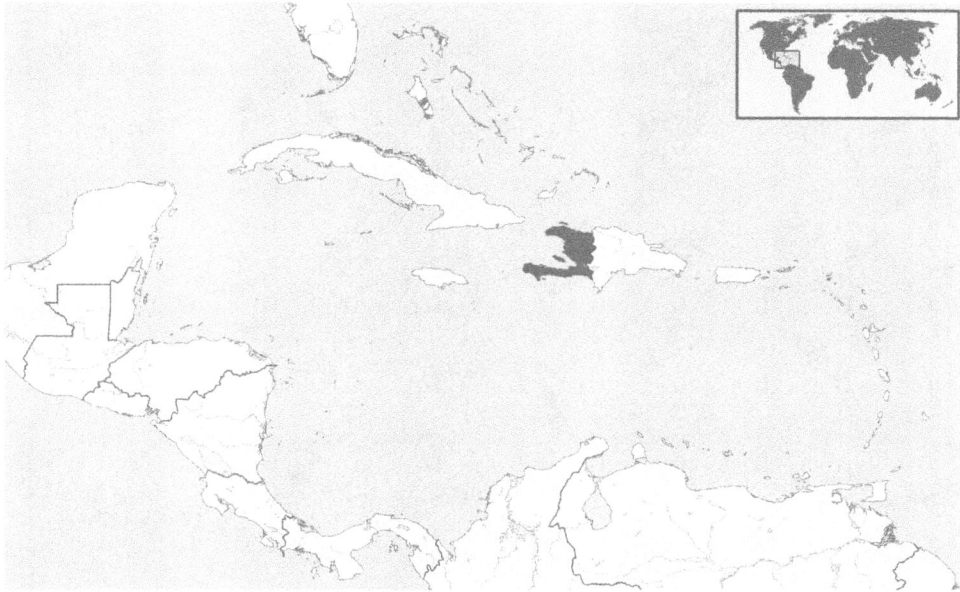

Figure 6.5 Location of Haiti in the Caribbean Sea. *Source:* **Stasyan117**

affect them and how they might influence foreign affairs elsewhere (e.g., Cuba lending military support to other nations and criticizing the impact of the American trade embargo). The identity of these islands is often defined and constructed as a counterpoint to the dominant power's ideology, as in Cuban socialism versus American capitalism, or Taiwan's democracy versus Chinese autocracy.

There tend to be fewer current examples of a **Settlement** island, the fifth governance type defined by Warrington and Milne (2018). They give the examples of Iceland and Greenland during the period from 800 to 1400 AD and New Zealand during the eighteenth century. Geopolitically, these islands are the outposts of imperial reach and have served as emigration "safety valves" from overpopulation or opposition groups. Since the migrating population is relatively homogenous it often leads to the adoption of the political system from the homeland. The dominant policy concern during this period on these islands is the orderly, rapid, and effective settlement of the new territory. As for identity, the settlers still retain a strong cultural association with their homeland and have not yet developed a unique island-based national identity.

Most of the Caribbean islands from the sixteenth century to today could be grouped into the sixth category, a **Plantation** governance system. Geopolitically, these islands have depended on the production and export of a narrow range of primary products back to their colonial or neocolonial metropoles (e.g., sugar, tea, timber, spices, cotton, bananas, minerals). Although this may create wealth for a minority of the population, the specialized economy means that they are vulnerable to rapid price changes. The land is also more vulnerable to environmental damage and depletion. A good example of an island that fits this category is the Pacific island nation of Nauru, where incredible wealth was generated for a short period of time at the end of the twentieth century as a result of the mining and export of phosphates for fertilizer. Today, the resource has been depleted, the landscape has been devastated, much

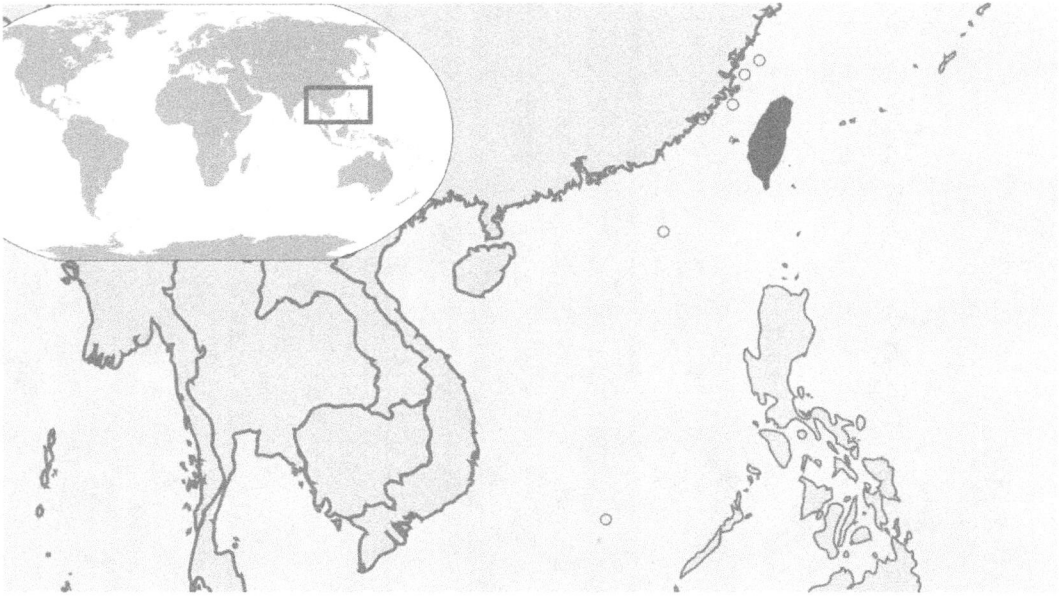

Figure 6.6   Location of Taiwan. *Source*: **JOSH tw**

of the population has left, and the island country has been referred to as "the first failed Pacific state" (Connell 2006). The political system of plantation islands is oriented around land-holding and the dominant policy concern is maintaining economic dependence and mitigating local class tensions. The wide differences in social and class standing make it difficult for the island to develop a national identity.

The last of Warrington and Milne's governance types is the **Entrepôt** island. As was the case with Fortresses, entrepôts enjoy a central location. However, their physical location as a gateway is less important than their strategic economical location. In other words, these islands play important roles as service-based sites for other regions of the world, in offering investment, finance, legal, regulatory, and dispute resolution services. The judicial and legislative system often mirrors that in the former colonial power and encourages trade and investment. Although there may be an entrepreneurial economic system, the political system values conservatism and political stability. The dominant policy concern is making sure that the government and public administration can support the economy and the centralized government can react quickly to social needs. The island population has a higher proportion of expatriates and identifies with conservative values, supporting strong, hierarchical leadership. As you may have guessed, many of the current offshore financial centres (or OFCs) would be members of this category. A lesser known example might be the small Indian Ocean island nation of Mauritius, east of the much larger island of Madagascar. Although not considered among the top OFCs, nonetheless the island has developed from an agricultural base into a more balanced economy with income distribution. It welcomes foreign capital investment, has transparent and well-defined legal and investment systems, and an efficient and competitive system of taxation. Government tends to play an important role, for example in controlling utilities and the imports of rice, flour, petroleum, and other products. It has been described as an African success story (Frankel 2016; Zafar 2011)

A QUESTION OF POOR GOVERNANCE?

The conventional wisdom is that poor governance and corruption is to blame for many of the problems of developing jurisdictions (Elahi 2009; Kaufmann, Kraay, and Mastruzzi 2003). Moreover, it has been stated that one of the best ways to achieve development goals is to improve governance, including law and order, regulatory barriers, property rights, and controlling corruption (Australian Agency for International Development 2006; Gani 2009a; Pacific Islands Forum Secretariat 2007). These principles have now become enshrined in many development policies. At the same time, there is also an assumption that smaller jurisdictions are, by virtue of their size, more democratic (D. Anckar 2010; Clark 2009; Diamond and Tsalik 1999). The reality, at least for some small islands, is more complex. For example, Baldacchino (2012a) suggests that democratic practices and pluralism may exist, but only if you conform to the prevailing, homogenous values and standards. Those who do not conform are more likely to be ostracized or, in Baldacchino's words (2012a, 112), it may be "hell for dissidents."

At the same time, democracy seems to prevail in some cases against all odds. For example, Corbett and Veenendaal (2018) describe the case of São Tomé and Príncipe, a small island state off the west coast of Africa. Despite being surrounded by autocracies and having a colonial past, it has been touted as having one of the most stable democracies in Africa. Moreover, small states in the Caribbean and the Pacific "are among the most stubbornly and disproportionately democratic countries in the world" (Corbett and Veenendaal 2016, 432). As to corruption and poor governance, there are many examples where this has stymied the development hopes of islanders, including on Nauru, Tonga, Vanuatu, the Turks and Caicos, and Fiji. However, in a biographical analysis of the Pacific's political elite, Corbett (2015) reminds us that politicians are not anonymous; they are people who get into politics with hopes, dreams, and the best intentions, only to find themselves caught up in the dirty, messy business of political life.

CONCLUSIONS

This chapter has provided many examples where islands have been treated as little more than inconveniences by larger political powers. When it has been convenient, they have been incredibly important in expanding the regional and global ambitions of larger, more powerful nations (Ratter 2018). The focus on islands in the larger political arena should not be minimized. The most recent and potentially dangerous geopolitical situation concerns the tensions associated with the various claims to a set of tiny islands in the East China Sea (Baldacchino 2016; Emmers 2009; Valencia 2007). Referred to as the Senkaku Islands in Japan, the Diaoyu Islands in China, and the Tiaoyutai Islands in Taiwan, these islands are now at the heart of a major territorial and resource dispute. Administered by Japan since the late nineteenth century, both China and Taiwan started to make claims after a survey in the late 1960s suggested that there may be oil and gas reserves in the surrounding waters. Since that time all three claimants have been involved in various military encounters in the territorial waters surrounding the islands.

The rhetoric of island vulnerability has been used when it best serves the interests of islanders. For example, the plight of islands in the Pacific facing sea-level rise has been strategically identified by the Small Island Developing States (SIDS) to galvanize world attention to their needs. Most subnational island jurisdictions (SNIJs) have also used their ambiguous political relationship to their advantage, and in the process, rejected opportunities for independence declared for them by external agencies of the United Nations. We seem to be entering an era where creativity and negotiation skills that have often been a characteristic of island life are increasingly being practiced and recognized on the world stage.

Key Readings

Corbett, Jack. 2015. *Being Political: Leadership and Democracy in the Pacific Islands.* Honolulu: University of Hawai'i Press.

Ratter, Beate. 2018. *Geography of Small Islands: Outposts of Globalisation.* Cham, CH: Springer.

Stuart, Kathleen. 2009. "A Listing of the World's Sub-National Island Jurisdictions." *In The Case for Non-Sovereignty: Lessons from Sub-National Island Jurisdictions,* edited by Godfrey Baldacchino and David Milne, 11–20. London, UK: Routledge.

Warrington, Edward, and David Milne. 2018. "Governance." In *The Routledge International Handbook of Island Studies: A World of Islands,* edited by Godfrey Baldacchino, 173–201. New York: Routledge.

Watts, Ronald. 2009. "Island Jurisdictions in Comparative Constitutional Perspective." In *The Case for Non-Sovereignty: Lessons from Sub-National Island Jurisdictions,* edited by Godfrey Baldacchino and David Milne, 21–39. London, UK: Routledge.

Chapter Seven

Islands, Population, and the Movement of People

Move Iceland to the interior of a continent and it would be uninhabited in ten years, or less. (Holm 2000, 5)

INTRODUCTION

Earlier in this book, it was noted that at least 600 million people live on islands. Although this is a large number, it is worth asking what are the most important demographic characteristics of islands? Are island populations generally growing or declining, and are people moving away from or to islands? The short and simple answer to that question is that it depends on the island. This chapter looks more closely at the populations of islands, how and why they have grown and declined, and the implications of that growth and decline for both the islands and the rest of the world. In particular, we will examine the phenomena of migration to and from islands, and look at the role that islands have played as entry points or gateways for large-scale continental immigration, including the movement of refugees. Finally, we will see how island émigré populations, otherwise known as diasporas, may continue to play a role in the future of islands.

THE BASICS OF POPULATION CHANGE

To put it simply, the natural rate of population growth changes as a function of the relationship between the birth rate and the death rate. When the number of births exceeds the number of deaths over a period of time, then we will see populations increasing. Conversely, populations will decline when the death rate exceeds the birth rate. Of course, this natural change also needs to be combined with immigration and emigration. All other things being equal, when the level of immigration (i.e., the number of people moving into a place) exceeds the level of emigration (the number of people moving away from a jurisdiction) then the population will increase. Of course, the reverse is true when emigration exceeds immigration.

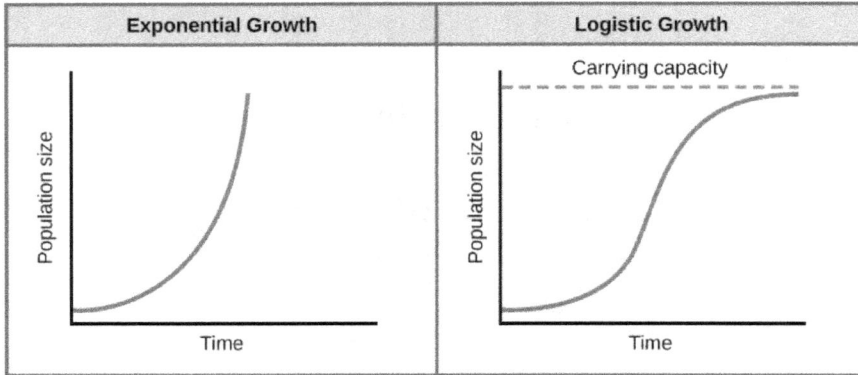

Figure 7.1   Population Change and Carrying Capacity. *Source:* **CNX OpenStax**

One of the earliest theories of population change was developed by the English clergyman Sir Thomas Malthus at the end of the eighteenth century. His *Essay on the Principle of Population* predicted a population catastrophe, including mass starvation, if populations continued to grow geometrically (e.g., 1, 2, 4, 8, 16, 32 …), as shown in Figure 7.1, while food supplies continued to increase at only an arithmetic rate (e.g., 1, 2, 3, 4, 5 …) (Malthus 1986). The dire consequences of these trends were taken up by other demographers and ecologists throughout the twentieth century, including by a group of scientists known collectively as the Club of Rome. Their research included the development of the concept of a population carrying capacity that might apply to specific regions as well as to the Earth as a whole (Ehrlich 1971; Meadows et al. 1972). These ideas were championed in a book by Meadows and colleagues titled *The Limits to Growth* that predicted dire consequences for the world if the trends in population growth, resource consumption, industrialization, and pollution continued at their current pace (1972). Although the predicted global collapse has not yet taken place, primarily because of technological innovation, expanding agricultural production, and a greater ability to match demand with supply in a more globalized world, there are still many astute and well-informed neo-Malthusians who believe that our current trends in consumption, production, and pollution (including human-induced global warming) are going to lead to a population correction regionally, if not globally. One of the neo-Malthusians was Fernand Braudel (1972) who wrote about the role of emigration, particularly from Corsica, in allowing those living around the Mediterranean to escape poverty and starvation. Where resources are particularly limited, such as on small atolls in the Pacific, reference to limits on carrying capacity have repeatedly been made (Bayliss-Smith 1974; Rapaport 2006; Williamson and Sabath 1982). Russell King (1993) provided a historical description of population change on islands. He said that when populations on small islands reach certain limits, one of three things is bound to occur: malnutrition or starvation, help from outside, or emigration. Moreover, the

WHEN POPULATIONS ON SMALL ISLANDS REACH CERTAIN LIMITS, ONE OF THREE THINGS IS BOUND TO OCCUR: MALNUTRITION OR STARVATION, HELP FROM OUTSIDE, OR EMIGRATION.

smaller the island, the earlier the point of saturation or, in Malthusian terms, the carrying capacity. King goes on to say that emigration has been an especially important element of the demographic histories of almost all of the islands of the Mediterranean and the Caribbean.

ISLANDS AND MIGRATION

Much has been written about the causes and consequences of migration. Although it has been criticized, behaviouralists have often tried to explain a particular migration flow as a function of a set of push and pull factors, some real and some perceived (L. Brown and Sanders 1981; D.S. Massey 1990; Portes and Böröcz 1989; J. Williamson 1986). So, for example, the civil war in Syria and the conflict in Iraq is currently pushing people away from their homes and forcing/encouraging them to move to safety. Several years ago, civil unrest in Tunisia and Libya resulted in the same processes of forced migration. As we have seen on the news, many of these emigrants and refugees have landed at least temporarily on gateway islands in the Mediterranean, including Malta, Lampedusa (Italy), and Lesvos (Greece), the latter located just off the coast of Turkey and one of the closest points of entry to the relative safety of the European Union. This is not the first time that islands have served as safe havens for people fleeing persecution and poor living conditions. Push factors may also include a lack of access to the basic necessities of life, including food and shelter. As will be discussed later in this chapter, in the mid-nineteenth century millions of Irish islanders were forced to leave their homes because of the failure of the potato harvest, brought on by environmental conditions and by human decisions.

Pull factors are also important motivators in the decision to migrate. Access to better economic and social opportunities, both for the immigrants and for their children, have led to massive relocations of people around the world. For example, population growth on some of the Caribbean islands is occurring partly as a function of attempts to access a pleasant physical environment and the leisure associated with this environment (i.e., sun, surf, sea, and sex) and partly to access the jobs associated with the tourism industry and the financial services sector in offshore financial centres (OFCs). Russell King and John Connell (1999, 9) have said that "migration is primarily a response to real and perceived inequalities in socio-economic opportunities and standards of living." But there seems to be something about islands that makes the movement of people especially important. In King and Connell's words (1999, 1), "islands have an unusually intense engagement with migratory phenomena." Perhaps the best way to understand this special relationship between migration and islands is to start with a few prominent examples, as raised below.

Migration and Population: Ireland in the Nineteenth Century

Arguably one of the most prominent stories of island depopulation in the western world in the last several centuries is that of Ireland during the second half of the nineteenth century. During the period from 1841 to the end of the nineteenth century, as a result of deaths exceeding births (i.e., natural decline) and emigration, the Irish population decreased by close to four million people, from 8.2 million to 4.5 million (see Figure 7.2). Even by 1961, the population of Ireland (both the Republic nation-state in the south and the United Kingdom's province

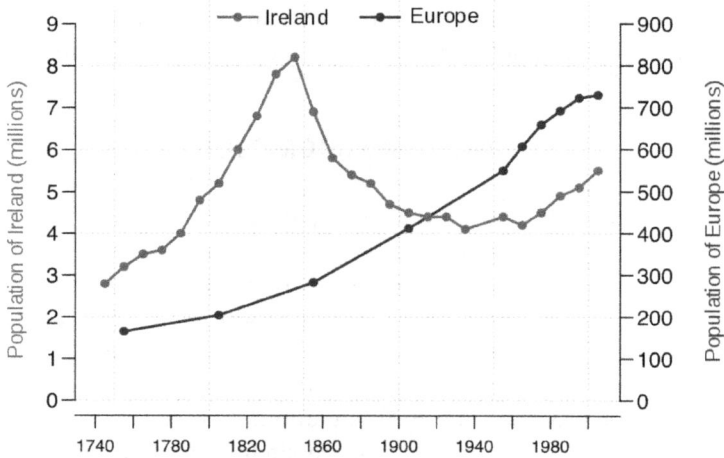

Figure 7.2 Ireland (the island) and Europe's indexed population between 1750 and 2006.
Source: **Ben Moore**

of Northern Ireland) had only recovered to about half of what it had once been in the early 1850s. To understand why this happened you need to understand the historical context, including the relationship between England and France leading up to the mid-1800s. During the first part of the nineteenth century (1793–1815), England and France were engaged in a series of military conflicts that led to the need for England to obtain food from sources other than the European continent. Ireland served this role, especially for the production and sale of potatoes. Both the average price of potatoes and subsequently the production of potatoes in Ireland for export increased significantly during this period (Woodham-Smith 1991). As a result, Ireland's birth rate and population increased (Mokyr and Ó Gráda 1984). As both a food and cash crop, potato production in Ireland crowded out other crops so that one-third of the population became dependent on potato production as a source of income and food (Kennedy et al. 1999). When hostilities ended, England once again had other options for food and the price of potatoes declined. Unfortunately, Ireland now had to support a much larger population with a much smaller revenue base. Moreover, the larger households meant that family farm holdings needed to be increasingly subdivided to allow the children, as adults, to have their own family farms. This meant that potatoes were grown more inefficiently on poorer quality land. A potato viral infestation in the 1840s wiped out much of the crop, with production declining from 14.8 million tons in 1844 to two million tons in only three years. The fact that Irish agriculture was now essentially a one crop industry (i.e., monoculture) contributed to the spread of the blight and the level of devastation. Combined with this was an increase in discrimination of the Irish by the English reflected in both the "Poor Laws" and the "Corn Laws," leading to higher duties on imported foodstuffs from Ireland and an unwillingness on the part of England to assist Ireland (Nally 2008). Effectively, the famine occurred more so because of the political economy of the region than as a result of a potato infection. Some have called the English response to the potato famine an act of genocide (Meierhenrich 2014).

An estimated one million Irish died of starvation and related illnesses in the following seven years (D. Ross 2002). From a Malthusian perspective, the population had greatly exceeded the carrying capacity of the land and, even with emigration, it crashed shortly after the decline in the production of food. You could also say that it was one of the strongest population migration push factors. From the 1840s to 1920, an estimated two million Irish emigrated from the island, primarily to the United States, Canada, and Australia, but also to New Zealand and South Africa (D. Ross 2002). At the end of the twentieth century, there were an estimated sixty million people of Irish descent living outside of Ireland (Royle 1999). This is a good example of population change and migration occurring as a function of a complex mix of economic, social, and environmental factors. The large population of Irish descent living off-island also allows us to introduce the word *diaspora* into this discussion, defined as "A dispersion of a people from their original homeland, (or) the community formed by such a people" (The Free Dictionary 2019a).

It should be noted that since the 1990s, immigration to Ireland has exceeded emigration away from the island. Before the global recession in 2008, Ireland was often touted as one of the "economic miracles," even being referred to in economic growth terms as the "Celtic Tiger" (Breathnach 1998; Gardiner 1994). The factors that led to this turnaround in the economy included a large, highly educated, relatively underemployed (female) labour force in Ireland and transnational corporations that were looking for sites for their service-oriented "back-office" functions, including data entry, customer relations, and accounting (Breathnach 1998). Although the post-2008 period has led to fewer jobs, higher levels of government debt, and austerity programs, there is still a net inflow of people to Ireland. Much of this net inflow was from other European Union countries, particularly Poland (McGinnity and Gijsberts 2018).

Migration and Population: Cuba in the Latter Half of the Twentieth Century

At 110,000 square kilometres and a 2017 population of approximately 11.5 million people, Cuba is the largest island in the Caribbean by total land area and the second largest (behind the combined populations of Haiti and the Dominican Republic on Hispaniola) in population. At its closest point it is also only one hundred kilometres from the Florida Keys of the United States (see Figure 7.3). For many years, the United States supported the Fulgencio Batista-led government of Cuba in order to protect the interests of American companies that had investments on the island. When Fidel Castro and his revolutionaries overthrew this government in 1959 and Fidel Castro came to power, the United States severed trade and diplomatic

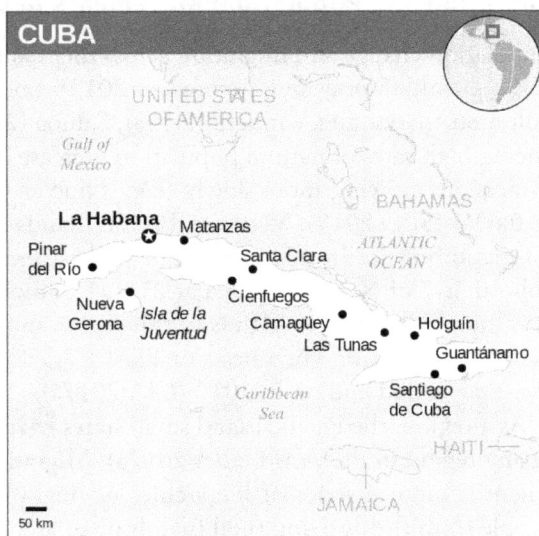

Figure 7.3 Cuba in the Caribbean. *Source:* **OCHA**

ties with Cuba. At that time, the population of Cuba was estimated to be about seven million. Especially during the Cold War period when Cuba was aligned with the former Soviet Union, the latter part of the twentieth century saw a number of incidents that brought about tension between the United States and Cuba, including a naval trade embargo, the Cuban Missile Crisis, and an attempted overthrow of the Cuban government by an American-backed opposition group (i.e., the 1961 "Bay of Pigs" incident) (Pérez 2002).

One of the more intriguing aspects of population-related changes on Cuba is related to emigration to the continental United States. In the years following the change in government, more than 250,000 political refugees and economic migrants left Cuba for the United States, settling largely in southern Florida. In the eight-year period from 1965 to 1973, a further 300,000 Cubans followed. There was another spike in emigration in 1980 known as the "Mariel boatlift" when an estimated 125,000 Cubans left for the United States. It is reputed that Castro used this as an opportunity to rid Cuba of criminals, undesirables, and the mentally ill (Larzelere 1988). Following this blip, the United States and Cuba came to an immigration agreement that included an annual quota of 20,000 immigrants from Cuba, the reunification of direct relatives and a "wet foot, dry foot" policy, wherein only those migrants who actually reached the shores of the continental US (including the Florida Keys) were allowed to stay and seek permanent residency. If boats carrying Cubans were intercepted before they reached the shore, they were sent back to Cuba or to a third country. As an example of the significance of this policy, the US Coast Guard annually intercepted between 500 and 3,000 Cubans at sea from 1995 to 2008 (Wasem 2009). It was only in late 2016 that this wet foot, dry foot policy was changed. As one of former US President Obama's final acts, he repealed the wet foot, dry foot policy. In return, Cuba agreed to take back illegal or deportable immigrants (Roig 2019). A harsher Cuban foreign policy by the Trump administration has once again made the repatriation of Cubans uncertain.

Migration and Population: Movements in Oceania

Population change and migration across the islands of Oceania is a centuries-old tradition and varies greatly (Storey and Steinmayer 2013). Looking only at crude birth and death rates, the Solomon Islands and Vanuatu (2.5%), Samoa (2.2%), and PNG (2.1%) have all been experiencing high rates of natural population increase (Secretariat of the Pacific Community 2019). Others are growing more slowly (e.g., Niue at 0.2%, Palau at 0.5%, and Wallis at Fortuna at 0.6%) (SPC 2019). Very few Pacific islands are experiencing rapid population growth as a function of net in-migration. The highest level of net migration is in the French "special collectivity" of New Caledonia at 21.9 /1,000 or +2.2% (Central Intelligence Agency 2017). It is more common for islands in this region to be losing people through migration, as in the Federated States of Micronesia or FSM (-7.9%), Tonga (-7.7%), and Samoa -6.7%) for the five-year period ending in 2012 (CIA 2017a).

As a region, the Pacific island small states have the greatest rates of outmigration (at -3.1%) of any region in the world (CIA 2017a). Migration in the islands of Oceania can be complex. There is often a series of hierarchies of migration, with one component the movement of people from the outlying rural islands of an archipelago to the more urbanized "main" island, as well as international emigration to metropolitan states on the edge of the region in places like New Zealand and Australia (Connell 2003a). Those living in territories associated with

THERE IS OFTEN A SERIES OF HIERARCHIES OF MIGRATION, WITH ONE COMPONENT THE MOVEMENT OF PEOPLE FROM THE OUTLYING RURAL ISLANDS OF AN ARCHIPELAGO TO THE MORE URBANIZED "MAIN" ISLAND, AS WELL AS INTERNATIONAL EMIGRATION TO METROPOLITAN STATES ON THE EDGE OF THE REGION.

the United States, such as the FSM, Palau, and the Marshall Islands, are more likely to use the Compact of Free Association to move to the US, while those in the Cook Islands, Tokelau, and Niue can more easily access NZ through the Pacific Access Category. By the 2013 census, it was estimated that 7.4 percent of New Zealand's population consisted of people who identify ethnically with the Pacific islands, sometimes referred to as Pasifika People (Pasifika Futures 2017). The number and percentage of ethnic Pacific Islanders in New Zealand has increased consistently over the past generation and, given that the median age of this group is sixteen years younger than the median age of the New Zealand population, it will likely form a much larger share of the NZ population in the future. In fact, it is now not uncommon for the NZ population that originated from a Pacific island to greatly exceed the population that still resides back on their home island.

The reasons for migrating are as varied as the contexts themselves. Topping this list would be the rapid population growth on some islands combined with relatively few employment opportunities to support this younger population and existing labour and mobility agreements that allow islanders to take advantage of economic opportunities elsewhere. Therefore, migration serves as a safety valve to better match labour supply with labour demand. Family reunification is also an important factor in migration decisions. Bertram and Watters (1985, 499) refer to Pacific migration patterns as "transnational corporations of kin," reflecting the ongoing family ties to home and communal land as well as employment prospects. With a growing attention to the impacts of human-induced climate change and sea-level rise, more recent research has suggested that the growing frequency and severity of storms combined with higher sea levels will become a more important reason to emigrate and indeed may force entire island populations to relocate (Ash and Campbell 2016; Kelman 2015).

Two of the most powerful voices for islanders in the Pacific Ocean are the writer Albert Wendt and anthropologist Epeli Hau'ofa. Wendt (1976, 49), originally from Samoa but part of the Aotearoa/New Zealand diaspora, coined the term Oceania and, in the first line of his article titled "Towards a New Oceania," states that, "I belong to Oceania—or, at least, I am rooted in a fertile portion of it—and it nourishes my spirit, helps to define me, and feeds my imagination." Much of Wendt's writing speaks to the movement of Pasifika people in the region (J.M. Wilson 2018). Given his mobility between Samoa and New Zealand, he speaks to the rootlessness of islanders and even refers to himself as living in exile as a condition of his life. Born in Papua New Guinea (PNG), and of Tongan descent, Hau'ofa worked and went to school in PNG, Fiji, Australia, and Canada. In describing the evolution of population movements by Pacific Islanders, Hau'ofa (1994, 155) stated that:

> They have since [independence] moved, by the tens of thousands, doing what their ancestors did in earlier times: enlarging their world as they go, on a scale not possible before. Everywhere they go,

to Australia, New Zealand, Hawai'i, the mainland United States, Canada, Europe and elsewhere, they strike roots in new resource areas, securing employment and overseas family property, expanding kinship networks through which they circulate themselves, their relatives, their material goods, and their stories all across their ocean, and the ocean is theirs because it has always been their home.

Hau'ofa is saying that the current mobility and reach of Pacific Islanders is nothing new. They have always viewed their world as a network of connected places. In fact, the "circulation" of islanders in the Pacific is often more complex than is first apparent from published census statistics including those referenced above. Islanders may move multiple times seasonally and throughout their life cycle and may have two or more residences in different locations (Bedford and Hugo 2008). In particular, the Samoan population in the twenty-first century has been described as truly transnational, with diasporas throughout the Pacific and in the United States, remittances that make up a significant portion of the Samoan economy, and tourism that consists largely of Samoans returning from abroad (Shankman 2018).

It might be useful to think of this population migration complexity as a kind of "churn," a term that Graham and Peters (2002) employed to describe the complex back and forth movement of Canadian Indigenous peoples moving between rural reserves and urban metropolises. In addition to making it more difficult to measure, this complex mobility also makes it challenging to develop a sense of identity or even "home" that is linked to one place. Instead, especially with the second generation of mobile Pacific Islanders, what might be emerging are multiple (Bedford and Hugo 2008) or hybrid identities (J.M. Wilson 2018), or a pan-Pacific identity (MacPherson 1997). At the same time, and perhaps more so among Pacific Islanders than in other regions, there seems to be a longer and stronger connection between descendants of migrants and their home communities and islands, partly to reaffirm an island culture in the children and partly to maintain a claim on family and kin-based land rights (Connell and Voigt-Graf 2006).

International immigration statistics also fail to capture the migration that is taking place internally within a given archipelago/jurisdiction. A common pattern has emerged recently where people have been moving from less populated, peripheral islands in the archipelago to more populated "urban" and centralized islands. In many respects, this movement of people mirrors the rural to urban migration that has taken place in mainland regions around the world and can be attributed to a lack of employment opportunities and access to adequate public services such as health and education. These intra-jurisdictional moves create problems for both origin and destination islands. As populations decline on peripheral islands, transportation access also deteriorates and public services are more difficult and costly to provide to

NOT ONLY DOES EMIGRATION BECOME THE NORM WITHIN FAMILIES AND COMMUNITIES, IT ALSO BECOMES INSTITUTIONALIZED IN GOVERNMENT POLICY. FOR EXAMPLE, GOVERNMENTS WILL REGULATE RECRUITMENT AGENCIES, SUPPORT THE RIGHTS OF MIGRANTS ABROAD, AND FUND UNIVERSITY PROGRAMS THAT PRODUCE MANY MORE GRADUATES IN HIGHLY SKILLED OCCUPATIONS THAN CAN POSSIBLY BE EMPLOYED DOMESTICALLY.

and from the rural islands. In the larger destination islands, it becomes increasingly difficult for the urban public infrastructure to accommodate large increases in numbers of people. It is also difficult for the labour force to absorb so many new workers seeking meaningful employment. Unfortunately, the result is often a poor, impoverished class living in shanty towns and no longer able to grow their own food to sustain their families.

A Culture of Migration

Many of the circumstances that might warrant emigration, such as ethnic conflict and oppression, natural disasters or a downturn in the economy, might be understood as short-term unique events that can be overcome. But what happens when the circumstances that encourage this movement of people continue for decades or generations? When economic or social deprivation, or the perception of a better life elsewhere, becomes the norm and when the values of migration become a community's values? This is when a "culture of migration" or a "migration ideology" takes root (Connell 2008, 2013a; D.S. Massey et al. 1993). Russell King (1993, 23) describes this as a situation where "Emigration may become institutionalized as part of island society, and necessary for its stable survival."

There are many examples where this has emerged. Connell (2008) describes it as existing on Niue where the remaining island population is just a tiny fraction of the total population of Niueans living in New Zealand and elsewhere. These remaining islanders are primarily public sector employees (Connell 2016a). Russell King (1993), Gailey (1959), and Royle (1999) suggest that islanders living on the Aran Islands off the west coast of Ireland have experienced this for more than a century, and the inevitability of migration is described in an article in *History Ireland* as "leaving the dreadful rocks" (Royle 1999). The cultural anthropologist Niko Besnier (2011) describes the migration decision among Tongans as desirable, possible, and inevitable. Finally, Asis (2006, n.p.) notes that millions of Filipinos are eager to work abroad. Quoting two surveys as evidence, she reports that 33 percent of adult respondents agreed with the following statement in a 2005 survey: "If it were only possible, I would migrate to another country and live there." Even more indicative of the pervasiveness of a culture of migration, in a 2003 national survey of Filipino children aged ten to twelve, Asis found that 47 percent of the children interviewed wished to work abroad someday.

Not only does emigration become the norm within families and communities, it also becomes institutionalized in government policy. For example, governments will regulate recruitment agencies, support the rights of migrants abroad, and fund university programs that produce many more graduates in highly skilled occupations than can possibly be employed domestically. In Tonga, the government has allowed Tongans living abroad to retain their Tongan citizenship and land rights even when they become citizens of their adopted host countries (Small and Dixon 2004). Even when there appears to be locally based employment and quality of life opportunities that would encourage islanders to stay home, the imperative to migrate becomes so embedded as a part of society that the migration continues. As discussed in an earlier section of this chapter, not only does this have significant consequences for the development of the island, it may also change one's sense of identity and perception of what constitutes home.

Refugee Movements in the Mediterranean Sea and the Role of Islands

Islands have frequently been used as offshore gateways to regulate the movement of people. Small islands such as Grosse Île in the province of Québec, Canada, Angel Island in San Francisco Bay, and Swinburne and Hoffman Islands in New York Harbour were used primarily to detect and quarantine those who might be bringing illnesses from abroad, in the hopes of preventing the spread of infectious diseases to new populations. The boundedness or separateness of islands was a feature that was thought to reduce the risk of contagion spreading to the mainland.

In the past several years, some of the most significant movements of people involving islands have taken place on the southern edge of Europe. Civil strife and economic hardship in the African states bordering the Mediterranean Sea (e.g., Libya, Tunisia), countries in the Middle East (e.g., Syria and Iraq), and even those places farther away (e.g., Sudan, Ethiopia, Afghanistan, Mali) have led to the mass movement of people attempting to escape by making their way north and west. This is not an easy voyage. For example, the United Nations Refugee Agency (UNHCR 2019) has estimated that 360,000 migrants crossed the Mediterranean in 2016 and almost 5,000 have died in the attempt. This is up from 3,771 deaths in 2015 when more than a million migrants attempted the crossing (Quinn 2016). The photograph in Figure 7.4 shows refugees packed into an inflatable boat in the Mediterranean. Kouremenos and Dierksmeier (2019, 2) suggest that in the middle of the sea "the boat itself looks like an island." Although the total numbers attempting to cross the Mediterranean had declined in 2019 (i.e., 16,000 between January and March and 503 deaths in 2018), this is still a significant movement of people and a tragic loss of life (UNHCR 2019).

For many of those migrants, islands such as Lampedusa, Malta, and the Greek islands in the Aegean Sea have become safe havens or gateways. As is the case with SNIJs, the particular rules

Figure 7.4   Refugees on an inflatable boat in the Mediterranean. *Source:* **Alejandro Carnicero/ Shutterstock**

and regulations often influence the patterns of these population movements. In this case, the European Union's regulations governing the movement of refugees between one EU member and another are critical. One of the important features is a provision called the Dublin Rules that normally require refugees to apply for asylum in the first EU state in which they arrive. They could be deemed inadmissible in another country if they gained protection in the first country of asylum. What this means is that if refugees find themselves on the island of Lampedusa, part of a province of Italy and one of the closest points to the main departure points in Libya and Tunisia along the coast of Africa, or Syrian and Afghani refugees make their way to the Greek island of Lesvos, one of the closest points to the non-EU member Turkey, these European governments are able to move them along elsewhere in Italy or Greece, respectively and, once on the mainland, the refugees may be able to move elsewhere within the European continent once their asylum case is heard. However, should these refugees find themselves on the shores of the island of Malta, an EU member that does not have any land on the continent, the government cannot legally pass them along to another EU country and the refugees have little means to move from Malta to another EU country. This has created tensions in Malta, as most of the refugees do not really want to be there and the government and the people of Malta have not shown a great deal of enthusiasm about their presence (Nimführ, Otto, and Samateh 2020). This humanitarian crisis has been inflamed as some European governments patrolling the Mediterranean have been turning these migrant flotillas back before they can land on their soil and declare that they are seeking asylum.

MIGRATION, REMITTANCES, AND SKILLED MIGRANTS

Throughout human history, islanders have been among the most mobile of people seeking opportunities to support themselves and their island families. As noted earlier, the migration of islanders seeking work away from home has become a common feature of island life. An inadequate supply of well-paying jobs or the seasonality of island industries has often forced one or more members of the household to leave home (Connell 2013). Much of the pay earned while working away is sent back by the migrants to the island households in the form of remittances. These funds are a major component of the world economy. According to World Bank estimates, global remittances totalled USD 689 billion in 2018 (World Bank Group 2019a), far exceeding the USD 150 billion in Official Development Assistance (ODA) provided by the thirty members of the Organisation of Economic Co-operation and Development (OECD 2019).

Although remittances are a part of the economy of every country, the relative importance of island diasporas and, arguably, the close social ties within islander families and communities, make remittances an especially important component of the social and economic development of many islands and island families. In 2018, the list of the top ten countries by absolute value of migrant remittances included only one island nation, the Philippines archipelago, at USD 33.8 billion (World Bank Group 2019a). However, these aggregate amounts underestimate the role of these remittances to the economies of island countries. When these same migrant remittances are measured as a share of total national Gross Domestic Product (GDP), three islands are in the top ten; Tonga in 1st position at 40.7 percent, Haiti ranked 4th at 32 percent and Bermuda ranked 7th rank at 22 percent (World Bank Group 2019b). Although data is not available for all of the subnational island jurisdictions like Bermuda, one might expect the role of remittances to be even more significant for these jurisdictions, where there are fewer restrictions on labour mobility to the metropole country.

The stereotype of migrants, and especially temporary migrants, may see them as under-educated and lower-skilled, filling menial, poorly-paid jobs at their destinations. In fact, the proportion of migrants who are skilled has been increasing and many island jurisdictions are now educating skilled graduates as an export, with the expectation that remittances would flow back to their home countries (Cabanda 2017). For example, "Filipino nurses make up the single largest group of foreign-born nurses in OECD countries" (Abarcar and Theoharides 2017, 3) with between 3,000 and 8,000 leaving the Philippines on permanent visas every year and over 100,000 leaving between 1997 and 2009 (M. Thompson and Walton-Roberts 2019). The major destinations for these nurses were Saudi Arabia, the United States, and the United Kingdom (Lorenzo et al. 2007). As further evidence, many island countries are seeing a substantial share of their trained physicians travelling abroad to work. Using data from the late 1990s/early 2000s, Mullan (2005) calculated emigration factors for countries sending doctors to four destination countries (Canada, the US, the UK, and Australia). The emigration factor was essentially a ratio of the number of doctors practicing in these four destination countries relative to the total number practicing at home and abroad. A value of fifty would mean that just as many physicians from the home jurisdiction are practicing abroad as there are physicians practicing at home. Mullan found that six of the ten countries with the highest emigration factors in the world were islands, including five of the top six (i.e., Jamaica (41.4), Ireland (41.2), Haiti (35.4), Sri Lanka (27.5), and New Zealand (22.6)). The role of remittances has become so important to the long-term economic health of island nations that an entire model of island economic development has emerged that incorporates remittances as a key component. Often referred to as the MIRAB model, standing for **MI**gration, **R**emittances, **A**id, and **B**ureaucracy, this model will be discussed in more detail later (Bertram 2006; Bertram and Watters 1985, 1986).

Despite the obvious importance of remittances, there may be a "tipping point" such that, as the proportion of the labour force living abroad continues to increase, there may be less need to send remittances. A report produced by USA International Business Publications (2011) found this to be the case for the south Pacific island of Niue. In the 1970s and 1980s remittances were one of the most important components of Niue's foreign exchange. However, by the late 1990s, as entire families had relocated to New Zealand, less was being sent back to Niue. In the case of Niue, remittances are no less important because almost every household on the island has at least one person working in the public sector. Households on these "government islands" may not need the same level of remittance support than elsewhere (Connell

THE OTHER FEATURE OF REMITTANCES IS THAT THEY CAN BE HIGHLY VARIABLE OVER TIME, BOTH IN ABSOLUTE TERMS AND IN THE ROLE THEY PLAY IN THE GDP OF A COUNTRY. THIS MAY MAKE IT DIFFICULT FOR ISLAND GOVERNMENTS TO RELY ON FEES OR TAXES BASED ON REMITTANCES AS A LONG-TERM, STABLE REVENUE STREAM.

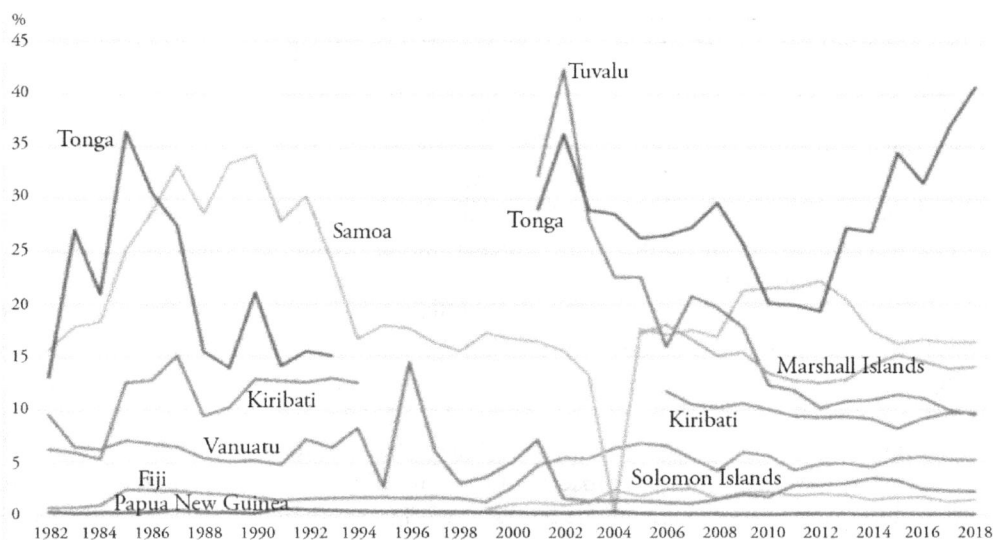

Figure 7.5   Share of Gross Domestic Product from Remittances to Pacific Ocean Islands, 1982-2018. *Source:* **World Bank Open Data**

2008, 2016a). Despite the evidence presented above, we need to be cautious about making broad generalizations. Cultural traditions among Pacific island families are sometimes strong enough to ensure maintenance of stable levels of remittances lasting several decades (Connell and Brown 2005).

The other feature of remittances is that they can be highly variable over time, both in absolute terms and in the role they play in the GDP of a country. This may make it difficult for island governments to rely on fees or taxes based on remittances as a long-term, stable revenue stream. Figure 7.5 shows the proportion of the GDP attributable to remittances for a set of seven Pacific island states, from 1982 to 2008. Not only are there variations in the role of remittances among these seven islands, but there are also wide fluctuations from one year to the next for individual places. This and other fluctuations may in part be explained because modest changes to the trade and tourism of a very small island economy can lead to large proportional changes from one year to the next.

The benefits of remittances when a member of that household moves away can be felt both in the household and the larger community. For example, in addition to helping individuals and extended families, it encourages consumption and investment and generates tax revenues for local and regional governments. It also supports the health and education sectors by allowing families to pay for their health care and their children's school fees, thereby replacing social spending while alleviating poverty (R. Brown, Connell, and Jimenez-Soto 2014). At the level of the community it may contribute to community projects such as building or renovating community halls, schools, and clinics, and is donated to those in need (Connell and Conway 2000).

Remittances can also cause problems for these same households and communities. First, they can magnify income inequities (especially in the short term) and possible resentment between households and communities (R. Brown, Connell, and Jimenez-Soto 2014; Koechlin and Leon 2007). Although these income inequities exist in any community, the differences between the "haves" and the "have nots" seem to be exaggerated when remittances are injected into a few households in a community in need. As well, there is a perception that remittances are spent on the "conspicuous consumption" of luxury goods such as new vehicles or appliances that normally have to be imported (Appleyard 1989; Connell and Conway 2000). Not only does this contribute to relative inequality, but these purchases may not lead to further locally based indirect consumption or employment.

Within households, the loss of a "head of household" for a long period places an additional social burden on the rest of the family, sometimes straining the family bonds. At an aggregate level, the loss of a large number of community members reduces the overall social capacity of a place. Especially in small communities, individuals may play multiple roles as ongoing volunteers (e.g., firefighters, local governing councils, churches) and in times of crisis (e.g., rebuilding houses or public buildings lost as a result of natural disasters). The absence of a significant number of potential volunteers means that the remaining community members have to do more, leading to volunteer burnout. These absences may also lead to a reduction in the economic and entrepreneurial capacity of communities. The remaining businesses and households do not have the services of the temporary migrants such as plumbers, electricians, and accountants. Perhaps more critical for the future of the communities and islands, the ability to apply new ideas and create new businesses is diminished. Ultimately, temporary employment elsewhere may become permanent, leading to the loss of entire households. Although the ties to one's former home may remain strong through kin and friends, over time (and especially for the children of the migrants) the ties become more tenuous, and the level and even the need for remittances to the remaining few family members decreases (Harling-Stalker and Phyne 2014).

RETURN MIGRATION

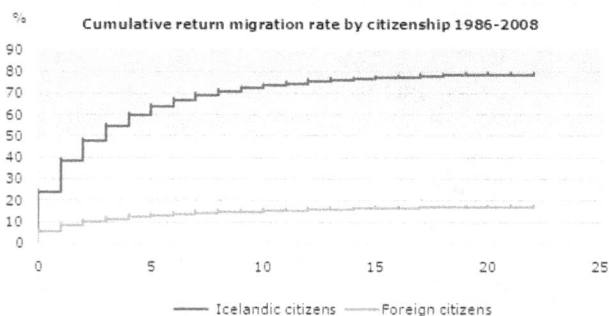

Figure 7.6 Cumulative Return Migration Rate of Icelandic and foreign citizens, 1986-2008. *Source:* **Statistics Iceland**

Ernst Ravenstein (1834–1913), one of the earliest population researchers, is credited with the idea that each current of migration produces a compensating counter-current (Ravenstein 1885). It is not surprising then that some islands that have experienced waves of emigration are now experiencing a returning echo of this emigration, as larger numbers of islanders return home, either temporarily or permanently. Many places that have been facing population loss have looked at this return migration as a development panacea and have been devising policies to encourage the return of this segment of the population as part of a broader economic development strategy (Byron 2000). Much less research has been undertaken to assess the success of these initiatives or the social and economic impact on the home communities. One island that seems to have been quite successful in seeing high levels of return migration has been Iceland. Figure 7.6 shows that almost 80 percent of Icelanders who moved away during the period from 1986 to 2008 returned within twenty years of their departure after staying away an average of 2.4 years (Statistics Iceland 2009). The factors that contributed to this counter-movement in migration were as varied as the reasons for emigration in the first place, including reconnecting with family and kin, a specific employment opportunity, and retirement. Lower-skilled emigrants may also return because of unsatisfactory experiences abroad. It should be noted that the level of return migration in Iceland following the recession of 2008–09 appeared to be lower than in previous years (Júlíusdóttir, Skaptadóttir, and Karlsdóttir 2013).

Another island where return migration has been studied is Tonga. Here, return migrants were drawn from many different age and social classes. They brought with them skills, capital, and experience that allowed them to contribute to the village economy (Maron and Connell 2008). In the Caribbean, Conway and Potter (2007) suggest that although the number of return migrants may be small, they are no longer the stereotypical retirees. Instead, they cover a range of ages, classes, and migration histories and can be effective agents of change. As discussed earlier, the more common pattern of migration appears to be messy. Rather than moving away as youth and returning as retirees, the movement of islanders is now more likely to be transnational, with multiple trips across multiple "homes" over multiple time periods.

EXAMINING FAST-GROWTH ISLANDS: A TALE OF TWO ISLANDS

Much of the discussion and the examples provided so far in this chapter have been about islands that are experiencing population decline. Islands are not near the top-ranked world jurisdictions in overall population growth. However, it is still useful to look at several places that have grown relatively quickly to better understand the underlying causes for that growth. As of 2017, the fastest-growing island over the previous year was Madagascar, ranked 18th in the world at 2.5 percent (CIA 2017a). Bahrain grew by 2.4 percent (ranked 35th), the British Virgin Islands grew by 2.3 percent (35th), and the Turks and Caicos grew by 2.2 percent (38th) (CIA 2017a). Other than higher-than-average population growth, these places do not appear to have much in common. They are spread around the globe and vary considerably in per capita economic growth and human development.

Turks and Caicos

This island chain in the Bahamas and Northern Antilles, located north of Hispaniola (see Figure 7.7), was once a dependency of Jamaica. When Jamaica gained independence in 1962, it became an overseas territory of Britain and the residents retained British citizenship. The Turks and Caicos Islands (T&C) had a population of just under 54,000 in 2018, having more than doubled since 2000 (CIA 2017a). Natural population growth in the T&C is relatively high at 1.2 percent, compared to neighbouring islands such as the Bahamas (0.8%), as well as the British Virgin Islands and the Cayman Islands (0.6%). However, much of the overall growth on the T&C is attributable to a higher level of legal and illegal immigration from Haiti and Cuba. More than half (57%) of the resident population of T&C are immigrants and, as is the case in many British Overseas Territories, many of these individuals are referred

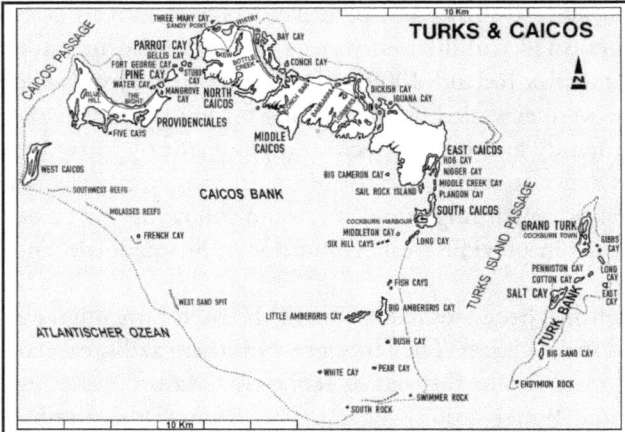

Figure 7.7   Map of Turks and Caicos. *Source*: **Jamaicajoe**

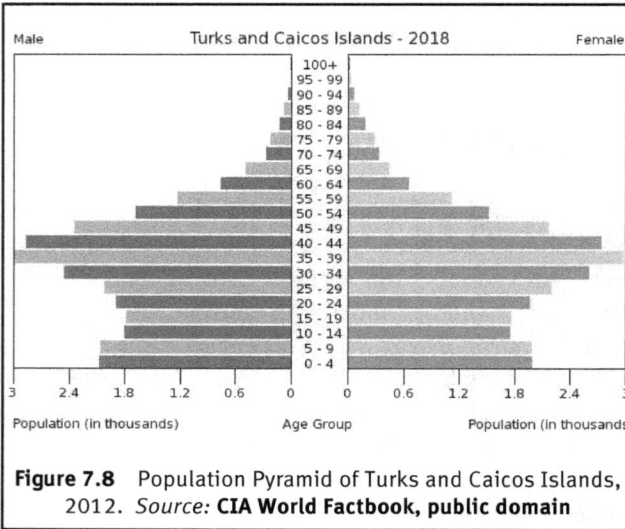

Figure 7.8   Population Pyramid of Turks and Caicos Islands, 2012. *Source*: **CIA World Factbook, public domain**

to legally as "non-belongers" (Carr and Liu 2016). Despite having limitations on the right to vote and own property, non-belongers play a key role as labour in the tourism and related service sectors. Population pyramids often provide a good visual snapshot of the current and future demographic structure of a jurisdiction. T&C's pyramid (see Figure 7.8) suggests that population growth from natural increase might be slower in the future as there are fewer individuals currently aged ten to twenty-nine years old. Unfortunately, this does not account for the impact of migration.

The economies of most of the SNIJs in the Caribbean, including the T&C, have consistently outperformed that of the independent nation-states (Clegg et al. 2017). The Turks and Caicos economy is dependent on two sectors: tourism and financial services. Agricultural production and export is negligible. Although the economy was adversely affected by the 2007–08 global recession and by hurricanes in 2017, in 2018 its GDP still grew by 4.6 percent and stayover arrivals (a proxy for tourism) had increased by almost 6 percent, a strong showing despite damage from extreme weather events (Turks and Caicos Statistics Department n.d.). As a SNIJ, responsibility for the maintenance of good governance in the T&C ultimately rests with Britain. In response to allegations of corruption at senior levels of island government, in 2009 Britain suspended the constitution and imposed direct rule on the territory through the Governor (Clegg and Gold 2011). After an investigation and implementation of reforms, a revised constitution was approved and new local elections were held. Although stability has been restored, this has undoubtedly limited its continued growth as an offshore financial centre relative to other islands in the region like the Caymans and the British Virgin Islands. However, it does not seem to have limited its appeal as a tourist destination. The number of international tourist arrivals in the T&C increased consistently from 2008 to 2014, reaching 1.33 million, before declining to 1.24 million in 2017 (Montanez 2019b). The government has made a conscious effort to expand the scope of the tourism industry to encompass cruise ship traffic as well as the niche markets of heritage and cultural tourism (Cameron and Gatewood 2008). It has also done what other Caribbean islands are doing: selling or signing long-term leases with cruise ship companies for most or all of the entire islands to allow them to develop their own terminals and beach facilities (Klein 2018).

Madagascar

Located off the southeast coast of Africa in the Indian Ocean, Madagascar has been discussed earlier in the context of a high level of biological diversity, endemism, and ethnic diversity (see Figure 7.9). A former colony of France, this second largest island nation in the world gained its independence in 1960. It had an estimated population in 2016 of almost 25 million people and has grown by one-third in the past decade. Unlike the Turks and Caicos, a graphic of the age structure of Madagascar resembles a classic pyramid shape, with a much higher share of the population in younger age cohorts. Therefore, unless there is a high level of emigration of adolescents or a high death rate, the population will likely continue to increase rapidly.

Unlike the T&C, the economy of Madagascar has a much greater reliance on agriculture and other primary sectors such as mining. Also unlike many other jurisdictions in the region, the total GDP of Madagascar has been increasing steadily throughout the twenty-first century, including +4.2 percent from 2015 to 2016 (Plecher 2019). Although the GDP/capita is remaining stable, because of the large and growing population, a lack of investment, and reliance on small-scale agriculture, it is much lower than the T&C at USD 450 per person in 2017 and has been stagnating or declining in real terms since the early 1970s (Razafindrakoto, Roubaud, and Wachsberger 2018). As was the case in the T&C, the governments of Madagascar have been accused of mismanagement and corruption, and there have been several power struggles between different groups that have called into question the stability of democracy and the rule of law (Connolly 2013; Ploch and Cook 2012; Razafindrakoto, Roubaud, and Wachsberger 2018). Despite abundant natural resources and cultural homogeneity, the economy of Madagascar has not performed as well as one might expect.

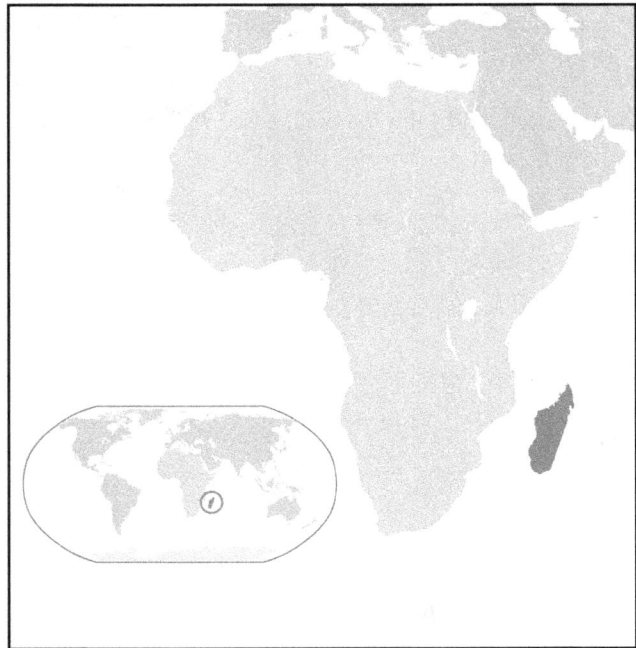

Figure 7.9 Map of Madagascar in the Indian Ocean. *Source:* **Quizimodo**

CONCLUSIONS

Because islands are not connected by land to continents, there may be a perception, at least on the part of mainlanders, that islanders are as disconnected from the rest of the world as their islands. This is not necessarily a belief that is shared by islanders. As the examples provided above show, there are many historic and current examples where islanders have become very mobile and connected, not only within the well-travelled water routes among island regions and archipelagos but also over much longer distances. Islanders have moved to escape poverty and hardship and to take advantage of opportunities to gain skills, experience, and income. More recently, a growing literature suggests that future moves may be tied more closely to the many impacts of global warming, including sea-level rise. Whether this takes the form of short-term mobility or signals the emergence of transnational populations that force us to rethink the concepts of migration and diaspora, the economic and social impacts are profound and mixed. However, it does mean that many islands are better connected to the world economy and culture than their physical distance might suggest. Despite rapid advances in transportation technology, islands also continue to serve as gateways or entry points for the flow of larger numbers of people moving across oceans and seas, either to escape danger or with the hope of building a better life. As such, they will remain an important component in many international issues related to migration and population change.

Key Readings

Bertram, Geoff. 2006. "Introduction: The MIRAB Model in the Twenty-First Century." *Asia Pacific Viewpoint* 47 (1): 1–13.

King, Russell, and John Connell, eds. 1999. *Small Worlds, Global Lives*. London, UK: Pinter.

Chapter Eight

Island Health and Epidemiology

In many oral and written traditions authored by Hawaiians, venereal diseases are portrayed as the "curse of Cook." (Jolly 1996, 203)

INTRODUCTION

This chapter discusses the health of islanders in very broad terms. In part, it identifies the health challenges that appear to be linked more closely with island populations, including the consequences of obesity. It provides an overview of individual and social determinants of health, and most importantly, it looks at islands as unique places that have allowed us to better understand issues related to health care epidemiology and the transmission of diseases, not only as they are applied to islands, but also for the lessons they may teach us about human health in all contexts.

EPIDEMIOLOGY AND ISLANDS

We begin by introducing a branch of the health sciences known as epidemiology. It has been defined by the Centre for Disease Control and Prevention as "the study (scientific, systematic, and data-driven) of the distribution (frequency, pattern) and determinants (causes, risk factors) of health-related states and events (not just diseases) in specified populations (neighbourhood, school, city, state, country, global)" (CDC 2016, n.p.). As a geographer by training, I have an affinity with epidemiology because it is so closely linked to space and place, where health factors and outcomes can be mapped so that we might better understand the relationships. One of the earliest and most celebrated epidemiologists is the physician John Snow (1813–1858) from London, England. He and other physicians at the time were trying to understand the causes of cholera in nineteenth-century Europe. While cholera was originally thought to be a function of "bad air," Snow found that in one of the outbreaks of the disease, the greatest cluster of cases was found around a public well on Broad Street in the neighbourhood of Soho, London. He was able to convince the local government to remove

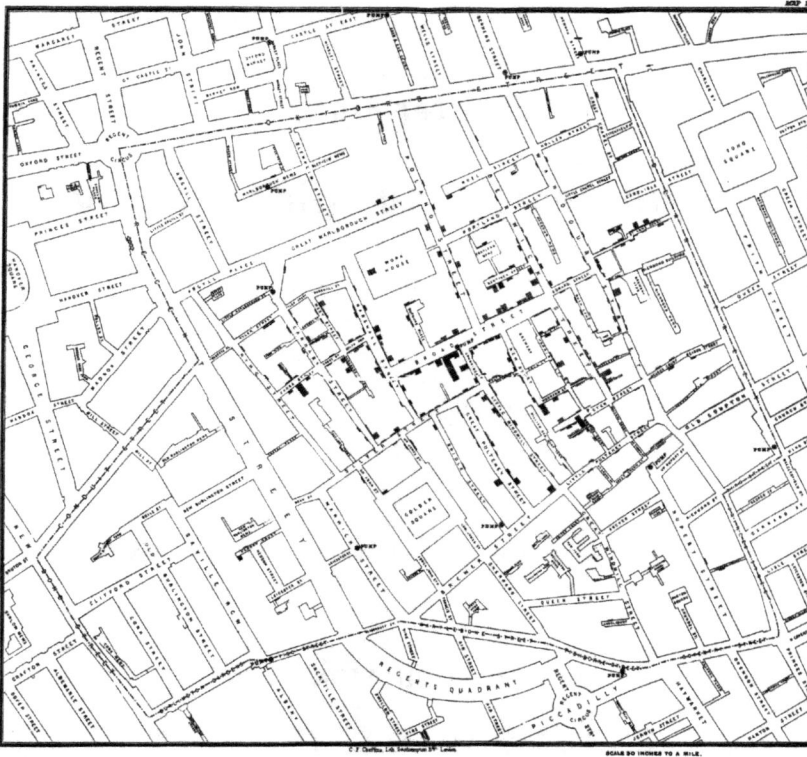

Figure 8.1   Original map made by John Snow in 1854. Cholera cases are highlighted in black. *Source:*
public domain.

the pump handle from the well. Afterwards, the number of new cases of cholera decreased significantly (McLeod 2000; Paneth 2004). Figure 8.1 shows the location of the homes of those who had contracted cholera in relation to the location of the public well. It is a powerful and clear example of the value of epidemiology. Even though physicians at the time may not have understood that it was the micro-organisms in the water that caused the transmission of cholera, or that the absence of new cases was at least partly a function of people moving away from the area to escape the outbreak, this example shows that seeing the spatial distribution of a characteristic of the population in relation to a health problem is often critical in arriving at a health solution.

This example of Snow and cholera happened to take place on the very large island of Great Britain. However, it does not really show the importance that islands have played in the field of epidemiology, and especially in communicable diseases. The best examples of this are provided in a body of research by geographers Andrew Cliff, Peter Haggett, and others (Cliff and Haggett 2004; Cliff, Haggett, and Smallman-Raynor 2000; Cliff, Haggett, Ord, and Versey 1981; Haggett 2000). In their 2000 book *Island Epidemics*, Cliff, Haggett, and their colleague Smallman-Raynor suggest that island populations have been important in allowing us to better understand health in five ways: 1) seeing a relationship between disease and the range and diversity of islands, 2) how we measure "disease thresholds" on various islands, 3) the impact of the changes in accessibility to islands over time on the likelihood of outbreaks, 4) the ability of discovering "index cases" on islands, and 5) the use of islands as sites for the creation and evolution of quarantine policies. The following section will examine each one of these features.

Range and Diversity of Islands and Health

Using islands to study the characteristics of diseases may be useful if you approach islands as closed systems or living laboratories. Despite the criticisms of this approach, there is something about the relatively precise boundary between water and land that makes people and researchers think that islands are ideal sites for social and biological experimentation. Despite the fact that every island is unique, it may be possible to better understand the underlying nature of certain diseases and health outcomes by comparing two or more islands that differ on a few fundamental features, such as population size. Cliff, Haggett, and Smallman-Raynor (2000) provide several historical cases where islands have been used in this way. For example, they describe the research on the spread of measles on the Faroe Islands in 1846 (Panum 1988), and rubella/German measles on the island continent of Australia (New South Wales) in the first half of the twentieth century. Although these kinds of outbreaks also took place on mainlands, the small scale of island communities and the more accurate record-keeping associated with migration by sea to and from islands in the nineteenth century allowed researchers to discover the carrying agents or vectors of the illnesses.

Disease Thresholds on Islands

The second way that islands have been useful to epidemiologists is in measuring the disease threshold necessary to maintain the existence of an infectious disease in a community. Sometimes called the "critical community size" or "endemism" (not to be confused with the endemism associated with flora and fauna on islands), it refers to the threshold population size necessary for a communicable disease to be constantly present. In other words, the community has reached a minimum population size so that it does not have to be reinfected from an outside source for the disease to remain. Although cities have often been used to describe the community, the high level of mobility of people between cities on continents means that there are very ambiguous or blurred boundaries when analyzing the transmission of infectious diseases. The defined boundaries between water and land associated with most small islands, and especially small islands prior to the proliferation of mass air travel, have allowed researchers to delimit and track the mobility of diseases. What Cliff and others found was that for those islands below the critical community size, the outbreak episodes or waves of diseases are farther apart in smaller, less populated islands than they are in larger islands, either above

ALTHOUGH THESE KINDS OF OUTBREAKS ALSO TOOK PLACE ON MAINLANDS, THE SMALL SCALE OF ISLAND COMMUNITIES AND THE MORE ACCURATE RECORD-KEEPING ASSOCIATED WITH MIGRATION BY SEA TO AND FROM ISLANDS IN THE NINETEENTH CENTURY ALLOWED RESEARCHERS TO DISCOVER THE CARRYING AGENTS OR VECTORS OF THE ILLNESSES.

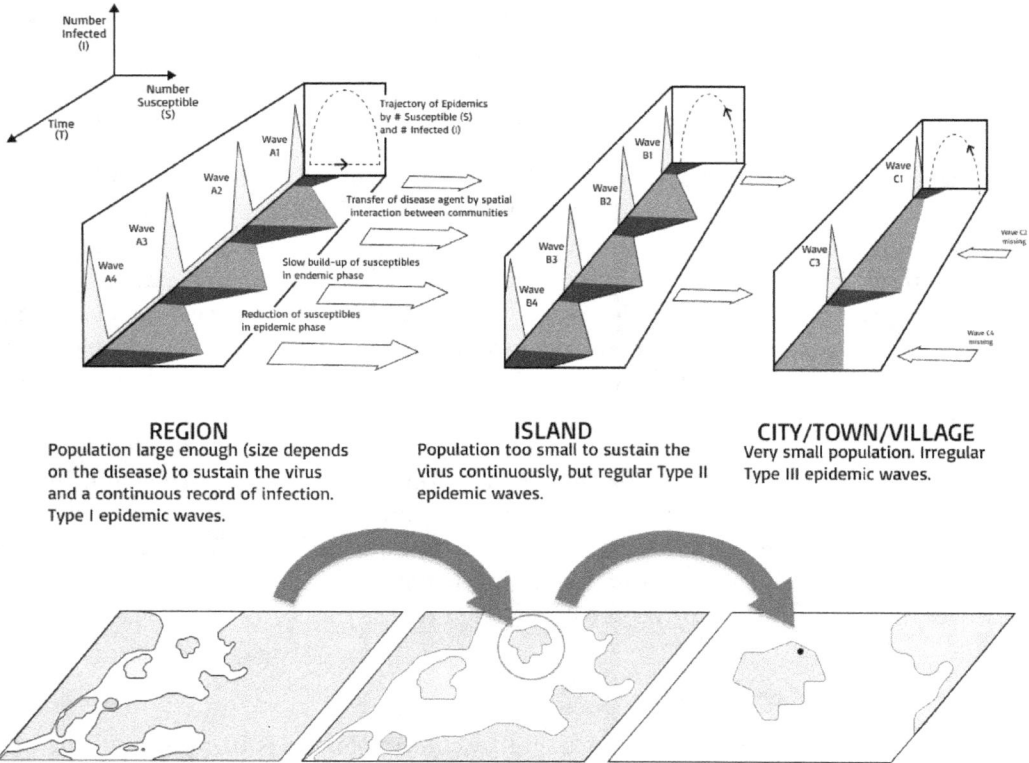

Figure 8.2 Population thresholds and the transmission of communicable diseases. *Source*: **Author, adapted from Cliff, Andrew, and Peter Haggett. 1995.**)

or below the critical community size (Cliff, Haggett, Ord, and Versey 1981). Although the base population for endemism was found to be between 250,000 and 345,000 people, this threshold is somewhat arbitrary. Its actual level depends on the kind of infectious disease and on whether a vaccination exists. For example, the existence of a vaccination might mean that the population threshold level would have to be greater, perhaps one million people instead of 350,000, for a disease to be continuously present in a community. Figure 8.2, adapted from Cliff and Haggett (1995), illustrates this for a hypothetical disease across three increasingly smaller, more isolated communities. As you move from a relatively large community (A) with a "susceptible" population large enough to continuously sustain a virus (i.e., it is endemic to that place) to a much smaller, relatively remote island community (C) where a virus needs to be reintroduced from elsewhere, the epidemic waves or outbreaks become much farther apart or are more sporadic. Much of the research on epidemic waves has been associated with the measles virus prior to the introduction of mass vaccinations (Bartlett 1957).

Changes in Island Accessibility

Earlier, it was pointed out that islands suffered a rapid loss of life after being exposed to the diseases brought to their islands by European explorers (Stannard 1993). In fact, Indigenous Hawaiians had a saying, "Lawe li'ili'i ka make a ka Hawai'i, lawe nui ka make a ka haole"

> AS THE AVERAGE TIME IT TAKES TO TRAVEL BETWEEN PLACES HAS DECREASED, THE LIKELIHOOD OF EXPOSURE DURING THE CONTAGIOUS PERIOD OF A VIRUS INCREASES.

which roughly translates as "Death by Hawaiians takes a few at a time; death by foreigners takes many" (Hodgetts et al. 2010). However, the relative inaccessibility of some small islands, and the change in this accessibility over time as transportation technology has evolved, has also proven to be useful in allowing us to better understand the relationship between technology and the transmission of diseases. In general, as the average time it takes to travel between places has decreased, the likelihood of exposure during the contagious period of a virus increases. Historically, and based on the prevailing mode of transportation at the time, some islands were so remote that island Indigenous populations were initially shielded from certain infectious diseases. Cliff and Haggett (1985, 2004) use the example of the island of Fiji during the latter part of the nineteenth century to show this technology/transmission relationship. Fiji required the labour of indentured Indians to work in the sugar cane plantations at the beginning of the nineteenth century. Early in this period, sailing ships were the primary mode of transport, taking about seventy days for the voyage. Although measles was detected onboard on about one-third of these voyages, the measles virus was not able to survive a voyage of this duration. Therefore, those on Fiji were not infected with the disease. However, by the early twentieth century, faster steamships took only thirty days to travel to Fiji. Because of this shorter time period, the measles virus was able to thrive on arrival on Fiji, infecting a larger susceptible population.

The other characteristic associated with a greater level of connectedness is the frequency of infectious outbreaks. For example, in Iceland prior to the era of jet travel (1896–1945), the intervals between the waves of measles epidemics were approximately five years and the severity of those epidemics was much greater. With a larger, more connected population travelling primarily by air (1946–1982), the frequency between outbreaks dropped to only 1.5 years. However, since the susceptible population was smaller, the severity of the outbreaks was much lower (Cliff, Haggett, Ord, and Versey 1981).

It may seem hard to believe now, but the reduction in travel time with an improvement in transportation technology was often quite significant and one of the early examples of globalization (Pascali 2017). For example, the average sailing time to travel around the world may have been about one year prior to the mid-nineteenth century. However, with the building of strategically placed canals (e.g., Suez, Panama), the use of more accurate charts and the use of coal and steamships, this travel time decreased to about one hundred days by the start of the twentieth century and further declined to about sixty days by 1925 (Rodrigue, Comtois, and Slack 2016).

Today, when most transoceanic passenger travel is by air, the era when most passengers travelled between continents by ship seems ancient. However, it was only in the late 1950s that the number of passengers travelling across the Atlantic Ocean by air started to exceed the number travelling by ship (Lawton and Butler 1987). In 1970, just over 310 million passengers were transported by air (World Bank 2019). This had increased to 4.2 billion in 2018 (ibid).

> AN INDEX CASE IS AN
> INITIAL PATIENT IN
> A POPULATION WHO
> EXHIBITS SYMPTOMS
> OF AN INFECTIOUS
> DISEASE. BY STUDYING
> THE CHARACTERISTICS
> AND ENVIRONMENT OF
> THIS INITIAL PATIENT,
> EPIDEMIOLOGISTS MAY
> BE ABLE TO TRACE THE
> SOURCE AND VECTOR OF
> THE DISEASE.

A good example of the relationship between the evolving risk of infection and transportation technology as it pertains to a metaphorical island occurred in 1989 in and around the Swiss airport of Geneva-Cointrin. Surprisingly, several malaria cases were recorded in the local population within two kilometres of the airport. This was despite the fact that these victims had never travelled to malarial areas. It was concluded that infected mosquitoes had inadvertently hitched rides aboard jet aircraft arriving from malarial areas and had then escaped and caused the outbreaks after the jet aircraft had landed in Geneva (Haggett 1994).

From recent outbreaks of infectious diseases such as severe acute respiratory syndrome (SARS) in 2003, the influenza A (H1N1) pandemic of 2009, the Ebola virus of 2014, and now the coronavirus/ COVID-19, we have seen that there are now very few communities, on islands or on continents, that cannot be reached during the contagious period of most infectious diseases. One island that is a notable exception might be the British overseas island territory of Saint Helena in the South Atlantic, 4,000 kilometres from Rio de Janeiro, Brazil and 3,130 kilometres from Capetown, South Africa. Although a major international airport has recently been constructed on this small island of approximately 4,000 people, previously the only access with the mainland was via the RMS (Royal Mail Ship) *St. Helena* to/from Capetown, South Africa, typically a five-day voyage. Despite improvements in transportation technology, Saint Helena may be one of the few places in the world that is effectively less accessible now than it was in the 1850s when over 1,000 ships visited the island annually (Hogenstijn and van Middlekoop 2005; Royle 2001). Coincidentally, one of the first uses of the new airport, even before it officially opened, was for a medical evacuation.

There is no simple relationship between the degree of isolation, scale, or population size of the island and health care. Once again, the South Atlantic British Dependency islands of Saint Helena, the Falklands, and Tristan da Cunha are useful examples. In his research, Royle (1995) points out that the small size and isolation of these islands has quite likely contributed to genetic inbreeding and higher levels of some genetic illnesses such as diabetes. At the same time, because these islands remain the political responsibility of Britain (i.e., they are SNIJs), the quality of healthcare is superior to most small island developing states (SIDS).

Islands and Index Cases

The concept of a "Patient Zero" or an "Index Case" with respect to communicable diseases has also been considered an important component of epidemiology on small islands. An Index Case is an initial patient in a population who exhibits symptoms of an infectious disease. By studying the characteristics and environment of this initial patient, epidemiologists may be able to trace the source and vector of the disease. As discussed above, it was not uncommon for the passenger manifests of ships to record significant amounts of information regarding passengers, including names and ages of all family members, as well as their geographical origins and destinations. This is more information than would normally be available on someone migrating by land across a continent, especially if their travel kept them within the same country. Therefore, if an outbreak of a disease does occur, it may be easier to trace those individuals who have arrived or departed from an island and to determine the links between them, as well as the characteristics of the particular disease and setting.

Islands as Places of Quarantine

Islands have frequently been used as sites for quarantining passengers suspected of carrying contagious diseases, and policies regarding these quarantines have emerged as a function of these sites. The origin of the word quarantine comes from the Italian word *quaranta* or forty, apparently in reference to the number of days needed to keep infectious patients in quarantine. Of course, the actual length of time that patients are able to infect a susceptible population varies on the basis of many factors, including the type of disease. It has been suggested that this number is derived from Christian biblical scripture, in reference to the time Jesus supposedly spent in the wilderness (Matthew 4: 1–11 New Revised Standard Version). Tognotti (2013) has also speculated that the term is derived from one of Hippocrates theories regarding acute illnesses.

There have been a number of islands just offshore of mainlands that were used in the past as quarantine sites in an attempt to isolate infected immigrants arriving from other parts of the world. These include Grosse Île, Quebec, Canada, Swinburne and Hoffman Islands in New York Harbour, and Angel Island in San Francisco Bay. Two of the earliest recorded examples of quarantine islands included a plague hospital (or *lazaretto*) on the small island of

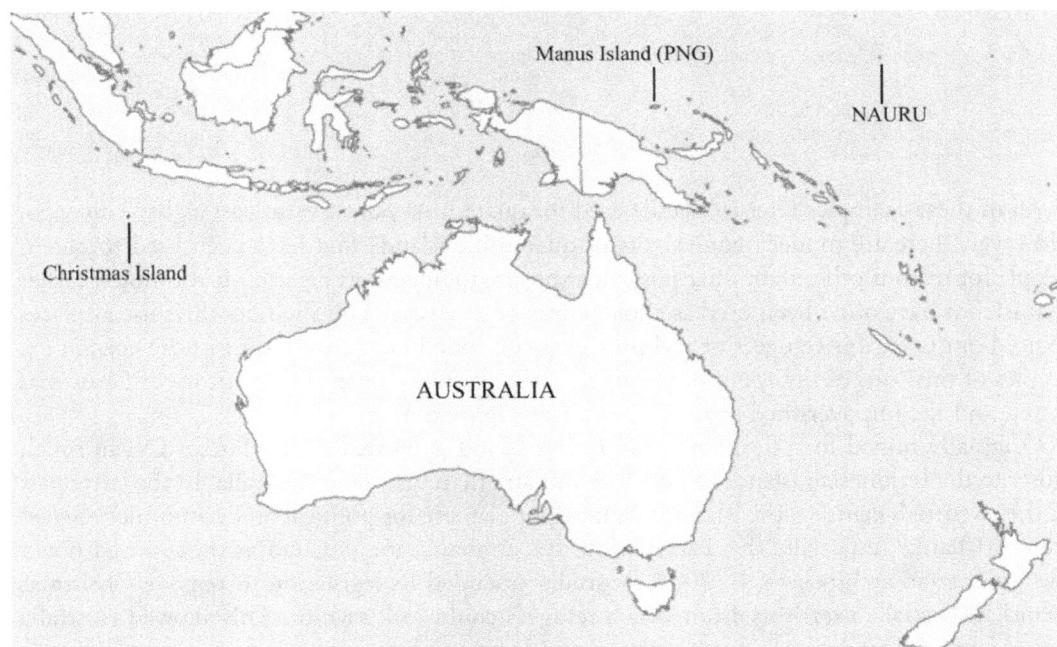

Figure 8.3   Map indicating locations of Christmas Island, Manus Island (PNG), and Nauru in relation to Australia. *Source:* **public domain**

Santa Maria di Nazareth just off the coast of the Republic (and island) of Venice to prevent the spread of bubonic plague at the end of the fourteenth century (Doty 1897), and the city state of Genoa that applied this same strategy in 1476 (Tognotti 2013). The artificial islands of Swinburne and Hoffman in New York Harbour are not as well-known as entry points as the immigrant checkpoint of Ellis Island, but they played important roles during a period of massive immigration to the United States. Over the course of fifty years (1892–1954), more than twelve million immigrants passed through Ellis Island. If any of the immigrants exhibited symptoms of illness, or if boarding officers who came aboard the ships before they landed suspected that anyone might be carrying a communicable disease (especially yellow fever, typhus, smallpox, or cholera), these passengers were sent directly to one of these two quarantine islands. Although there is no definitive record of the total number of people detained on these two islands, in 1901 there were reported to be 7,801 people on Hoffman Island and, during a cholera outbreak in a six-week period in 1892, there were 5,788 people confined on this island (Seitz and Miller 2011).

It is noteworthy that although the intent of this process was to provide an effective barrier between those infected and the larger American population, in reality, the process was neither systematic nor effective. In many cases, only those ships originating from ports that had a history of infections were boarded and it was not unusual for only the steerage class passengers to be inspected, with the implicit rationale that first- or saloon-class passengers were not susceptible to infectious diseases. Finally, despite the inspections, someone could still be a carrier of an infectious disease but may not yet have entered a stage where the symptoms were visible to the health inspector.

> ISLANDS HAVE OFTEN BEEN USED AS CHOKEPOINTS OR GATEWAYS. THIS INCLUDES THEIR USE AS PLACES TO HOLD OR QUARANTINE REFUGEES OR ASYLUM SEEKERS FOR INDEFINITE PERIODS.

All of these examples refer to islands used for quarantine purposes at least eighty years ago. However, there are modern equivalents of quarantine islands that have been used to detain people for reasons other than their possible exposure to infectious diseases. It was noted earlier that islands have often been used as chokepoints or gateways. This includes their use as places to hold or quarantine refugees or asylum seekers for indefinite periods. One of the current examples of this "out of sight, out of mind" policy is Australia's use of its territory of Christmas Island and the impoverished state of Nauru (see Figure 8.3).

Originally mined for phosphates, Christmas Island is located in the Indian Ocean much closer to the Indonesian islands of Java and Sumatra than mainland Australia. In the latter part of the twentieth century, the island became a popular site for political and economic refugees from Sri Lanka (especially the Tamil minority), Afghanistan, and Iran as they moved down the Indonesian archipelago. In 2001, Australia amended its legislation to remove Christmas Island as Australian territory from which refugees could seek asylum. This allowed Australia to hold arriving asylum seekers indefinitely without hearing their case for refugee status, or to transship them to other sites, including Papua New Guinea and Nauru. The detention

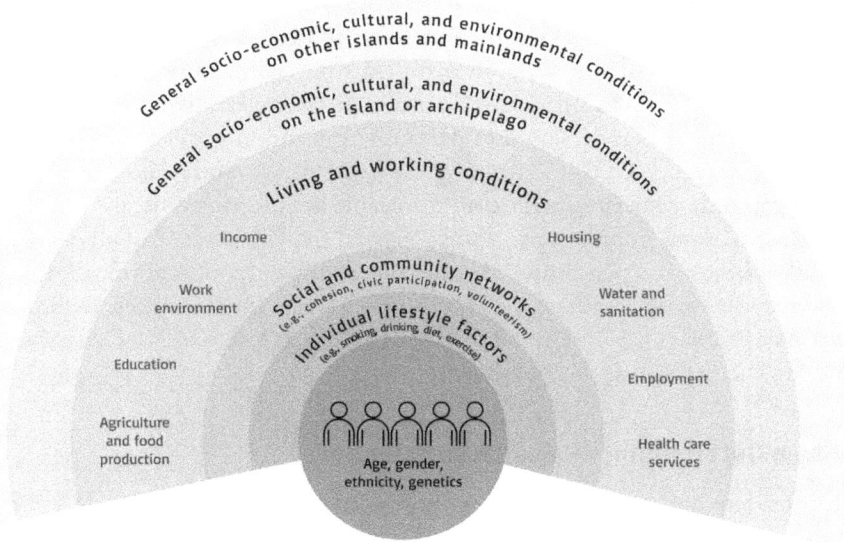

Figure 8.4 Social determinants of health. *Source:* **Author, Adapted from Anja Heilmann**

facilities established in 2001 were part of the Australian government's "Pacific Solution." Although as of late 2019 the detention centre was virtually closed, at one point there were more than 3,000 detainees held on Christmas Island (Refugee Council of Australia 2019). Two other islands have been used by Australia for the same purpose: Manus Island (part of Papua New Guinea) and Nauru. The formerly "rich" but now impoverished nation-state of Nauru was paid twenty million Australian dollars to establish a detention camp in 2001 for migrants seeking asylum who may have landed at other locations or were intercepted on the high seas. This detention facility was closed in 2008 but was reopened in 2012 after the election of a new federal government. Although the cases of many of the asylum seekers brought here were eventually approved and they moved on to Australia, there were still 562 asylum seekers in detention camps on Nauru and PNG as of September 2019 (Refugee Council of Australia 2019).

DETERMINANTS OF HEALTH AND ISLANDS

Much of the discussion regarding the factors associated with health outcomes is now being considered within a determinants-of-health framework. There are many definitions of "determinants" or "social determinants" of health but the following definition found on the Wikipedia site (n.d., n.p.)is one of the clearest and most concise. Social determinants of health "are the economic and social conditions that influence individual and group differences in health status. They are health-promoting factors found in one's living and working conditions (such as the distribution of income, wealth, influence, and power), rather than individual risk factors (such as behavioural risk factors or genetics) that influence the risk for a disease, or vulnerability to disease or injury." Figure 8.4 shows how factors or influences on health outcomes at the level of the individual, such as obesity, genetics, race, gender, age, and the consumption of tobacco products, are nested within the larger context of that person's social, economic, and physical environment. Even though this diagram implies a clear distinction between individual and social determinants of health, in reality, many determinants involve both spheres. A focus on health determinants coincides with a shift in the kinds of health challenges facing islanders, and in particular the rise in problems associated with non-communicable diseases (NCDs) (Tolley et al. 2016). While island health problems in the past were most commonly associated with communicable or infectious diseases, using 1990s data on sixteen Pacific Island populations, it was reported that 75 percent of all deaths and 40–60 percent of all health care expenditures are now associated with non-communicable or lifestyle-related health problems (I. Anderson 2013; World Health Organization, Regional Office for South-East Asia 2008). One determinant of health linked closely to NCDs that is particularly worrisome for Pacific Islanders is obesity.

Obesity as a Determinant of Health on Islands

Although considered by some to be too simplistic, one of the most common indicators of excessive weight is the Body Mass Index or BMI, a measure of the weight of individuals relative to their height. A person is defined as overweight if their BMI is between 25.0 and 29.9 and obese if their BMI is 30.0 or higher. According to an article in the British Medical Association's journal *The Lancet*, of the thirteen countries where the average resident has a BMI that is greater than thirty (i.e., obese), eleven of them are islands (Ng et al. 2014). In 2014, seven of the eight most "obese countries" are islands in the Pacific, seventeen of the twenty-nine most obese countries are islands, and more than 60 percent of the population is obese in the Cook Islands, the Federated States of Micronesia (FSM), Nauru, Niue, Palau, Samoa, and Tonga (Dillinger 2018; Gani 2009b).

Figure 8.5   Cans of SPAM. *Source:* **Michael Mozart**

> **OBESITY IS AS MUCH A PROBLEM OF THE INTERRELATIONSHIP BETWEEN THE LAND TENURE SYSTEM, TRADE POLICIES, AND THE ROLE OF TRADITIONAL VERSUS MODERN AGRICULTURE AS IT IS A MATTER OF INDIVIDUAL PREFERENCE AND LIFESTYLE CHOICES.**

Obesity is a health factor or determinant because it leads directly and indirectly to so many other health problems. For example, many of the island populations experiencing the highest levels of obesity are also experiencing a greater prevalence of diabetes, the highest levels of cardiovascular problems, and hypertension (Hawley and McGarvey 2015; World Health Organization, Regional Office for South-East Asia 2008). It has been estimated that more than 138 million people in the Western Pacific are living with diabetes (Nanditha et al. 2016). Although not all health problems are related to obesity, adverse health outcomes from obesity are particularly severe in children and pregnant women suffering from iodine deficiencies and anemia (i.e., an iron deficiency that leads to fatigue and a higher level of maternal deaths and cognitive problems in children). Robert Hughes and Geoffrey Marks (2009) and the WHO (2008) report found higher levels of vitamin A deficiencies on many of these islands, leading to higher rates of death from malaria, measles, diarrhea, and maternal deaths.

As is often the case with the determinants-of-health approach, the high level of obesity among islanders is the result of a complex set of interrelated factors. As recently as the first half of the twentieth century, many Pacific Islanders lived in relative isolation that was associated with a subsistence farming and fishing lifestyle and a diet consisting largely of traditional root crops, vegetables, fruits, and seafood. An economy that relied on physical labour and a diet that was low in processed foods, sugars, fats and sodium meant that there was a lower level of obesity in the population. During and after the Second World War, native islanders became more connected to the outside world and were exposed to the refined or processed Western diets that were high in fats and sodium, such as rice, sugar, flour, soda, beer, and fast food, brought to the island by the military (Cassels 2006; Cheng 2010; Davis et al. 2004). This diet is perhaps best illustrated by the popularity of two products: mutton flaps, consisting of scraps of meat and bone taken from sheep that are up to 50 percent fat, and the canned meat product SPAM, a mixture of precooked processed pork and ham, as seen in Figure 8.5 (McCartan 2010). In Hawai'i, this is sometimes referred to as "The Hawaiian Steak."

The economy of many Pacific Islands also changed during this period, becoming dominated by sedentary jobs considered high-status by the island population (Hawley and McGarvey 2015; Hodge et al. 1995). Moreover, in many socially connected societies dominated by feasts, a large physique is considered a mark of high social status (Brewis and McGarvey 2000; Hardin, McLennan, and Brewis 2018). It is not uncommon for large quantities of food to be consumed in these frequently occurring feasts. Some early research has even suggested that the ethnicity, and specifically the genetic make-up, of Pacific Islanders is one in which the body stores higher levels of energy in order to better adapt to feast-famine cycles (Gosling et al 2015; McGarvey et al. 1989; Neel 1962). Finally, McLennan and Ulijaszek (2014) suggest that a history of colonialism that is associated with lower self-esteem and a lack of affordable health services has also contributed to obesity.

In looking at the situation in Fiji and Tonga, Snowdon and her colleagues (2010) point to many inappropriate policies or gaps in policies as contributing to obesity-related problems on islands. For example, island consumers have very little control over food quality (e.g., healthy versus unhealthy foods), food advertising, or even the duties on imported foods that tend to be lower for unhealthy food choices. They point to the fact that much of the arable land on tropical islands is used to produce monoculture export crops rather than healthier, more nutritious foods that could be sold locally. Therefore, obesity is as much a problem of the interrelationship between the land tenure system, trade policies, and the role of traditional versus modern agriculture as it is a matter of individual preference and lifestyle choices.

There are many more examples where it appears that islanders exhibit higher levels of poor health outcomes or where illnesses or diseases are endemic to specific islands, regardless of whether the causes are communicable/infectious or genetic. For example, the Indigenous peoples on the Torres Strait Islands north of Australia exhibit greater levels of obesity, early onset diabetes, and hypertension than similar populations on mainland Australia (Leonard et al. 2002). In the Maldive Islands in the Indian Ocean and in Cyprus and Sardinia in the Mediterranean Sea, there is a much higher rate of beta-thalassemia than in the general world population (Banton 1951; Firdous, Gibbons, and Modell 2011; Galanello and Origa 2010). This genetic condition can lead to anemia and other health problems. On the island of Tristan da Cunha in the South Atlantic, more than half of the population has a certain degree of asthma (Citron and Pepys 1964; Zamel 1995). Research that stretches back to the 1960s suggests that there appears to be a genetic predisposition that is separate and apart from any of the environmental factors that might influence a diagnosis of asthma. On Guam and the other Mariana Islands, a form of the neurological Parkinson's disease, also known as amyotrophic lateral sclerosis or ALS, are reported to contribute to deaths at a rate that is one hundred times greater than that which is reported for any other population (Hirano et al. 1961). Finally, a number of health researchers have reported that small tropical islands in the Indian Ocean, Pacific, and the Caribbean seem to be emerging as the most significant incubators and transmitters for a wide variety of mosquito-borne viruses, including dengue, chikungunya, and Zika (Cao-Lormeau 2016; Cao-Lormeau and Musso 2014; Nahn and Musso 2015). The implication is that these islands, and small island developing states in general, are more likely to experience natural disasters, often lack a safe supply of water and sanitation, are at a greater distance from major health providers, and have local governments with limited resources to manage outbreaks (Cao-Lormeau 2016).

AS A COMPOSITE INDICATOR, THE HUMAN DEVELOPMENT INDEX IS STILL A FAIRLY SIMPLE MEASURE, CONSISTING OF AN AGGREGATION OF THE AVERAGE LIFE EXPECTANCY, LEVEL OF EDUCATIONAL ATTAINMENT, AND AVERAGE INCOME PER CAPITA FOR A POPULATION IN A GIVEN JURISDICTION.

HUMAN DEVELOPMENT INDEX AND HEALTH

The determinants-of-health model discussed earlier is holistic and multidimensional. Therefore outcome indicators should not consist solely of the physical health of an individual or group. Unfortunately, the lack of comparable data across small islands, and especially small islands that are not independent nation-states, makes it very difficult to describe, understand, and compare patterns of health and disease, and we are often left to discover the differences using data on the prevailing causes of death (N. Lewis and Rapaport 1995).

One of the most frequently cited measures of the health and development prospects of populations is the Human Development Index (HDI). As a composite indicator, it is still a fairly simple measure, consisting of an aggregation of the average life expectancy, level of educational attainment, and average income per capita for a population in a given jurisdiction. As such, it is supposed to reflect the goals of a long and healthy life, knowledge, and the resources needed for a decent standard of living (Stanton 2007). Although this is still fairly simplistic, it does allow for simple comparisons across the many different kinds of jurisdictions in the world, each with their own data collection systems. The HDI ranges from 0.0 to 1.0; the higher the value, the better the level of human development. If we looked at all of the small island developing states (SIDS) as a group, the average HDI has consistently been in the medium rank in comparison

Year		Country or Territory	HDI
Publication	Data		
Very high human development			
2019	2017	Macau	0.914
2018	2017	Taiwan	0.907
2017	2016	Martinique	0.863
2017	2016	Guadeloupe	0.850
2017	2016	Reunion	0.844
2017	2015	Puerto Rico	0.845
2009	2008	Jersey	0.985
2009	2008	Cayman Islands	0.983
2009	2008	Bermuda	0.981
2009	2008	Guernsey	0.975
2009	2008	Gibraltar	0.961
2009	2008	Faroe Islands	0.950
2009	2008	Isle of Man	0.950
2009	2008	British Virgin Islands	0.945
2009	2008	Falkland Islands	0.933
2009	2008	Aruba	0.908
2009	2008	Guam	0.901
High human development			
2017	2016	Mayotte	0.789
2015	2010	Greenland	0.839
2009	2006	U.S. Virgin Islands	0.894
2009	2008	Northern Mariana Islands	0.875
2009	2008	Turks and Caicos Islands	0.873
2009	2008	Anguilla	0.865
2009	2008	Cook Islands	0.829
2009	2008	American Samoa	0.827
2009	2008	Monserrat	0.821
2012	2010	New Caledonia	0.789
2012	2005	Wallis and Futuna	0.763
2012	2010	Saint Pierre and Miquelon	0.762
2012	2010	French Polynesia	0.737
2012	2000	Saint Martin	0.702
Medium human development			
2009	2008	Saint Helena	0.797
2009	2008	Niue	0.794
2009	2008	Tokelau	0.750
2012	2000	Saint Barthélemy	0.688

Figure 8.6 Human Development Index for many subnational island jurisdictions.
Source: **Adapted from Wikipedia**

to other nations. It also appears that human development on these island nations has been improving but only slightly. In 1990, the average HDI for all SIDS was 0.574. This had increased to 0.607 by 2000 and to 0.660 by 2014 (Jahan 2015).

In an earlier chapter, it was argued that subnational island jurisdictions (SNIJs) may be better off in many respects than small islands that are politically independent. A comparison of the Human Development Index between these two groups shows that this may be the case, at least by this one measure. Using 2017 data from the United Nations Development Program in Figure 8.6, the average HDI for all SIDS (most of which are politically independent) was 0.676 (UNDP 2018).

From a selection of thirty-five jurisdictions, most of which are island territories, thirty-one of them are in either the High or Very High HDI categories and all but nine of them have HDIs greater than 0.800 (UNDP 2018). Beyond the Human Development Index, there is an increasing number of indicators that are attempting to measure and compare a holistic interpretation of health, as defined as quality of life, well-being, and even happiness. The background rationale surrounding many of these measures is that there is a causal relationship between attitudes and perceptions of well-being or happiness and the mental and physical health outcomes of a population (Post 2005; Scheier and Carver 1985, 1987). To use just one of these indicators as an example, the "Happy Planet Index" has emerged and been used as a measure that incorporates life expectancy, perception of well-being, inequality, and nations' ecological footprints (i.e., the average amount of land needed to sustain a country's consumption patterns). In 2016, only Vanuatu (in 4th place) ranked among the top ten "happy places," and only four islands or archipelagos (Jamaica in 11th, Indonesia in 16th, and the Philippines in 20th) were among the top twenty happiest places in the world.

CONCLUSIONS

Islands are not simply other places or contexts that experience health challenges. Historically, they have played critical roles in allowing us to better understand the transmission of diseases, factors associated with obesity and other non-communicable diseases, and the influence of the determinants of health. In Cliff and Haggett's (1995, 199) words, "island epidemiology has implications both for practical questions of disease control and for academic questions about the origin and persistence of infectious diseases." These roles are not just limited to the past, where islands were often used to quarantine migrating populations carrying communicable diseases; some islands are now quarantining other kinds of "undesirables," including those seeking asylum from war, persecution, and poverty.

As with their unique ecosystems, small islands appear to be associated with some diseases and "lifestyle-related" health characteristics such as obesity that are not present to the same degree on continents. This creates unique challenges and forces the adoption of unique solutions. At the same time, and despite a lower level of access to many health services, the strong social bonds that exist within the communities on many islands may encourage a level of resilience and sustainability that may not be present in comparable mainland communities. In the next chapter, we will look at the ways in which small islands may be vulnerable to economic change but also how they and their residents have proven to be quite resilient and adaptable to evolving circumstances in the larger global economy.

It is premature for us to understand all of the dimensions and consequences of the relationship between islands and the 2020 COVID-19 coronavirus pandemic. In the past, and at least partly as a result of their ability to control entry at air and seaports, some islands did not experience the impacts of communicable diseases as early and to the same degree as on mainland communities. As Figure 8.2 in this chapter shows, if a disease is less infectious, has fewer people who can contract it (i.e., the susceptibles), and the community is relatively small, it is less likely to emerge and remain permanently on the island. However, our world is much more highly connected now than in the past. This feature, combined with the infectious nature of this illness while people are asymptomatic and the fact that most people are susceptible to it, has meant that it has emerged on many islands. The following factors might allow some islands to escape the worst public health consequences of the pandemic. If they are relatively small and remote and have few points of entry, they may be more likely to identify and isolate any travelers, including those who may be infected. However, this means the island governments would have to move quickly and aggressively to enact public health measures, including testing and tracing the contacts of those who are infected. Unfortunately, most small islands do not have the testing capabilities of larger places making it difficult to quarantine those infected and prevent community spread. Individual islanders must also be willing to self-isolate for at least several weeks, something difficult to do for many socially active island cultures.

The island city of Singapore is an interesting case study in this pandemic. Although it is not remote and is highly connected to the rest of the world, the government moved quickly to ban flights from the initial region of the epidemic in Wuhan, China. It placed those coming from outbreak sites in mandatory quarantine, and aggressively tested and traced the contacts of all those infected. As a relatively affluent jurisdiction, it had a highly developed health infrastructure and testing capabilities. Moreover, it is not uncommon for the Singaporean government to intervene quickly and decisively in enacting public policies, including in the area of economic development. For the most part, the citizens of Singapore accept this intervention in order to achieve a greater common good including, in this case, accepting the inconvenience of temporary self-isolation.

Unfortunately, even if islands have been successful in limiting the local public spread of these kinds of infectious diseases, they will not escape the economic consequences. A decline in tourism arrivals, revenues, and employment are the most obvious initial outcomes. This is especially problematic for islands that are heavily dependent on tourism. In addition, remittances from the island diaspora populations may decrease as other jurisdictions experience recessions. At least in the short term, it may even be more difficult for temporary island workers to move internationally if there are continued restrictions on foreign travel. Those island nations such as Papua New Guinea that are dependent on the royalties from the extraction of minerals will also suffer economically during a global recession, as will places that export fossil fuels such as Trinidad and Tobago. Even the offshore finance centres may suffer if the banking industry contracts. Although the full story of this pandemic has yet to be told, it appears that islands will play an important role.

Key Readings

Cassels, Susan. 2006. "Overweight in the Pacific: Links Between Foreign Dependence, Global Food Trade, and Obesity in the Federated States of Micronesia." *Globalization and Health* 2 (1): 10.

Cliff, Andrew, Peter Haggett, and Matthew Smallman-Raynor. 2000. *Island Epidemics*. Oxford, UK: Oxford University Press.

Cliff, Andrew, Peter Haggett, John Keith Ord, and G. R. Versey. 1981. *Spatial Diffusion: An Historical Geography of Epidemics in an Island Community*. Cambridge, UK: Cambridge University Press.

Haggett, Peter. 2000. *The Geographical Structure of Epidemics*. Oxford, UK: Oxford University Press.

McLennan, Amy K., and Stanley J. Ulijaszek. 2014. "Obesity Emergence in the Pacific islands: Why Understanding Colonial History and Social Change is Important." *Public Health Nutrition* 18 (8): 1499–1505.

Chapter Nine

Economic Change, Development, and Islands

For small island states and territories, life continues to be a scramble to exploit one niche or opportunity, then another, moving as nimbly as possible from one to the next or from one crisis to the next as one dries up and (hopefully) another presents itself. (Baldacchino 2015a, 6)

INTRODUCTION

Prior to the 1970s, when a politician or a researcher talked about development, they were almost always speaking about economic development. Moreover, this version of development was usually synonymous with a change in employment, income, or production for a country or region. If social characteristics or indicators such as the level and quality of education, health outcomes, rights and freedoms of citizens, or equality were raised at all, they were considered as indirect outcomes of a growing economy. Islands, and especially small islands, were marginal players within the global economy. Other than being mysterious and remote, their economic role was limited to providing mainlanders with food, such as sugar cane, pineapple, and bananas, or minerals, such as phosphates used in fertilizer. Even earlier, when most small islands were colonial outposts, they served as way stations for the slave trade or were exchanged as if they were pawns in a global geopolitical chess match (Pocock 2005). Now, many small islands are said to be among the most creative, entrepreneurial places in the world (Baldacchino 2015a). For example, some tropical islands capture enormous sums of money from the millions of tourists escaping to vacation paradises. Other islands have become politically astute or destitute enough to secure large sums of money in aid. And there are yet other islands that have become so closely associated with services that the phrase "offshore financial centre" has become synonymous with islands, both as centres of commerce and international finance as well as with allegations of larceny.

This chapter examines the economy and economic development of islands. It does so by first providing a description of the measures or indicators of development that have been applied to small islands. It then builds on this by reviewing three of the most prominent models of island economic development: 1) resource exploitation, 2) remittances and aid, and 3) strategic flexibility. Although these models may constitute idealized and oversimplified representations of island economies, they still tell us much about the economic nature of many small islands. This chapter also approaches the subject with the understanding that island economies are very diverse and are constantly evolving to fit their changing circumstances. It also recognizes that an island's scale, location, culture, and external political relationships have been important features in shaping the nature of their economies.

> ISLANDS, AND ESPECIALLY SMALL ISLANDS, WERE MARGINAL PLAYERS WITHIN THE GLOBAL ECONOMY. OTHER THAN BEING MYSTERIOUS AND REMOTE, THEIR ECONOMIC ROLE WAS LIMITED TO PROVIDING MAINLANDERS WITH FOOD, SUCH AS SUGAR CANE, PINEAPPLE, AND BANANAS, OR MINERALS, SUCH AS PHOSPHATES USED IN FERTILIZER. . . . NOW, MANY SMALL ISLANDS ARE SAID TO BE AMONG THE MOST CREATIVE, ENTREPRENEURIAL PLACES IN THE WORLD (BALDACCHINO 2015A).

ECONOMIC INDICATORS

It was not that long ago that the development of all countries was measured against the economic history of America and Europe after the Second World War. This model implied a consistent, positive growth in Gross Domestic Product (GDP) and stability within the democratic political system that supported this growth. It implied that there was a natural evolution in the economic structure of a country, with specialization first in the production and export of primary goods, then a shift to the production of manufactured goods and, eventually, to an economy based on the production and export of producer or intermediate services, such as finance and banking (Rostow 1959). Finally, it was thought that although this path to growth might start in one or more key industrial sectors and regions, over time growth was expected to "trickledown" or "spill over" to other sectors and to other regions of a country (Darwent 1969; Moseley 1974; Richardson 1976). For example, investment by private companies and the government in automobile manufacturing and assembly in the northeastern United States or in the information technology sector in California's Silicon Valley was coordinated with government investment in public infrastructure to support these sectors, such as transportation, communications networks, and education. Not only did this result in growth in the region where these sectors were clustered, but it also generated growth among the companies that served as suppliers and customers, not only in that region but throughout the country and abroad.

Since the absolute level of Gross Domestic Product is influenced by the size of the economy, it could be made comparable across a range of countries by dividing GDP by the total population (i.e., GDP per person). This model of development and these indicators have come under increasing criticism as a method to assess social welfare (O'Neill 2014; Philipsen 2015; Van den Bergh 2007, 2009). Among other things, it was felt that the GDP did not pay enough attention to the relative distribution or equity of that wealth across the total population and did not measure informal work, thereby marginalizing the roles played by women and those living in rural regions. Despite these criticisms, it is still useful to look at how islands fare according to this economic indicator. In general, many islands appear to be doing quite well relative to mainland jurisdictions.

Using mostly 2017 data, twelve out of the world's top thirty wealthiest jurisdictions (as measured by GDP/capita) were islands (Central Intelligence Agency 2017). This includes Bermuda (6th), Singapore (7th), Isle of Man (8th), Brunei (9th), Ireland (10th), the Falklands/Malvinas (12th), Sint Maarten (14th), Hong Kong (17th), Jersey (21st), Guernsey (24th), Iceland (25th), and Taiwan (28th). Although these places are different in many ways, they appear to share some characteristics. For example, several of them are "island city-states" such as Singapore, Hong Kong, and Brunei. As Grydehøj and others have pointed out, large cities on small islands have been an overlooked aspect of island studies research and teaching (Grydehøj 2014, 2015; Grydehøj et al. 2015). In addition, more than a few of them are subnational island jurisdictions that specialize in offshore banking and international finance and are also characterized by having relatively stable governments and institutional structures, such as the judicial system. Very few of the islands listed above still rely on agriculture, other primary activities, or even manufacturing.

At the other extreme, and once again using the measure of GDP/capita from the most recent CIA World Factbook, six of the thirty jurisdictions with the lowest levels of GDP per capita are islands (CIA 2017b). These include Tokelau (226th), Madagascar (218th), Comoros (216th), Haiti (212th), Kiribati (208th), and the Solomon Islands (203rd). Once again, these places differ in many respects. However, with few exceptions, the economies of most of these islands are dominated by the production and export of primary goods and they experience relatively higher rates of population growth. The inclusion of several islands on this list points to the challenge of using GDP per capita as a summary measure of development. For instance, Tokelau may be a good example of what has been referred to as "subsistence affluence" (Conroy 2012; Fisk 1982). Most of Tokelau's inhabitants have New Zealand citizenship and are therefore able to move freely between the islands. Although their GDP per capita value may be low, this does not include the substantial government subsidies and revenue paid out to Tokelauans in fisheries royalties. Countering this, there are other island researchers who say that "subsistence affluence" may be a mirage in the goal of achieving real long-term development (Chand and Duncan 1997; Connell 1991). As such, seeing smiling natives performing for tourists masks a deeper struggle for cultural and economic survival.

In the last chapter, we used the Human Development Index (HDI) as an alternative measure of the health of islanders, primarily because it incorporated life expectancy but also because the adult literacy rate, school enrollment, and GDP per capita are often directly correlated with health outcomes. Therefore, it can also be used as a substitute or a complement to GDP. As is the case with the simple GDP measure, the HDI can also be calculated and compared relatively easily for most jurisdictions, including small island developing states and many subnational island jurisdictions. In its simplest form, the HDI can range from 0.0 to 1.0. Convention suggests that if a country or region has an overall value of more than 0.8 it is considered highly developed, while an HDI of lower than 0.5 is considered a low level of development.

If you ranked all jurisdictions by their 2017 HDI, you would find that ten islands are in the top quartile (25%) of places, with the Republic of Ireland being the highest at 0.938 in fourth place and Bahrain at 0.846 ranked in 43rd position. Of the 25 percent of jurisdictions in the lowest quartile, only six of the forty-seven jurisdictions are islands. Haiti is consistently the lowest-ranked island using the Human Development Index measure with a 2017 value of 0.498. The HDI is not without its own critics (Bagolin and Comim 2008; Kovacevic 2010). Some suggest it really isn't much better than purely economic measures in assessing the true human condition (McGillivray 1991; Ogwang 1994; Ravallion 1997), while others are opposed to it on methodological grounds (Lind 1992; Ogwang 1994; Srinivasan 1994). In response, advocates of the measure note that the HDI was never intended to "be a comprehensive measure of human development or well-being—but rather a summary alternative to economic measures" (Kovacevic 2010, 1).

> ONE OF THE CHALLENGES OF USING INDICATORS TO ASSESS THE STATE OF AN ISLAND'S ECONOMIC DEVELOPMENT IS THEIR INHERENT BIAS.

One of the challenges of using indicators to assess the state of an island's economic development is their inherent bias. For example, two of the eight indicators used by the United Nations Committee for Development Policy to assess economic vulnerability are population size and remoteness. Since islands tend to be smaller and more remote than mainland jurisdictions, they are more likely to be defined as vulnerable. Also, despite urban island nations such as Singapore consistently having one of the highest levels of GDP/capita and HDI among any country in the world, the Commonwealth Secretariat's Composite Vulnerability Index ranks them as being among the most vulnerable countries in the Commonwealth, primarily because of the volatility of economic production. What really matters in economic vulnerability is income, not production. During downturns in the economy, remittances and aid may make up for reductions in formal income. Government spending in places like Singapore and Tokelau, as noted earlier, make these places less vulnerable than these measures might suggest.

If we relied solely on these two measures (Gross Domestic Product and the Human Development Index) we might be tempted to conclude that islands have performed quite well when they are compared to other mainland jurisdictions. However, since the mid-1990s, arguments have been made that these indicators do not accurately reflect the nature of small island economies (see, for example, Armstrong and Read 2003; Briguglio 1995; Read 2004). Briguglio (1995) summarized this perspective by suggesting that small size (i.e., lack of economies of scale), remoteness/isolation (i.e., higher transportation costs), dependence on the outside world for markets and materials (i.e., greater exposure to external economic shocks), and even a greater likelihood of experiencing natural disasters all combined to make small islands more vulnerable to economic crises. Other researchers (Baldacchino and Bertram 2009) have criticized this approach as not recognizing the advantages of small size and the active role of these islands in a globalized world. Despite the flaws of the vulnerability index, this early work has allowed organizations like the Association of Small Island States (AOSIS) to have a more powerful voice on the international stage, including at United Nations meetings on the sustainable development of small islands, especially during a climate crisis.

MODELS OF ISLAND ECONOMIC DEVELOPMENT

Resource Exploitation as Economic Development

One of the most common and more traditional roles played by small islands has been as a producer and exporter of primary resources, including agricultural products, seafood, and minerals. Depending on the region, this role existed well before contact with Europeans (Bellwood 1978; I. Campbell 1987; D'Arcy 1998). Once under colonial rule, many islands were forced to adopt a plantation-based economy, and this model was perpetuated even after islands gained their independence and became more integrated into the global economy (Beckford 1969). Some islands, such as New Guinea (copper and gold) and New Caledonia (nickel), continue to have an abundance of natural resources on and offshore; others are less fortunate (Opeskin and MacDermott 2009). As is often the case when colonial or corporate entities engage economically with small (island) states that have a finite and specialized resource base, short-term wealth is usually followed by longer-term political destabilization, resource depletion, and environmental degradation (Connell 2007). Another challenge facing places with a specialized economy around natural resource exports is the so-called "Resource Curse," where an overemphasis on this one economic activity combined with poor governance hinders the long-term development of other sectors (M. Ross 1999, 2003). Some believe that this is what took place in Papua New Guinea, an island developing country that has specialized in the export of minerals and petroleum, while also experiencing the 1988 to 1998 Bougainville civil war (Banks 2008; Togolo 2006). In this latter case, concern about foreign gold and copper mining operations, a lack of compensation for repossessed land, environmental degradation brought about by the mining operation, and ethnic conflict all contributed to an armed con.

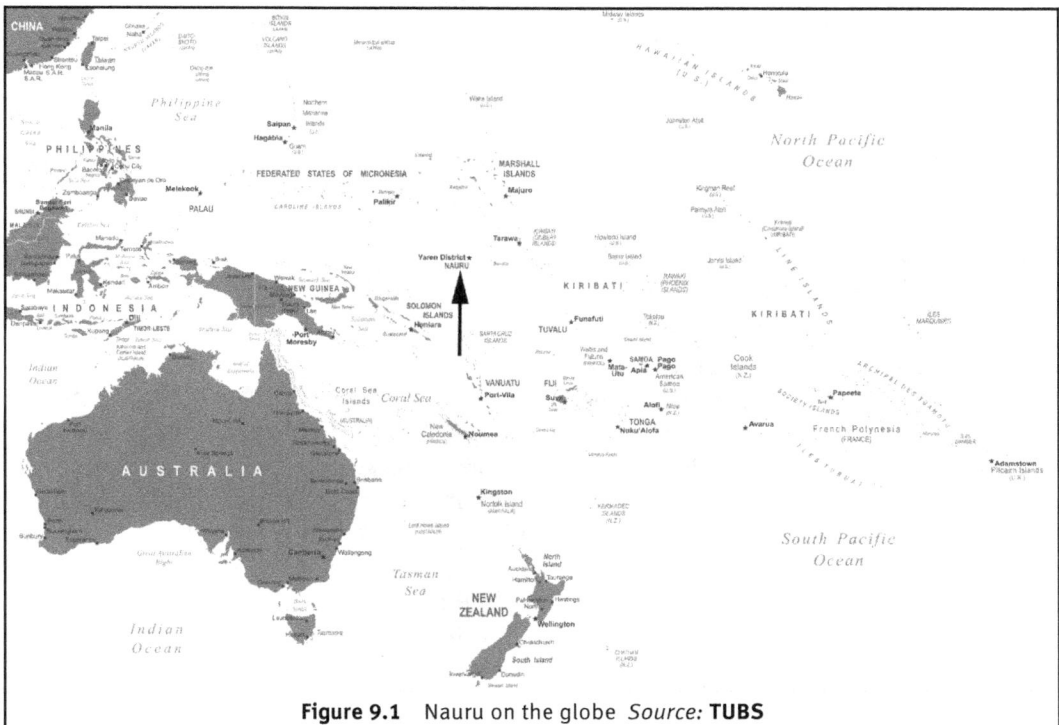

Figure 9.1   Nauru on the globe *Source:* **TUBS**

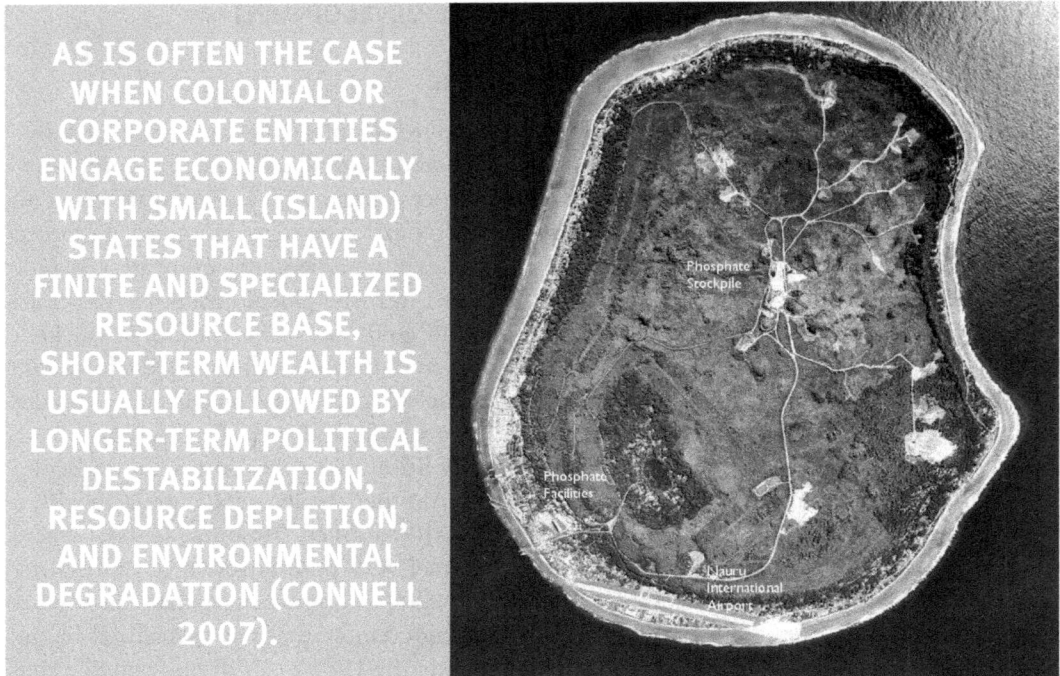

AS IS OFTEN THE CASE WHEN COLONIAL OR CORPORATE ENTITIES ENGAGE ECONOMICALLY WITH SMALL (ISLAND) STATES THAT HAVE A FINITE AND SPECIALIZED RESOURCE BASE, SHORT-TERM WEALTH IS USUALLY FOLLOWED BY LONGER-TERM POLITICAL DESTABILIZATION, RESOURCE DEPLETION, AND ENVIRONMENTAL DEGRADATION (CONNELL 2007).

Phosphate Stockpile

Phosphate Facilities

Nauru International Airport

Figure 9.2   Satellite map of Nauru. *Source:* **ARM Image Library,** public domain

flict between the Bougainville islanders and the PNG government that resulted in the deaths of up to 20,000 people (Adomo 2017). Secessionist pressures remain to this day, with 98 percent voting in favour of independence in a November 2019 referendum.

A good example of an island that has relied primarily on the production and export of primary resources has been the Republic of Nauru in Oceania. Located roughly on the equator and about 2,600 kilometres east of Papua New Guinea, Nauru has a land area of twenty-one square kilometres and a 2017 population of just less than 14,000 people (see Figures 9.1 and 9.2). Although it has been occupied by tribal peoples for almost 3,000 years, non-Indigenous colonial states have controlled the island since the 1880s: first Germany (1888–1914), then Australia (1914–1919), and finally a trusteeship of Australia, New Zealand, and the United Kingdom (1919–1942, 1945–1968). Nauru achieved independence in 1968. The mining and export of phosphates, used primarily in making fertilizers, has been a major generator of export revenues on the island since the late nineteenth century, first under colonial rule and then under the state-owned Nauru Phosphate Corporation. Although official government data is not complete, in the 1970s and 1980s the popular press estimated that Nauru was one of the richest countries in the world, as measured by income or GDP per capita, with almost all of this wealth from the royalties associated with phosphate sales (Connell 2006; McQuade 1975). In 1991, Howard (50) wrote of Nauru, "Imagine an island nation where the typical family of five has an income of $100,000 a year without even working; where there are no taxes or duties; where all medical and dental care is free, and even medicine, band-aids and aspirin are dispensed without charge."

An estimated seventy-seven million tons of phosphate was mined from the island from the early 1950s until it was exhausted in the mid-2000s (Déry and Anderson 2007; S. Gale 2019; H. Hughes 1964). Although a government trust (i.e., the Nauru Phosphate Royalties Trust) was established so that the profits from the resource could be reinvested in other economic activities to benefit future generations of islanders, and this Trust at one point exceeded AUD 1.3 billion, mismanagement and a series of poor investment decisions left the trust with only AUD 300 million in 2004 and it is now virtually insolvent (J. Cox 2009; Toatu 2004). Nauru was never a stereotypical jewel of the tropics (H. Hughes 1964). However, as a consequence of the mining, much of the landscape of Nauru has also been denuded of trees and consists of heaps of waste material.

The depletion of this resource and the failure to diversify into other sectors during this period has also led to higher unemployment and a wholesale emigration of Nauruans to Australia, New Zealand, and other Pacific countries. It has also led the Nauru government to embrace a number of creative schemes to generate other sources of revenue, including establishing itself as a tax haven (it was blacklisted for purportedly engaging in money laundering), trading their vote on United Nations motions in return for aid from Taiwan, Russia, and China, selling Nauruan passports, and as a detention centre for asylum seekers to Australia who were intercepted on the high seas (Herr and Potter 2006; Kendall 2009; Morris 2019). Remittances from the diaspora of Nauruans living elsewhere now represent one of the largest sources of national income. The most recent schemes include greening the landscape and marine areas (Baines 2014) and engaging in tourism (Fagence 1999). Some, however, are not so optimistic. The economic prospects for Nauru are now so bleak that it has been referred to as the first failed Pacific state and a fiefdom of Australia (Connell 2006).

Migration, Remittances, and Aid as Economic Development

> WHEN YOU COMBINE REMITTANCES AND INTERNATIONAL DEVELOPMENT AID, IT IS NOT UNCOMMON FOR THESE SOURCES OF INCOME TO PLAY DOMINANT ROLES IN THE ECONOMIES OF SMALL ISLAND STATES.

Island studies scholars have also suggested that the economies of many small islands now rely primarily on a combination of remittances generated from migrants living and working abroad, international aid, and public-sector spending. This has been referred to as the MIRAB model, where the **MI** is migration, the **R** is remittances, the **A** is aid, and **B** represents bureaucracy. Migration leads to flows of remittances and international aid leads to a greater presence of the public sector, in the form of bureaucracy and employment (Bertram and Watters 1985). In an earlier chapter, we examined remittances in relation to the migration of islanders. Here we look at it as an integral part of the economies of small islands. As with almost all of the theories of island economic development, this model relies on a high degree of connectivity with the outside world. The initial empirical examples that were used to support this nationalistic development model were from the formerly colonial small island states in the Pacific. Bertram (2006) suggests that a dependence on these activities in places such as the Cook Islands, Tokelau,

Kiribati, and Tuvalu was understandable since they were emerging from a long period of colonial welfarism and had a history of public-sector-led investment in long-term infrastructure projects and short-term development projects.

One measure of the significance of remittances to the economies of small islands comes from the World Bank (2016) in its Migration and Remittances Factbook. When remittances were measured as a share of a country's 2014 GDP, nine of the top thirty countries were island nations. This included Tonga ranked fourth at 27.9 percent, Bermuda ranked seventh at 23.1 percent, and Haiti ranked eighth at 22.7 percent. These numbers may be a little misleading. For example, most of the remittances flowing into the offshore financial center (OFC) of Bermuda are really deposits as a function of its status as a low tax financial centre.

With respect to international aid, small islands are also prominently represented. For example, using World Bank figures showing Official Development Assistance as a share of total 2017 Gross National Income, thirteen of the thirty developing countries or territories with the highest values were islands, including Tuvalu with 65 percent of its GDP coming from international development aid. When you combine remittances and international development aid, it is not uncommon for these sources of income to play dominant roles in the economies of small island states. Bertram and Poirine (2018) note that of the fifty island states and territories in their research, nineteen of them relied on remittances and official transfers for over 40 percent of their funding, and for fourteen of these, the figure rises to more than 60 percent of their funding. Connell (2016a, 468) has referred to some islands of the Pacific as "government islands," where almost every household has at least one person employed in the public sector, including health and education, thereby enabling a level of prosperity and population stability that would otherwise be impossible.

Strategic Flexibility as Economic Development

An earlier chapter discussed islands as subnational political jurisdictions (SNIJs), and the sometimes favourable arrangements they may be able to negotiate with their metropoles or parent nations, including freer trade, special tax concessions, less spending on the military, and access to a larger labour market and a pool of investment capital. Many small islands that find themselves in this ambiguous relationship, and even other politically independent small island states (SIS), appear to be able to use their size, location, and political negotiating skills to their advantage. Island studies scholars now suggest that this is not a coincidence and that, at a broader conceptual level, there is something unique taking place here. The acronym used to describe this model of economic development is PROFIT, which loosely translates to encompass **P**eople, **R**esource management, **O**verseas engagement, **FI**nance, and **T**ransportation. Baldacchino and Milne (2000) refer to this as the "resourcefulness of jurisdiction," defined by McSorley and McElroy (2007, 142) as "the ability particularly for non-sovereign dependent territories to manipulate metropolitan links for local benefit."

This concept is discussed in much more detail in Baldacchino's (2010b) *Island Enclaves*. The contributors to this volume were looking at the circumstances of SNIJs in the North Atlantic and the Caribbean, such as the Faroes, Canary Islands, Azores, Bermuda, and the Channel Islands of Jersey, Guernsey, and the Isle of Man, and found that their relatively strong economies were not really MIRAB-related. Instead, these islands had become economically and politically entrepreneurial and were especially successful in negotiating and gaining

THESE ISLANDS HAD BECOME ECONOMICALLY AND POLITICALLY ENTREPRENEURIAL AND WERE ESPECIALLY SUCCESSFUL IN NEGOTIATING AND GAINING CONTROL OVER A VARIETY OF ECONOMIC LEVERS WITH THEIR METROPOLITAN PATRONS. IN THIS SENSE, THEIR POLITICAL ACUMEN AND RELATIONSHIPS HAD BECOME AN ASSET, IN MUCH THE SAME WAY THAT EXPORTS OF NATURAL RESOURCES OR A DIASPORA POPULATION PROVIDING REMITTANCES COULD BE CONSIDERED ADVANTAGES.

control over a variety of economic levers with their metropolitan patrons. In this sense, their political acumen and relationships had become an asset, in much the same way that exports of natural resources or a diaspora population providing remittances could be considered advantages. Ferdinand, Oostindie, and Veenendaal (2019) suggested that it is their constitutional status with their metropoles that largely explains their economic success. When you look more closely at some of these islands, you find that one of the main topics dominating internal political discourse is the nature of the relationships the islands have with their metropoles or other larger mainland nations. This often took place regardless of which political party might hold power on the island. Even those parties elected on a platform of island independence often appeared to use the threat of independence as a means to negotiate concessions from the parent nation. This has led to a higher level of economic specialization on some of these islands, including serving as tax and insurance sector havens, offshore banking centres, and duty-free manufacturers.

So what are some of the policies or regulatory relationships that some of these small islands are able to negotiate? For the **P**eople part of the model, this might be preferential immigration pathways to and from the patron states. Although some of these policies are now becoming more restrictive, inhabitants of many island territories have held passports to their metropoles of the United States, France, Denmark, Netherlands, and New Zealand for decades, giving them the right to travel, work, and study abroad relatively easily (Veenendaal 2016). Not only has this resulted in modest levels of remittances flowing back to the islands from those living abroad but it may be used as a labour market and political "safety valve" on those occasions where there may be too few local jobs for a growing island population.

The **R**esource management component of this model refers to the ability of an island government to gain favourable regulatory powers over the management and revenue associated with island natural resource development. Some examples of this might include the ability of the provincial government of Newfoundland and Labrador, Canada, to retain higher levels of royalties from the federal Canadian government for the extraction of offshore oil and gas deposits (Blake 2015; Fusco 2007). In the same manner, royalties from oil and natural gas extraction in the North Sea are supporting local governments in the Orkney and Shetland Islands of Scotland (Johnson, Kerr, and Side 2013). These same island governments have built wind turbines with the support of capital provided by Scottish or British governments and are then able to retain a portion of the revenues from the sale of electricity back into the grid to support local public services (Kerr, Johnson, and Weir 2017). Another example may be the ability of many small islands such as the Falklands, Kiribati, Tuvalu, Tokelau, and others to retain greater control over the revenues generated from foreign fishing within the 200-mile exclusive economic zones surrounding their islands (Bertram 2013; Seidel and Lal 2010).

Overseas engagement refers to several kinds of abilities. In some cases, island jurisdictions are able to negotiate greater control over local economic and social activities not available to other jurisdictions in the larger mainland country, such as policing, education, health services, and even immigration policy. Moreover, island territories may participate in regional and international associations or organizations that they would normally be excluded from as non-sovereign jurisdictions. This is referred to as "para-diplomacy" and, because of their membership in these supranational organizations, either in a voting or observer capacity, they may obtain benefits that are not available to other mainland regions (Bartmann 2006; Tavares 2016). For example, Greenland, a Danish territory, has played an increasingly prominent role in the geopolitics of the Arctic, especially as it relates to hydrocarbon and mineral deposits (Ackrén 2014; Jacobsen 2019). The Faroe Islands (Denmark) and New Caledonia (France) have signed trade treaties with other nations (Prinsen, Lafoy, and Migozzi 2017), and the Cook Islands and Niue (territories in free association with New Zealand) have been permitted to join United Nations agencies such as UNESCO (Bartmann 2006).

Another example of overseas engagement or strategic management as practiced by small island nations is the sale of passports (van Fossen 2018). In some cases, such as in Tonga, the Marshall Islands, Nauru, and Vanuatu, these ventures have been short-lived and been accompanied externally by concerns about corruption and internally about an influx of outsiders (Eggenberger 2018). On other islands, such as in Malta, Antigua and Barbuda, St. Kitts and Nevis, and Cyprus, passport sales programs are more integrated internationally and internally, within the tourism sector (Přívara 2019)

The FInance part of the PROFIT model refers to the powers devolved to island jurisdictions to develop and implement their own finance, insurance and taxation policies, and laws. For example, the British island crown dependency (or "bailiwick") of Jersey located between England and France has a unique constitutional relationship with both the United Kingdom and the European Union which, when combined with the entrepreneurship of its residents, has allowed its financial services sector to thrive (Entwistle and Oliver 2015; Hampton 1994; 1996). It is noteworthy that, as is the case with MIRAB-based islands, this may lead to a higher level of public-sector bureaucracy and government employment. However, the nature of that bureaucracy is quite different. In this model, the state facilitates the relationships to trade financial services between external private and public-sector actors such as companies and other governments, whereas in the MIRAB model the bureaucracy is focused on either providing internal services or seeking ongoing international aid.

Lastly, the Transportation component of the model refers to policies and regulations that pertain specifically to public control over air and marine infrastructure and routes. It is not uncommon for island governments to negotiate guarantees from their federal or metropole governments to maintain and pay for transportation access to and from the island and between islands within an archipelago. The provincial Prince Edward Island government was granted this right when it agreed to join the federation of Canada, and federal funding continues to flow to the province to support ongoing ferry service to the mainland. At an estimated cost of almost £300 million, the British government has built a major international airport on the tiny island of Saint Helena, located in the South Atlantic. Even small island states are proving to be entrepreneurial in their management of transportation pathways. For example, Iceland's Icelandair explicitly advertises their location as a convenient and inexpensive stopover for travellers moving between Europe and North America, while at the same time encouraging travellers to stay a few days to enjoy the sites of Iceland (Lund, Loftsdóttir, and Leonard 2017).

These examples make us rethink the stereotypical image of small islands as powerless, dependent, and vulnerable to the economic and political winds of globalization. Instead, they appear to be able to seek out and take advantage of specific economic and business niches. Baldacchino and Bertram (2009, 156) use the analogy of the finches' beaks to paint a picture of the adaptability of small islands. Just as naturalists and biologists discovered that the beaks of finches in various parts of the Galápagos Islands adapted to the circumstances associated with their environments, so too are many small islands able to adapt to their own economic circumstances. Moreover, they suggest that "island economies have the added capacity to refashion their beaks in response to actual or potential environmental opportunity."

Offshore Financial Centres: An Example of Strategic Flexibility

> THE TERM OFC REFLECTS THE FACT THAT THERE IS USUALLY A GEOGRAPHICAL SEPARATION BETWEEN THE SOURCE OF THE WEALTH AND THE REGION THAT IS PROVIDING A FINANCIAL SERVICE FOR THESE CUSTOMERS

Offshore financial centres (OFC) or tax havens have generated a considerable amount of media attention over the past few years, not least because some governments and advocacy groups see them as places where wealthy individuals and companies can escape paying their "fair share" of domestic taxes. The very fact that these jurisdictions have been given this label implies that islands are represented prominently within this group. This, despite the fact that some OFC jurisdictions such as Switzerland and the American State of Delaware are only metaphorical islands. The term reflects the fact that there is usually a geographical separation between the source of the wealth and the region that is providing a financial service for these customers, and they have been formally defined by the International Monetary Fund (IMF) as "a country or jurisdiction that provides financial services to non-residents on a scale that is incommensurate with the size and the financing of its domestic economy" (Zoromé 2007, 7). Hampton (1994, 237) refers to them as centres "that host financial activities that are separated from major regulating units by geography and/or legislation." In other words, the primary purpose of island companies in these sectors is not to serve the domestic population but rather to serve foreign corporations and individuals. The term "tax haven" is often used more pejoratively to suggest that the client companies and the jurisdictions are somehow skirting laws (Hampton and Christenson 2007; Hampton and Levi 1999). Hines (2010, 103) defines tax havens neutrally as "countries and territories that offer low tax rates and favourable regulatory policies to foreign investors." The OECD uses slightly more sinister wording, describing them as places that impose no or only nominal taxes, have very little transparency (including having laws that prevent the effective exchange of information with other governments), and that are trying to attract investment and transactions that have little or no activity (OECD 1998). It is not surprising that the OECD, through its

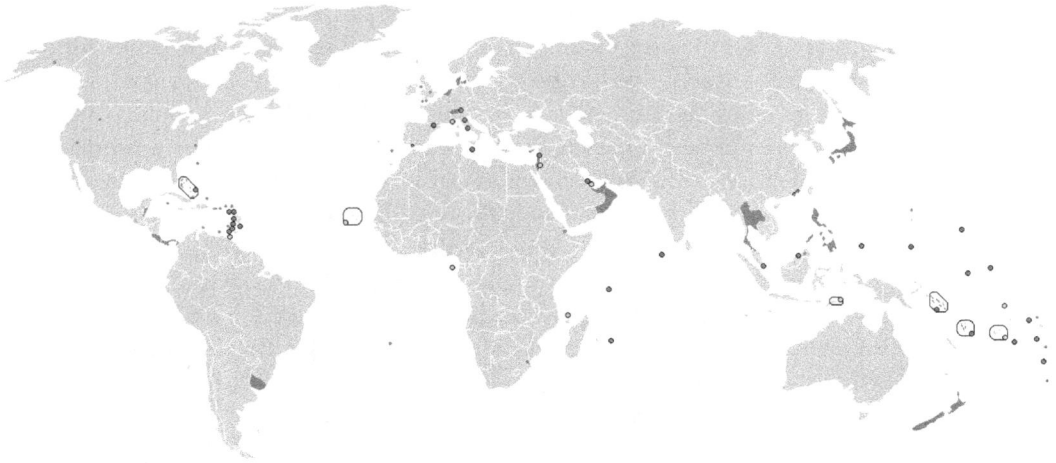

Figure 9.3 Locations of Offshore Financial Centers, circa 2013. *Source:* **Alinor**

Financial Action Task Force, is also tasked with evaluating these tax regimes for criminal or unethical behaviour.

There is a range of activities and services offered in OFCs, including banking, insurance, and investment or hedge funds being designated as the headquarters location for corporations, and serving as a repository for the capital of wealthy individuals and companies at relatively low or no rates of taxation. They are also used by the reinsurance industry, i.e., insurance that is purchased by insurance companies to minimize their own risks. Estimates of the total value of offshore wealth vary widely, from almost USD nine trillion to as high as thirty-two trillion (Gaggero and Rua 2015; Henry 2012). This wide range speaks to the level of secrecy associated with measuring the stocks and flows of money in and out of OFCs. Of the eight leading OFCs identified by the IMF which collectively handle 85 percent of all investments, six of them are islands: Bermuda, BVI, Cayman, Hong Kong, Ireland, and Singapore (Damgaard, Elkjaer, and Johannesen 2018). Palan, Murphy, and Chavagneux (2013, 1) state that "If you think of tax havens as sun-kissed exotic islands reminiscent of the Garden of Eden where a few billionaires, mafiosi, and corrupt autocrats hide their ill-gotten gains, then think again." Although this quote suggests that OFCs may exist anywhere in the world, in fact, most OFCs really are located on small islands, sometimes in the tropics and sometimes in cold-water regions. Figure 9.3 illustrates that, as of 2013, OFCs were found in almost all of the coastal regions of the world, including the South Pacific, the North Atlantic/Mediterranean, and the Caribbean. This figure also shows that the political jurisdictions that serve as OFCs are not always subnational island jurisdictions, with some prominent examples being nation states (including the Bahamas, Cyprus, Philippines, and the Marshall Islands) as well as cities or city-states such as Dublin, London, and Hong Kong.

OFCs make their money in several ways. Some still charge a nominal tax rate on corporations. They also charge banking and service fees as well as fees to register or renew licenses for companies doing business in the OFC. Although each fee might be quite low, the volume of business can generate considerable revenues. For example, it has been estimated that 50 percent of the Cayman Islands GDP, and 20 percent of the GDP of the Bahamas is from offshore financial activity (McLean and Jordon 2017). The competitiveness of an OFC is not necessarily based on which jurisdiction can offer the lowest fee or tax rate. As important is the degree of flexibility in regulation, the level of transparency or confidentiality of the information (where less transparency is better), and the degree of economic and political stability of the jurisdiction (Cobb 2001). Stability is represented by an adherence to the rule of law, an independent judicial system, and a minimal level of interference in the economy by the local government. So companies may not prefer a place that has too little regulation because that may jeopardize the security of the assets held there and also draw the attention of international regulatory agencies like the Financial Action Task Force. Nor would they want a place with too much regulation, which might lead to excessive government interference in their affairs. This balance between stability and compliance has been described by Cobb (2001) as "place reputation."

One estimate by the Tax Justice Network (2019) is that up to USD 189 billion in potential government revenues are lost per year as a result of tax avoidance by wealthy individuals and corporations to offshore locations. In 2005, 220 of the 700 largest transnational corporations in the United Kingdom paid no corporate taxes largely as a function of the use of OFCs (Cobham and Klees 2016; Farnsworth and Fooks 2015). There have been sporadic protests from the citizenry in some mainland cities against the actions of their locally based companies and celebrities who have moved their wealth to offshore islands for the purpose of avoiding higher taxes. The Occupy movement that started on Wall Street in New York City in 2011 and spread to other major world centres is one of the best examples of a rebellion against the power of large corporations to avoid taxes in countries in which they operate (Fuchs 2014; Shaxson 2012). In 2016, the release of the Panama Papers led to the forced resignation of Iceland's Prime Minister, Sigmundur Davíð Gunnlaugsson, when it was revealed that he and his spouse had set up a company in the British Virgin Islands (Trautman 2017).

The term "tax evasion" might be a little misleading. First, there have always been differences in tax rates across different countries, and multinational companies have always used the existence of these differences to their competitive advantage. Second, the terms "tax avoidance" and "tax evasion" are often conflated in the public media. Tax avoidance may be viewed by some to be unethical but it is not illegal; accounting firms specializing in international taxation exist for the very purpose of using provisions in existing tax laws to the advantage of their clients. Tax evasion, however, is illegal and involves activities such as money laundering. Islanders see this controversy very differently, which is sometimes problematic. When the OECD's Financial Action Task Force released its Harmful Tax Competition Initiative report on tax havens in the late 1990s, many of the jurisdictions they designated as uncooperative tax

THE COMPETITIVENESS OF AN OFC IS NOT NECESSARILY BASED ON WHICH JURISDICTION CAN OFFER THE LOWEST FEE OR TAX RATE. AS IMPORTANT IS THE DEGREE OF FLEXIBILITY IN REGULATION, THE LEVEL OF TRANSPARENCY OR CONFIDENTIALITY OF THE INFORMATION (WHERE LESS TRANSPARENCY IS BETTER), AND THE DEGREE OF ECONOMIC AND POLITICAL STABILITY OF THE JURISDICTION (COBB 2001).

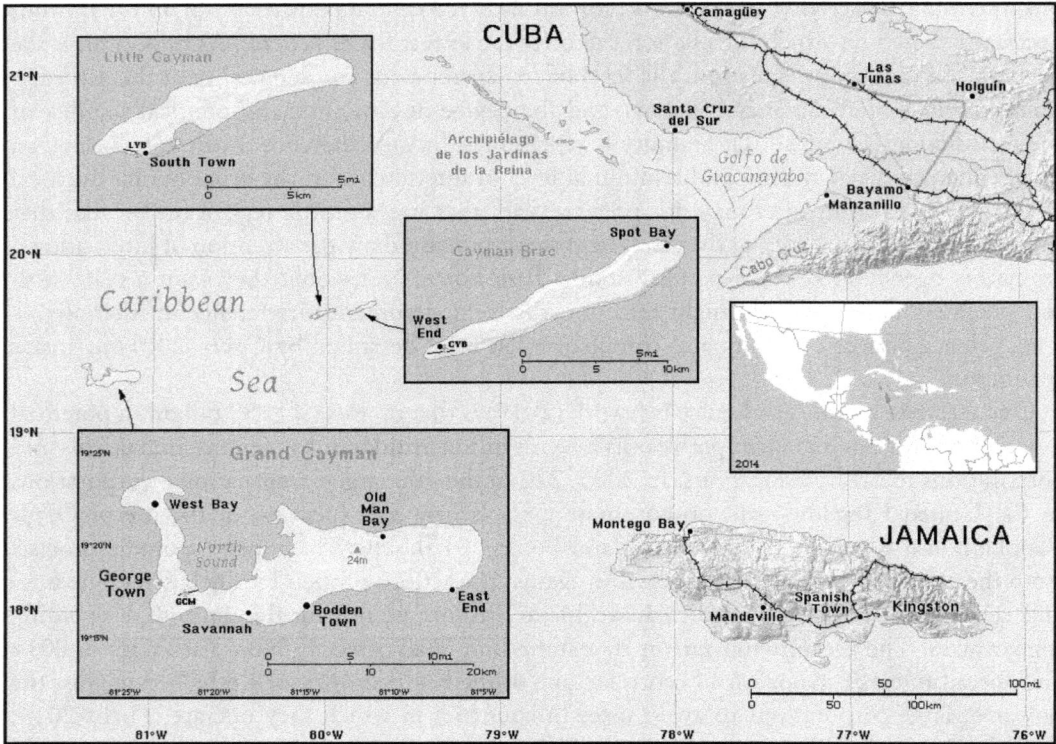

Figure 9.4   Location of Cayman Islands archipelago in Caribbean Sea. *Source:* **Ian Macky, public domain**

regimes were islands. Although almost all of these places have been forced to comply with the "soft law" standards established by the Task Force, some islanders still view this intervention by an agent of the wealthy developed world as just another form of fiscal colonialism and a continued threat to the islands' economic sovereignty (Bravo 2011; Sanders 2002). It is notable that since the initial focus on recovering lost tax revenues, the objectives of the OECD initiative have evolved to address much more subjective and value-laden issues such as money laundering and general political corruption (Larmour 2005).

Just as Nauru was a good example of an island that relied on resource extraction as a form of economic development, the Cayman Islands is a good case study of an island that has become one of the most important small island offshore financial centres. The Cayman Islands is a grouping of three islands located approximately 400 kilometres south of Cuba and about the same distance northwest of Jamaica (see Figure 9.4). Relatively little is known about the Indigenous groups that may have settled here prior to European contact, but they were undoubtedly an evolving mixture of Taíno and Carib peoples. Fewer than 1,000 people lived there when the first census was taken in 1803, of whom half were counted as slaves. Given its proximity to the wealth passing around the area on Spanish galleons, it was not uncommon for the islands to be used as a base by pirates and English smugglers and privateers (Murray 2010). It separated from Jamaica after that island gained its independence in 1962, preferring instead to remain a territory of Britain. Prior to this period the island was sparsely populated (8,000), had little infrastructure, and the main source of employment was as itinerant sailors in the international shipping fleet merchant marine. Given that a substantial amount of the island income at the time would have been derived from money sent back home by sailors and from the parent country of Britain, the Cayman Islands would have been a good example of a MIRAB-type economy.

During this period of British rule, the island was governed by the Historical Royal Charter. One of the provisions of this charter was that islanders were exempt from taxation so long as it was governed by Britain. One of the first boosts in population and investment occurred shortly after the separation from Jamaica, when Jamaicans opposed to independence moved to the smaller island. In the 1960s, politicians and business leaders on the Cayman Islands embarked on a deliberate strategy to develop the island economy on the dual pillars of tourism and offshore finance and, by 1972, more than 3,000 companies and 300 trust companies had been established (Roberts 1995). When the Bahamas gained independence in 1973, even more banks and trust companies shifted their operations from this island to the Caymans, fearful of increased and more strict banking regulations that might accompany Bahamian independence.

Cayman currently has a population of approximately 62,000 people, half of whom were not born on the island. It is ranked as one of the world's largest banking centres with foreign capital of over USD 4.1 trillion in assets, direct investment, and portfolio investments, an amount that is 1,500 times the value of its GDP (Fichtner 2016). It also is the home to more than 280 banks and 65,000 registered companies (CIA 2019), resulting in a bank for every 226 people (Gaggero and Rua 2015). The international focus of the banking industry is reflected by the fact that only nineteen of these banks are licensed to conduct business domestically.

Financial services account for approximately 50 percent of the territory's GDP, resulting in one of the highest values of Gross National Income (GNI) per capita in the world, at USD 53,271 (Clegg et al. 2017). Most of the rest of the Cayman economy and government revenue was generated from tourism-related activities, with an estimated 2.1 million tourists visiting per year (Montanez 2019a). On Cayman, as is the case on many tropical island financial centres, there is a close relationship between financial services and tourism, with much of the tourism serving the dual purpose of recreation and business, through conventions and meetings (Hampton and Christensen 2007). Now, remittances and international aid are almost non-existent and the country exports very few primary or manufacturing products. Therefore, the Cayman Islands has evolved from being a MIRAB island to being a PROFIT island (Baldacchino and Bertram 2009).

Researchers have criticized these models for their failure to translate into a long-term, sustainable base for island development and for not capturing the complexity of island economies and islanders' lives (Bertram 2004; Marsters, Lewis, and Freisen 2006). For instance, although the revenue streams associated with a MIRAB island may help families and communities, emigration still involves a loss of some of the brightest and most talented islanders to other regions of the world, often permanently. Remittances may continue for a number of years or even decades but as personal ties with a community fade and outmigration from a community intensifies, it is unlikely that remittances will continue at the same level. The same has been said about aid and donor fatigue (Bertram 2004). Moreover, it is not just the skills and capacity of the migrants that are lost, but also the social cohesion that these individuals had contributed to their families, their communities, and their islands. Long-term dependence on international aid may also leave islands in a tenuous position. Whether donors are inter-

> AS A PRESCRIPTION FOR LONG-TERM ISLAND PROSPERITY, THE PROFIT MODEL ALSO HAS ITS DETRACTORS. AS WITH MANY MIRAB ISLANDS, A GROWING PROFIT-BASED ECONOMY BRINGS NO GUARANTEES THAT ISLAND DEMOCRACIES ARE THRIVING (CLARK 2013) OR THAT, IN THE CASE OF SNIJS ESPECIALLY, ISLANDERS MAY HAVE GIVEN UP TOO MUCH CONTROL ABOUT MATTERS THAT DIRECTLY AFFECT THEIR LIVES (WYNNE 2007).

national organizations, governments, or foundations, maintaining a steady stream of revenue in the long term can be challenging and may also be accompanied by conditions that are not in the best interest for long-term development of the island. This makes it especially hard to commit to long-term infrastructure and human development initiatives, especially if the priorities of the donors do not necessarily match those of the islanders.

As a prescription for long-term island prosperity, the PROFIT model also has its detractors. As with many MIRAB islands, a growing PROFIT-based economy brings no guarantees that island democracies are thriving (Clark 2013) or that, in the case of SNIJs especially, islanders may have given up too much control about matters that directly affect their lives (Wynne 2007). For example, some on Aruba feel that the Dutch metropole has hampered cultural development on the island (Razak 1995), while Androus and Greymorning (2016) make the case that the language, land, and life stories of Indigenous Corsicans and Hawaiians have been lost with economic dependence on France and the United States. It also does not fully appreciate how poor governance and neocolonialism has just as often worked to the detriment of development goals instead of being a nimble stakeholder able to respond quickly to opportunities (Nadarajah and Grydehøj 2016).

The other significant approach to better understand the economies of small islands is the SITE (or **S**mall **I**sland **T**ourism **E**conomy) model. This model is premised on the fact that tourism for many islands is a pervasive element of island life, not only in generating employment and revenue, but also in influencing social and cultural development, politics, and environmental management. Although some island studies researchers (McElroy and Parry 2010; Oberst and McElroy 2007) view the SITE approach as a subset of the larger "strategic flexibility/PROFIT" approach, the larger question of tourism is such a significant part of the life of small islands that this discussion deserves its own chapter.

CONCLUSIONS

In concluding this chapter, it is important to ask if there are any other pathways to economic development emerging for small islands. Internationally, one of the concepts that has been gaining momentum appears to be economic development associated with the "Blue Economy" or the "Oceans Economy." At this point, most of the discussion on this approach seems to be associated with the United Nations (e.g., UNCTAD, WTO, SIDS) and less so from among island studies researchers. The origin of the Blue or Oceans Economy approach appears to be from the 2012 Rio de Janeiro Conference on Sustainable Development. In the outcomes document from this meeting, UN members agreed to "protect, and restore, the health, productivity and resilience of oceans and marine ecosystems, to maintain their biodiversity, enabling their conservation and sustainable use for present and future generations" (United Nations Conference on Trade and Development 2014, 1).

In this discussion, the terms "Oceans Economy" and "Blue Economy" are often used interchangeably and are said to have originated from an earlier Green Economy concept (Smith-Godfrey 2016). They also share similar desired outcomes, including the "improvement of human well-being and social equity, while significantly reducing environmental and ecological scarcities" (UNCTAD 2014, 2). However, unlike a green economy focussed primarily on the terrestrial or land-based environment, the Blue Economy is intended to encompass both the oceans and the atmosphere as an integrated whole. The term Blue Economy was coined by Gunter Pauli (2010) because it refers to the shared colours of three environments: the sky, the oceans, and the planet Earth or Gaia as seen from space. In this context, the concept has Indigenous parallels. For example, a Maori worldview on Aotearoa/New Zealand sees a strong connection between people and the environment, both on land and in the surrounding marine areas (Bargh 2014).

The premise behind this approach is that the lives and livelihoods of islanders and coastal residents have always been bound up in a connection between the sea and the land, and to separate the two of these environments is artificial, unsustainable, and shortsighted from an economic development perspective. Not only do 350 million people rely on fishing for their livelihood, but over 1 billion people rely on fish as their main source of protein. As was discussed earlier, with the implementation of the Convention on the Law of the Sea, the ability of small developing island nations to exercise greater control and responsibility over the space around them has also increased. For example, although the land area of the Cook Islands archipelago may only be 240 square kilometres, the area contained by their Exclusive Economic Zone extends a further 1.8 million square kilometres. There is also a growing realization of the value of oceans for the tourism sector, for shipping, for the generation of sources of non-renewable and renewable sources of energy, and for many other economic activities (Spalding 2016)

More importantly from the perspective of future island economic development is the utilization of marine areas for aquaculture, for mining of the seabed, and for marine "bioprospecting." This last item is especially intriguing in that it refers to the use of genetic materials from natural products found primarily in the shallow waters around islands and coastlines to derive new products, including food, biochemical, pharmaceutical, cosmetics, and bioenergy products. It has been estimated that by 2014, 18,000 natural products had been discovered and 4,900 patents of marine genetic origin had been filed, and this was increasing at a rate of 12 percent annually (Whelan, Annis, and Guajardo 2014). In 2011, thirty-six marine-derived drugs were in clinical development, including fifteen related to cancer treatments (OECD 2012). More significantly, the number of patents being published using marine genetic resources has increased dramatically and very few of these are covered under Access and Benefit Sharing (ABS) agreements with the island nations, meaning that any benefits accruing from these products could flow solely to the patent holders (UNCTAD 2014). Despite the establishment of international agreements such as the Nagoya Protocol that have been put in place to ensure that these environments are stewarded sustainably and the resources are shared in a fair and equitable manner with the local islands, it remains to be seen whether the harvesting of genetic materials and the mining of sea beds for minerals may merely be a twenty-first-century version of the extraction and exploitation of land-based natural resources on islands that took place throughout the twentieth century.

Key Readings

Baldacchino, Godfrey. 2010. *Island Enclaves: Offshoring Strategies, Creative Governance, and Sub-national Island Jurisdictions.* Montreal: McGill-Queen's University Press.

Baldacchino, Godfrey. 2015. *Entrepreneurship on Small Island States and Territories.* New York: Routledge.

Baldacchino, Godfrey, and Geoffrey Bertram. 2009. "The Beak of the Finch: Insights into the Economic Development of Small Economies." *The Round Table: Commonwealth Journal of International Affairs* 98 (401): 141–160.

Baldacchino, Godfrey, and David Milne, eds. 2000. *Lessons from the Political Economy of Small Islands: The Resourcefulness of Jurisdiction.* New York: St. Martin's and MacMillan Press.

Bartmann, Barry. 2006. "In or Out: Sub-National Island Jurisdictions and the Antechamber of Para-Diplomacy." *The Round Table* 95 (386): 541–559.

Clark, Eric. 2013. "Financialization, Sustainability and the Right to the Island: A Critique of Acronym Models of Island Development." *Journal of Marine and Island Cultures* 2 (2): 128–136.

McElroy, Jerome L., and Courtney E. Parry. 2010. "The Characteristics of Small Island Tourist Economies." *Tourism and Hospitality Research* 10 (4): 315–328.

Pauli, Gunter A. 2010. *The Blue Economy: 10 Years, 100 Innovations, 100 Million Jobs.* Taos, NM: Paradigm.

Smith-Godfrey, Simon. 2016. "Defining the Blue Economy." *Maritime Affairs: Journal of the National Maritime Foundation of India* 12 (1): 58–64.

Van Fossen, Anthony. 2018. "Passport Sales: How Island Microstates Use Strategic Management to Organise the New Economic Citizenship Industry." *Island Studies Journal* 13 (1): 285–300.

Chapter Ten

Island Tourism

The sustainability of island tourism depends on embracing the turbulence of tourism, the paradox of sustainability, and the duality of islandness. (Peterson 2011, 7)

INTRODUCTION

As an activity, tourism is arguably one of the most important components of economic development for many small islands. However, tourism is more than just economic development. It is also a cultural phenomenon, as the tourism industry markets images of islands and islanders to the world. Cultural symbols and practices such as island-specific music and dance are communicated and sold to tourists. Tourism also has implications for the physical environment given the pressures placed on destinations as a result of mass tourism. As we discussed in Chapter Three, island tourism is also intertwined with our images and perceptions of islands in literature, cinema, and advertising. Given the multifaceted connections between tourism and islands, this topic deserves special attention.

This chapter provides a better understanding of the role tourism plays in the lives of islanders and island institutions. There are many ways in which tourism has benefited islanders economically and socially. At the same time, tourism has also strained the fabric of island life, created tensions within island communities as well as between islanders and outsiders, and has led some islanders and island governments to question the ultimate value of tourism for their islands. We start by describing the growth of tourism as a sector and, specifically, its increasing presence on small islands. We then address the question of whether island tourism is really a unique type of tourism or whether islands are just one more context within which tourism takes place. Island tourism is then situated within a broader framework of economic development on islands, similar to the models presented in the previous chapter.

By focusing on the marketing and branding of islands and the fascination they appear to hold for tourists, we can perceive how islands are portrayed to the world and compare these stereotypes to the reality of island life. This "tourist gaze" is an important concept related to how tourists experience island destinations. One of the most significant aspects of the imagery of islands is that, despite the stereotypes of tropical paradises, there is considerable variety among island vacation sites. For example, the distinctions between "warm water" and "cold water" island destinations, or between cultural-historical motivations and "sun, sand, and sea" motivations, filter and modify the nature and impacts of tourism.

Perhaps the most important features of island tourism are the impact on islands, both positively and negatively, and across multiple economic, social, and ecological dimensions. This is especially significant given the finite space and boundedness of islands where impacts are concentrated and felt in smaller spaces. However, it is not enough to just understand the current impacts. It is also important to trace how the tourism industry is changing and what these changes might mean for islands in the future. For example, how will the growth of cruise ship tourism influence the future of small islands? The last section of this chapter emphasizes sustainability and sustainable development applied to tourism on islands. It also reflects on some of the harmful and contradictory impacts this approach can have on islands and island societies. For example, travellers may undertake an ecotourism holiday in an attempt to minimize their carbon footprint while at a destination. However, by travelling by jet to an ecologically sensitive destination, tourists may have contributed significantly to the production of greenhouse gases.

> THERE ARE MANY WAYS IN WHICH TOURISM HAS BENEFITED ISLANDERS ECONOMICALLY AND SOCIALLY. AT THE SAME TIME, TOURISM HAS ALSO STRAINED THE FABRIC OF ISLAND LIFE, CREATED TENSIONS WITHIN ISLAND COMMUNITIES AS WELL AS BETWEEN ISLANDERS AND OUTSIDERS, AND HAS LED SOME ISLANDERS AND ISLAND GOVERNMENTS TO QUESTION THE ULTIMATE VALUE OF TOURISM FOR THEIR ISLANDS.

THE MAGNITUDE OF TOURISM IN THE WORLD AND ON ISLANDS

Up until the middle of the twentieth century, long-distance tourism was an activity reserved primarily for the privileged class. In 1950, receipts for tourism that involved international travel totalled less than USD two billion and there were twenty-five million total international arrivals (Pantelescu 2012). Tourism continued to be a leading global sector until the rapid oil price increases in the early 1970s. In the past decade it has regained its position as the largest economic sector in the world (Crick 1989). The most recent statistics from the United Nations World Tourism Organization (UNWTO) suggest that as of 2018 there were 1.4 billion international tourist arrivals generating export earnings of USD 1.7 trillion (World Tourism Organization 2019). In the past several years, international tourism revenue numbers have increased between 4 and 6 percent annually. As large as they are, even these figures underestimate the total magnitude of tourism because they do not include domestic travel and tourism.

These global statistics do not allow us to measure the value and magnitude of tourism specific to islands. However, Figure 10.1 does provide one indicator of the significance of these activities. It lists the proportion of the Gross Domestic Product (GDP) attributable to the travel and tourism sector for those twenty-five countries with the highest share attributable to this sector in 2018. Nineteen of these twenty-five countries are islands or archipelagos with an average of 38.3 percent of their GDP associated with travel and tourism. Several of them, including the Seychelles, Maldives, St. Kitts & Nevis, and Grenada, derived more than half of their GDP from tourism. The economic importance of tourism on these nineteen islands also seems to be increasing. From 2010 to 2018, their share of GDP derived from tourism had increased by three percentage points.

COUNTRY	2018 (% of GDP)	2010 (% of GDP)
Macau	72.2	76.9
Seychelles	67.1	64.0
Maldives	66.4	69.8
St. Kitts & Nevis	62.4	48.0
Grenada	56.6	41.0
Vanuatu	48.0	51.1
Cape Verde	46.2	41.9
St. Vincent & the Grenadines	45.5	37.3
Belize	44.9	33.9
Antigua & Barbuda	44.1	52.3
St. Lucia	41.8	35.7
Bahamas	40.4	39.1
Fiji	38.9	40.0
Barbados	34.9	37.9
Jamaica	34.0	39.1
Georgia	33.7	14.2
Dominica	33.4	38.9
Cambodia	32.8	26.9
Iceland	32.6	19.4
Sao Tome and Principe	27.8	13.5
Albania	27.3	23.8
Croatia	24.9	20.3
Philippines	24.7	11.9
Mauritius	24.3	24.7
Cyprus	21.9	14.3

Figure 10.1   Travel and Tourism as a Percent of Gross Domestic Product, 25 Top Ranked Countries. (Island nations are in bold.) *Source:* **Author** (From statistics found at KNOEMA Data Atlas)

TOURISM ON ISLANDS OR ISLAND TOURISM

The statistics in Figure 10.1 clearly show that tourism plays an important economic role on islands. We shall see later that tourism is also closely connected to questions of island identity, and to social and physical environments on islands. However, this does not necessarily mean that tourism on islands is so different that it is distinct from the kinds of tourism that take place everywhere else. In other words, is there something that might be called "island tourism" or is this just tourism that happens to take place on islands? Should we be considering islands as special cases when we look at tourism? Sharpley (2012) speaks to this question by reminding us of what is different about many islands: their remoteness, their social insularity, their economic dependence on tourism, and their vulnerability. He notes that because they are surrounded by water, islands are logistically and psychologically different from mainland destinations. The water/land boundary makes it more costly to reach but, perhaps more im-

A Case Study of Tourism in Iceland

Located in the North Atlantic, midway between Europe and North America, Iceland is not a stereotypical warm-water tropical vacation paradise. Despite facing many of the same challenges shared by other cold-water islands, including costly transportation links with external markets, a small specialized economy and a fragile natural environment, in the early part of this century Iceland was experiencing modest increases in international tourism (Jóhannesson et al. 2010). After the 2008–09 global recession when Iceland's economy was devastated, tourist numbers and the subsequent economic, political, social, and environmental impacts on Iceland have accelerated. According to the Icelandic Tourist Board, the number of foreign visitors to Iceland increased from 488,622 in 2010 to more than 2.3 million in 2018, an increase of 380 percent (Icelandic Tourist Board 2019). This included a four-year period between 2014 and 2017 when the rate of increase exceeded 20 percent every year. Despite this exponential growth, there are more recent signs that the number of visitors is now levelling off and even declining (Icelandic Tourist Board 2019).

What were the reasons for this increase and what impact has this had on Iceland today? Most research suggests that following the collapse of the financial sector in Iceland in 2008–09, there was a concerted effort to rebrand and develop the economy through tourism (Karlsdóttir 2013). A marketing campaign that capitalized on a growing interest in adventure and nature tourism portrayed Iceland in the minds of potential travellers as "marauding financial Vikings from a desolate wilderness island at the edge of the habitable world" (Lund et al. 2017, 21). The phrases "Iceland Naturally" and "Inspired by Iceland" were also created as brands to convey purity, health, safety, and the country's beauty (Huijbens 2011). In addition, despite the perceived isolation of Iceland, it became central to trans-Atlantic air travel following the May 2010 eruption of the Eyjafjallajökull volcano. Although this natural calamity grounded many international flights between Europe and North America for up to a month, it ironically also focused attention on the island as a transportation hub and possible tourist destination (Gil-Alana and Huijbens 2018). The Icelandic government and the major domestic air carrier (Icelandair) started promoting the island as a stopover destination (Lund et al. 2017), offering inexpensive

portantly, it creates a "liminal" zone of transition that may not exist surrounding mainland destinations. Islands are also perceived by travellers to be different. The separation and the journey set them apart, both in the minds of the tourists and by the tourism marketing industry. Despite these characteristics, Sharpley concludes that islands are not collectively different enough from mainland destinations for island tourism to be a separate subfield.

Butler (2012) is also not convinced the research shows that tourists visit islands specifically because they are islands. However, others have arrived at a very different conclusion. Conlin and Baum (1995) argue that the special allure of islands occurs as a function of the less tangible characteristics of islandness that is reflected in tradition, distance, and insularity, while Baum (1997) says that islands provide an opportunity for tourists to escape. Similarly, the appeal of islands may be linked to their physical and political separation from the mainland, combined with a desire on the part of tourists to experience something different (Butler 1993). Poetically, Baldacchino (2012b, 55) notes that the allure of islands "speaks to a yearn-

flights between North America and Europe, the two most prominent sources of international tourists and business travellers, and stopover opportunities at little or no additional cost. Finally, the devaluation of the Icelandic currency (the króna) following the financial collapse made it even more competitive as a vacation destination (Gylfason and Zoega 2018).

The rapid increase in tourist numbers noted earlier has had significant impacts on the island. Economically, tourism is now seen, together with fisheries and heavy industry (e.g., aluminum smelting), as one of the key national industries (Jóhannesson 2015). Up to 30 percent of Iceland's GDP is directly or indirectly related to the tourism sector (Sutherland and Stacey 2017). The tourism boom has also contributed to an increase in employment, filled at least in part through migrant labour, an increase in new business formation, and in construction (ibid.). At the same time, it has now led to increased pressure on Iceland's currency, making it more expensive to visit and live on Iceland (Cook et al. 2019; Sutherland and Stacey 2017). With greater numbers of tourists, various stakeholder groups have started to voice their concerns about the negative impacts on Iceland, using terms such as a "gold-rush mentality" (i.e., rapid growth followed by rapid collapse), comparing this to what occurred in the fisheries and financial services, and a worry that tourism might be "spinning out of control" (Jóhannesson 2015, 186–187). Surveys of residents of Iceland show that they are becoming increasingly concerned about the sustainability of tourism growth (Helgadóttir et al. 2019).

There have also been increased environmental pressures on the major nature-based tourist sites on the island, especially those located in the popular Golden Circle route outside of Reykjavik (Cook et al. 2019). Subarctic regions like Iceland that are already vulnerable to extreme weather events, overgrazing, agriculture, and wood harvesting are now facing the impacts of overtourism (Hale 2018). Although Icelandic tourism policy may stress the importance of environmental issues on paper, in practice actions to preserve the natural environment have not kept pace with the policy rhetoric (Jóhannesson et al. 2010). A decline of about 20 percent in the number of visitors travelling by air to Iceland in 2019 points to the challenges faced by island destinations in implementing a stable, sustainable tourism plan while the destination is experiencing rapid changes in numbers of travellers (Icelandic Tourist Board 2019).

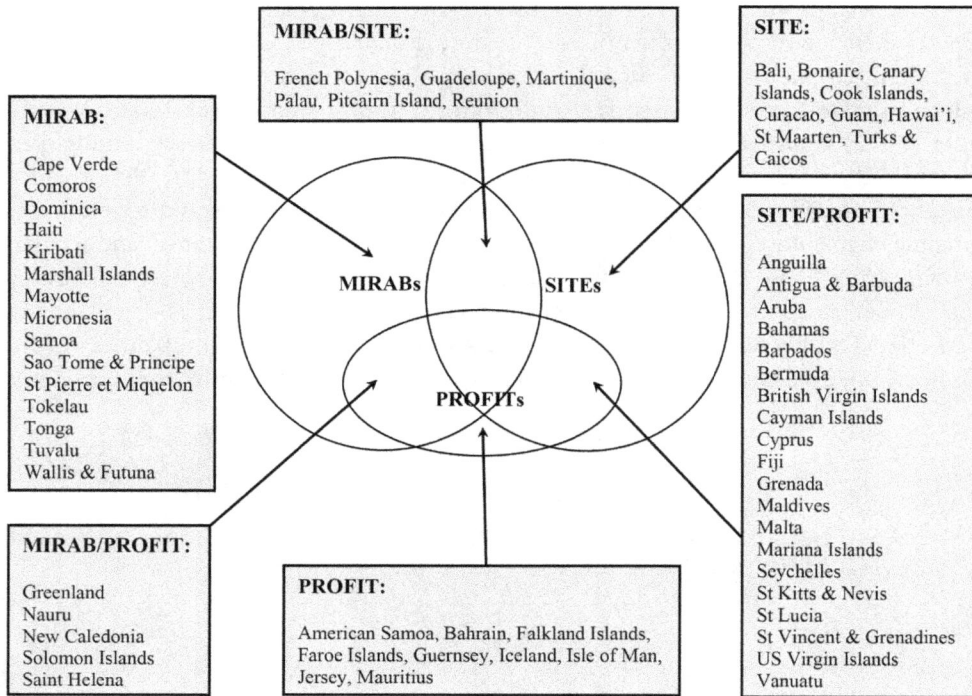

Figure 10.2 A three-fold taxonomy of small-island economies. *Source*: **Godfrey Baldacchino and Geoff Bertram**

ing for an island space and life that is part myth, part marketing hype, part reality … and not all continental or mainland driven."

A TOURISM MODEL OF ISLAND ECONOMIC DEVELOPMENT

The importance of tourism to islands has led to the creation of a model that places this sector at the centre of economic development. As was the case with the MIRAB and PROFIT models, this has also been given an acronym-type label—in this case, Small Island Tourist Economies or SITE places. Developed largely by McElroy and colleagues (McElroy 2003, 2006; McElroy and Hamma 2010; McElroy and Parry 2010; Oberst and McElroy 2007), it suggests that small islands specializing in tourism are likely to be more developed than other small island developing states (SIDS) across a range of indicators. These include higher incomes and GDP per capita, greater mobility of the labour force, longer life expectancies, lower infant mortality rates, and higher levels of literacy. This research found that islands specializing in tourism are more likely to be subnational island jurisdictions (SNIJs) than independent states and are more likely to share characteristics similar to the PROFIT-type islands than those relying primarily on remittances and aid.

Baldacchino and Bertram (2009) took this one step further in grouping many of the world's small islands into one or more of the three major economic models: SITE, MIRAB, and PROFIT (Figure 10.2). Although this figure suggests that there are many places relying primarily on remittances and aid (i.e., MIRAB islands), there are fewer islands that are pure PROFIT or SITE-dependent places. It is more likely to find islands sharing characteristics of strategic flexibility and tourism, such as the offshore financial centres in the Caribbean (e.g., Cayman, BVI) than it is to find islands that combine tourism (SITE) and aid/remittance (MIRAB) dependence. They also found a considerable level of change among islands. It was not unusual for islands to be specialized in tourism at one point, then evolve to be a mixture of SITE/PROFIT places, before returning to an emphasis on tourism.

MOTIVATIONS FOR ISLAND TOURISM OR ISLAND "FASCINATION FACTORS"

The utopian image of a tropical island where you can escape from your everyday life did not start with the rise of tourism marketing in the middle of the twentieth century. It had its beginnings in the encounters between Indigenous peoples and outsiders in the Caribbean and the Pacific, and how those encounters were portrayed in books, art, film, music, and even postcards (Connell 2003b; Lindstrom 2007; K. Thompson 2007). The images in these media portrayed islanders as welcoming and even uninhibited, and places that were exotic, mysterious, and different. Many travel writers have conflated this imagery, at least as it has been applied to tropical environments, to the three S's of sun, sea, and sand, with a fourth descriptor "sex" and even a fifth "shopping" sometimes added to the first three (Crick 1989; Morgan 2013). It has even been argued that this portrayal of islands was one of the principal features that transformed many islands from a dependence on primary industries to sites of mass tourism (Carlsen and Butler 2011).

This imagery is part of a much larger question of the emotional attachment and attraction that tourists have with islands. This has been called the tourist imaginary (DeLoughrey 2013) and the island lure (Baldacchino 2010a, 2012b). Some have even suggested that the very journey across water represents an escape from the mundane for tourists or a rite of passage (Carlsen and Butler 2011; MacDonald and MacEachern 2016). These motivations are important because they shape the behaviour of tourists (Iso-Ahola 1982). So what are, in Baum's (1997) words, the "fascination factors" that are established and reinforced by the island tourism industry and island governments to entice vacation travellers? The answer is that there are many factors and they vary by island. It may be the isolation or remoteness of a place that contributes to a sense of the unknown or mystery. The very fact that the destination is physically separated and can only be reached by ship or plane contributes to this allure. So the effort associated with the journey itself contributes to the appeal. Many people speak

> THIS SENSE THAT ISLANDS ARE CULTURALLY AND ECOLOGICALLY DIFFERENT, YET NOT SO DIFFERENT AS TO BE UNCOMFORTABLE FOR THE TOURIST, IS A PREVAILING THEME IN TOURISM MARKETING. VISITORS CAN EXPERIENCE A DIFFERENT CULTURE, LANGUAGE, OR FOOD BUT NOT BE TOO INCONVENIENCED OR CHALLENGED BY THOSE DIFFERENCES.

of taking a ferry as transformative, putting you in a different state of mind or, in MacDonald and MacEachern's (2016) words, "a visceral marker of 'otherness'" (289). Baum (1997) and Gössling and Wall (2007) refer to the island experience as being one where the total experience is greater than the sum of the parts. This suggests a sense of adventure in both the journey and at the destination that is less likely if your destination was just another stop in a network of connected highways. Consider, for example, the British Overseas Territory of Pitcairn Island in the Pacific that can only be reached by ship. Even then, there is no harbour, so cruise ships have to unload small numbers of passengers in dories and land only when calm sea conditions permit. This creates both a challenge and an allure for tourism marketing (Amoamo 2011, 2013). Of the sixteen cruise ships that visited Pitcairn in 2014, only four were able to land passengers (Amoamo 2017).

One of the arguments against building bridges, tunnels, or causeways connecting islands and mainlands is that these newfound connections will contribute to islands losing their allure and sense of difference (Hay 2006) to the point where they may no longer even be perceived as islands (Pigou-Dennis and Grydehøj 2014). Therefore, it is not uncommon to find an active local resistance to developing more efficient connections in order to preserve the cultural identity of islands (Vannini 2011). This sense that islands are culturally and ecologically different, yet not so different as to be uncomfortable for the tourist, is a prevailing theme in tourism marketing. Visitors can experience a different culture, language, or food but not be too inconvenienced or challenged by those differences. This may then become a learning experience. Not only are tourists able to relax, but by learning something perceived to be authentic directly from islanders, they gain knowledge. Gillis (2007) states it this way: "islands slake the modern thirst for that authenticity which seems in short supply on mainlands" (280). Some tourism marketing takes this theme even further, conveying the idea that your vacation will be transformative. For example, the 1991–92 Jamaican Tourist Board slogan was "Come back to romance. Come back to excitement. Come back to yourself. Come back to Jamaica" (Bolles 1992, 30). This approach harkens back to the fictionalized accounts of shipwrecked travellers marooned on tropical islands. From Gillis's perspective, the vacation becomes a search for sanctuary and renewal.

Many islands are explicitly marketed as places for relaxation; the sun, sand, and sea stereotype is paramount. For example, the Spanish Balearic islands in the western Mediterranean, including Mallorca, are one of the leading mass tourism sun and sand destinations (Aguiló et al. 2005). Also, in a content analysis of travel writers' articles on the Caribbean which appeared in newspapers in the United Kingdom, the recreational motivation dominated, catering to those tourists who may not care for the authentic so long as their experiences are entertaining or relaxing (Daye 2005). Relaxation on other islands may have a different interpretation. For example, the Greek islands of Ios and Cyprus are seen by tourists as party destinations where bars and nightlife are the most attractive features (Sharpley 2004; Stylidis et al. 2008).

Beyond this dominant stereotype of islands as places of relaxation, there are many other motivating factors linked closely to islands. One is associated with health or medical tourism (Connell 2011b), where people travel across borders for surgical procedures, dentistry, or more exotic procedures that might not be available at home (Connell 2013b). Although one might argue whether this is really tourism, many of the Caribbean islands, including Cuba, Barbados, Bahamas, the Cayman Islands, the Dominican Republic, and Jamaica have all embraced this as part of their economic development strategy (Bernal and Lowe 2019; Connell

2013; Ramírez de Arellano 2011). The link to tourism becomes stronger when you think of this more broadly as health and wellness tourism (Voigt and Pforr 2014). Rejuvenation in spas and the reputed therapeutic properties of geothermal and mineral waters has been a form of tourism since the Greek and Roman bathing cultures (Boekstein 2014). The Spanish island territory of Gran Canaria (Medina-Muñoz and Medina-Muñoz 2012) and Jamaica (Bernal and Lowe 2019; Valentine 2016) are trying to capitalize on this form of high-end, resort-based relaxation for the wealthy and increasingly middle-class tourist.

One of the more recent niche tourism marketing strategies, perhaps driven more by the news media than by the numbers of tourists, is what has been referred to as last-chance or voyeuristic tourism (Dawson and Lemelin 2015). This occurs in situations where tourists wish to visit a place or a feature before it vanishes completely. This might include the low-lying atolls such as Tuvalu and Kiribati that are especially threatened by sea-level rise (Farbotko 2010a; Huebner 2011), Indigenous communities in the Trobriand Islands of Papua New Guinea (MacCarthy 2012), or the wildlife ecosystems of the Galápagos Islands (Lemelin et al. 2010). As the label implies, the motivation for tourists is to be one of the last visitors to see a place, a people, or some element of wildlife. Ironically, for all of these examples, the increased tourism may also hasten their demise.

WARM-WATER AND COLD-WATER ISLAND TOURISM

As suggested from the section on tourism marketing and motivations, many of our perceptions of island-based tourist destinations are synonymous with warm, tropical locations. Much of the early research on island tourism focused primarily on these equatorial regions and the motivations associated with tourists choosing these destinations (Harrison 2001, 2004; R. King 1993). Although this is still very much the focus of much destination advertising, in the past twenty years we have seen an increased focus on tourism in temperate or cold-water

CHARACTERISTICS OF WARM VS. COLD WATER ISLANDS

Warm-Water Islands	Cold-Water Islands
• Extended season	• Short-limited season
• High temps (air & water)	• Rain & wind, potentially snow
• "Unspoiled" beaches	• Rocks, cliffs, patches of beach
• Lush vegetation, colourful flowers	• Few dishes (locally orientated)
• Fruits, spices, rich buffets	• Exploration/adventure
• Relaxation, quietness	• Social/historical/cultural is important (acquire knowledge & understanding of local society
• Loosening of inhibitions	• Three I's: Ice, Isolation, & Indigenous peoples (and maybe thoughtful leisure)
• The three S's: Sun, Sea, Sand (maybe a 4th—Sex, and a 5th—Shopping!)	

Adapted from Gössling and Wall, 2007

locations. In the northern hemisphere, examples include Hokkaido in northern Japan, Iceland, the Faroe Islands, islands off the coast of Scotland, the Channel Islands between Britain and France, and the Baltic Islands of Gotland and the Ålands. In North America, it would include Canada's Newfoundland and Labrador, Cape Breton, and Prince Edward Island, and in the southern hemisphere, it would include New Zealand, the Falklands/Malvinas, and both Kangaroo Island and Tasmania off the south coast of Australia. In fact, there are very few locations on the planet that have not experienced the reach of tourists to some degree. This expansion of the tourist gaze is only partly a result of the greater number of tourists and a decline in the relative cost of travel to more remote locations. It may also represent a greater fragmentation of the tourism industry—not necessarily away from mass tourism associated with sandy, tropical beaches, but now also incorporating options for those seeking a different and perhaps more authentic experience, based on motivations that do not include relaxing on the beach (Baldacchino 2006a). For example, in a 1999 survey of tourists to the temperate Prince Edward Island, Canada, only 10.5 percent said their primary motivation was relaxation (Bansal and Eiselt 2004). More than half (51.9 percent) of the tourists said they travelled there for adventure and another 31 percent said their main purpose was for personal reasons.

A growing body of research now exists describing the differences between tourism on warm-water versus cold-water islands (Baldacchino 2006f; Butler 2006; Gössling and Wall 2007). Typically, all of the characteristics associated with mass tourism, including large numbers of tourists concentrated in a small area, greater involvement by transnational corporations, and a more comprehensive built infrastructure catering to tourists, are more commonly associated with warm-water locales. Cold-water island vacation experiences are often linked with alternative tourism. In the early stages of the discovery of a destination, this normally takes place at a smaller scale over a more limited season, is more expensive, and is often organized by local companies or by the travellers themselves. Although tourism on cold-water islands may not be as visible to the public, tourists may actually have a greater intensity of experience in these communities because they are more likely to interact with local people and travel to places that have not received as many tourists. In fact, unlike the three S's of sun, sand, and sea, Gössling and Wall (2007) describe cold-water tourism using the three I's of ice, isolation, and Indigenous peoples. The sidebar on page 191, adapted from Gössling and Wall's work, contrasts the characteristics of tourism on warm- and cold-water islands. Understandably, this is a simplification of reality. Some islands, such as Prince Edward Island, Kangaroo Island, and New Zealand, have attempted to market themselves as locations that include both temperate and tropical features.

ONE PROBLEM WITH ISLAND TOURISM IS THAT, ALTHOUGH EMPLOYMENT AND INCOME MIGHT BE GENERATED IN OTHER LOCAL ISLAND SECTORS THAT SERVE THE TOURISM INDUSTRY THROUGH A MULTIPLIER EFFECT, IT IS NOT UNCOMMON FOR MANY OF THE INPUTS OR SUPPLIES NEEDED FOR TOURISM TO BE PURCHASED AND SHIPPED FROM ELSEWHERE (SHAREEF 2003).

Tourism and Prosperity in Small Island Developing States	Tourism Share of GDP%	GDP per Capita US$	HDI Rank
Countries with Tourism >25% of GDP			
Antigua and Barbuda	44.1	27,981	70
Cape Verde	46.2	7,316	125
Barbados	34.9	18,534	58
Cook Islands	70	16,700	n/a
Maldives	66.4	21,760	101
Palau	27.9	14,952	60
St Kitts and Nevis	62.4	29,820	72
St Lucia	41.8	14,355	90
Grenada	56.6	16,167	75
Belize	44.9	8,501	106
Dominica	33.4	9,886	103
Fiji	38.9	10,234	92
St Vincent	45.5	11,956	99
Sao Tome and Principe	27.8	3,324	143
Vanuatu	48	2,862	78
Seychelles	67.1	30,505	62
AVERAGE	47.5	15,303	105.29
Countries with Tourism 15-25% of GDP			
Tonga	20	6,111	98
Kiribati	18.8	2,086	134
Marshall Islands	15.6	3,697	106
Samoa	25	5,890	104
AVERAGE	19.85	4,446	110.5
Countries with Tourism <15% of GDP			
Comoros	10	1,632	165
Fed. States of Micronesia	7.9	3,482	131
Solomon Islands	12.5	2,242	152
Tuvalu	5.6	4,052	n/a
AVERAGE	9	2,852	149.3

Figure 10.3 Relationship between Tourism GDP and the Human Development Index on a Selection of Islands, 2017. *Source:* **Compiled by the author with assistance from Sarah Davison.**

IMPACTS OF TOURISM ON ISLANDS

Some of the most significant work on island tourism pertains to the kinds of impacts it has on islands and islanders, both positive and negative, and across economic, social, and ecological dimensions (Archer et al. 2005; Stylidis et al. 2008). Although it is a little misleading to consider these impacts as being mutually exclusive, each of these dimensions are discussed in this next section.

Economic Impacts

From Figure 10.1, we have seen that the travel and tourism sector is exceptionally important as a share of the total economy of many islands, and there is a considerable body of research that debates the advantages and disadvantages of tourism for islands (Carlsen and Butler 2011; Shareef et al. 2008). Island governments, often regardless of the circumstances facing their islands, also appear to inevitably gravitate towards tourism as a tool for economic development (Wilkinson 1989). However, does this automatically follow that tourism is "good" for islands economically? The economic arguments given for supporting tourism have been documented extensively (Briguglio, Archer, et al. 1996; Briguglio, Butler, et al. 1996; Lockhart and Drakakis-Smith 1997; Pattullo 1996). They include: (1) generating government revenue and foreign exchange including improving an island's balance of trade, (2) attracting foreign investment into places where very little local capital currently exists, (3) providing employment for islanders, either directly in the tourist industry itself, or indirectly through those companies supplying goods and services to the tourist sector, as well as those producing crafts and other island-specific souvenirs for the tourists who want to purchase authentic items (Healy 1994; Richards 2009), (4) diversifying the economy, especially away from a dependence on aid, remittances, and low-value-added agricultural and other primary sector exports, (5) discouraging the emigration of labour (i.e., the "brain drain") by providing opportunities for skilled and unskilled islanders at home, and (6) sharing in the costs of building and maintaining public infrastructure such as roads, airports, and seaports, and water and sanitation systems, that are used by both islanders and tourists.

Using 2005 data for a selection of small island states, Levantis (2010) found that there was a positive relationship between dependence on tourism as measured by the percentage of GDP, the average GDP per capita, and the Human Development Index (HDI) ranking of the islands. Using 2017 data, Figure 10.3 updates Levantis's work and the same general relationship exists. On those islands where tourism constitutes more than one-quarter of the GDP, the average GDP/capita was USD 15,303 and the average HDI rank of the places was 105th place. Conversely, on those islands where less than 15 percent of their GDP is associated with tourism, the average GDP/capita was only USD 2,852 and the average HDI rank of the group was 149th place.

Despite these reported benefits, it is also true that economic benefits to islands may be overstated, and even where islands benefit as a whole, there are going to be those who benefit more than others. One problem with island tourism is that, although employment and income might be generated in other local island sectors that serve the tourism industry through a multiplier effect, it is not uncommon for many of the inputs or supplies needed for tourism to be purchased and shipped from elsewhere (Shareef 2003). This leakage problem is especially apparent with resorts owned by transnational companies. Since many of the major airlines and hotel chains are headquartered elsewhere, it is also more likely that profits earned in these resorts are not reinvested locally but are instead sent back to the headquarters location. An extreme example of this is where cruise ship companies purchase or lease small islands or portions of islands, such as Labadee in Haiti and Coco Cay in the Bahamas (both owned by Royal Caribbean) that are then used as one of their ports of call (P. Jones et al. 2016; Rodrigue and Notteboom 2013). In these cases, most of the economic benefits associated with the purchase and consumption of goods or services by the tourists while "at port" are earned by the cruise ship company.

Another challenge is that the level and kinds of employment accessible to islanders may not be as beneficial as one might think. It is not uncommon for skilled and managerial positions to be filled by immigrants, leaving the local labour force with mostly lower-wage, lower-skilled jobs with little room for advancement. Although this employment may still be highly sought after in comparison to other local options, it is still likely to be seasonal or part-time with few benefits. There may also be more subtle leakages that lessen the local economic benefits of tourism. For example, just as profits are repatriated off the island, purchases by local employees on luxury items such as cars and electronics may also result in revenues leaving the island, reducing the benefits for local companies and employees (S. Pratt 2015). In addition, although there may be a demand for locally produced crafts, modernization may marginalize some of these vendors as less expensive, mass-produced replicas of the local crafts undercut the demand for locally-made authentic craft products (Paraskevaidis and Andriotis 2015).

Although competition for land with the tourist sector might benefit those who own their properties (Marjavaara 2007), those businesses and families who lease or rent their properties might see their rents increase as property values increase. This results in the gentrification of the most attractive properties and regions, where the construction or renovation of hotel com-

RANGE OF ECONOMIC IMPACTS ON ISLANDS

Positive	Negative
• Increases positive tax revenue (e.g., taxes and foriegn exchange earnings	• Low multipliers/"leakage" of benefits off-island
• Attracts foreign investment/capital	• Questionable impact on local employment (seasonal/part-time few benefits, largely unskilled jobs)
• Provides employment for Islanders	• Profits repatriated off-island
• Diversifies the economy	• Purchases by islanders from non-island sources
• Discourages labour emigration "brain drain"	• Competition for land and labour drives up local costs
• Assists in development of shared public infastructure	• New specialization in tourism "Dutch disease" and co-opting of government priorities

Source: Compiled by Author

plexes and summer homes squeezes out locals who may not be able to compete with wealthier tourists or property owners (Clark et al. 2007). The recent proliferation of short-term rental units (or STRs) such as Airbnb has this same effect of creating property winners and losers (Dodds and Butler 2019). One of the indirect outcomes of this is that many low-skilled service employees must travel longer distances to commute to their tourism jobs, decreasing families' quality of life (Cocola-Gant 2018). Another outcome is that the cost for local companies to hire local service providers, such as those in the trades, may increase with competition from the resorts. Finally, with the greater participation by non-local companies in the local tourism industry, decision-making by municipal, regional, and national governments may be co-opted or compromised by foreign investors or corporations.

The sidebar on page 195 summarizes the kinds of positive and negative economic impacts that may take place on islands as a result of tourism. However, the specific outcomes and the magnitude of these outcomes will vary considerably based on the circumstances facing each island. For example, Wilkinson (1989) suggests that small (island) states that exert greater control over the decision-making surrounding tourism, including integrating tourism into national and regional development planning and taking a small-scale development approach, are more likely to have positive economic outcomes. Scheyvens and Momsen (2008) provide a number of examples—from Samoa to the Cook Islands to Dominica—where local control over tourism has been viewed by the islanders themselves as beneficial for the overall development of their islands.

Social and Cultural Impacts

In the previous section, it was suggested that one of the possible positive economic impacts of tourism is the shared use of the physical infrastructure such as ports and roads that are built ostensibly to serve the demand by tourists and the tourism sector. A similar positive impact may occur in the social realm, where improvements in health and communications infrastructure serve both the local population and tourists. Tax revenues generated from tourist activities can also be used to support spending on social services and the education system. In a systematic analysis of twenty-eight SIDS from 2005 to 2016, Puig-Cabrera and Foronda-Robles (2019, 61) found that tourism development was "a driving force for enhancing the residents' quality of life" and that government expenditures had a four times greater positive effect on the population than private-sector investment.

Some have suggested that revenues generated from heritage tourism may allow for the recovery and preservation of culturally significant artifacts and sites (Found 2004). Such has been the case on Rapa Nui/Easter Island with respect to the ancient moa heads, iconic cultural artifacts endemic to that island (Barfelz 2011). Possible positive cultural outcomes of tourism include the preservation or even reawakening of local customs, traditions, symbols, and languages, especially if these features are the main motivations for tourism on that island (Schouten 2007). For example, tourism is thought to have helped preserve the traditional tivaevae art made by women of the Cook Islands (Ayres 2002), an enhanced sense of identity and community in the Turks and Caicos (Gatewood and Cameron 2009), and contributed to a sense of pride, place, and culture in Aruba (Croes et al. 2013). The argument is also made that exposing tourists to island societies serves as a form of education and enlightenment, so that the tourists come away with a greater appreciation of the society. Firat (1995, 118) notes

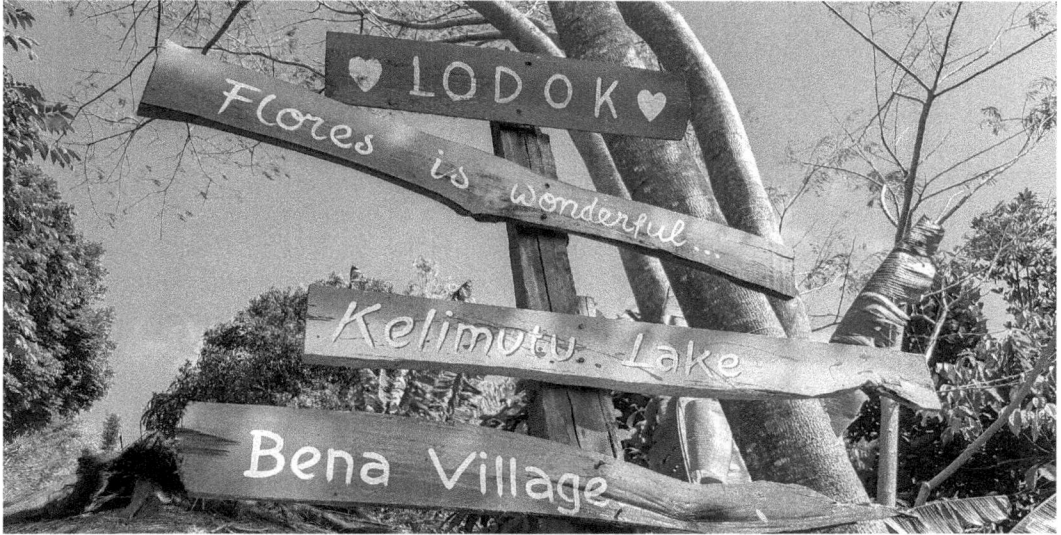

Signs on Flores, Indonesia. *Source*: Barbara Bednarz

that "cultures of all types—ethnic, national, regional, and the like—that are able to translate their qualities into marketable commodities and spectacles find themselves maintained, experienced, and globalized."

There appear to be greater concerns about the possible negative repercussions from tourism on an island's society and culture. For example, although tourism may allow for an expansion of health services and infrastructure, it may also lead to a two-tier health-care system, with a siphoning off of health-care professionals from the public system into the private medical system using superior facilities to satisfy the demand by vacationing tourists (Johnston et al. 2015). These kinds of inequities may also lead to a sense of relative deprivation by local islanders as they encounter, directly or indirectly, the differences in wealth between the tourists and themselves. This may lead to resentment and a loss of self-esteem.

Although funding from tourism may help to preserve various physical and cultural artifacts, it is more likely to lead to the dilution or commodification of these local, unique cultural characteristics, traditions, and symbols (MacLeod 2006). The globalization and commodification of culture, including the allure of Western cultural symbols and material consumption, may contribute to a shift in value systems within families, communities, and organizations

WE SHOULD NOT ASSUME THAT ISLANDERS HAVE NO AGENCY IN THIS PROCESS. FOR EXAMPLE, IN THE EASTERN INDONESIAN ISLAND OF FLORES, VILLAGERS RECOGNIZE THE DIFFERENCE BETWEEN THE AUTHENTIC AND THE INAUTHENTIC AND BELIEVE THAT THE CULTURAL TOURISM IN THEIR COMMUNITIES PROVIDES THEM WITH "A SENSE OF PRIDE AND IDENTITY AND ALSO POLITICAL RESOURCES TO MANIPULATE" (COLE 2007, 943).

(van Rekom and Go 2006). It has been suggested that traditional cultural practices for tourists become a form of staged authenticity (MacCannell 1973) and the places themselves have become "Disneyfied" (Baldacchino 2012b, 57). In these cases, cultural expressions such as the Hawaiian hula dance and music have been co-opted and reproduced in an inauthentic manner for tourists (Imada 2004).

The younger generation, in particular, might be more likely to see tourists as role models or view the cultural traditions as outdated in a globalized world (Stylidis et al. 2008), and in so doing may lose touch with their own cultural history and language. It is not unusual for locals to start mimicking the cultures of tourists, thereby marginalizing their own culture (Conlin and Baum 1995). However, we should not assume that islanders have no agency in this process. For example, in the eastern Indonesian island of Flores, villagers recognize the difference between the authentic and the inauthentic and believe that the cultural tourism in their communities provides them with "a sense of pride and identity and also political resources to manipulate" (Cole 2007, 943). In Bali, a sculptor may realize that his finished products have little to do with the traditional Balinese culture, but the act of sculpting preserves his cultural identity as a Balinese (Schouten 2007). And in Roatan, off the coast of Honduras, the Garifuna people have cut out the non-Indigenous middlemen and taken control of their *punta* dance performances to ensure that they are the ones who will benefit from these cultural performances for tourists (Kirtsoglou and Theodossopoulos 2004).

Ecological or Physical Environmental Impacts

Tourism, and especially the mass tourism that is often associated with warm-water island destinations, is rarely considered to have positive impacts on the local physical environment. However, as is the case with the built environment in the form of airports, roads, and hospitals, revenues generated from tourism may be used to preserve and protect fragile natural environments. This is especially the case if these physical structures or landscapes are drawing tourists to the island. The incorporation and care of sensitive land in national parks or coral reefs in marine parks may be made possible because of the revenue generated by tourists. One example is the Galápagos Islands, defined as a UNESCO World Heritage site in 1978 and a Biosphere Reserve in 1985. The Galápagos National Park was formed in 1959 and covers almost all (97.5 percent) of the land area of the islands, while the Galápagos Marine Reserve, formed in 1986, incorporates 70,000 square kilometres surrounding the archipelago. Entrance fees charged to tourists as well as fees charged to tour operators contribute to the cost of preserving the natural environments and funding conservation programs (Benitez 2001), although it is rare for these kinds of fees to cover the full cost of managing and restoring fragile habitats damaged by tourists (Goodwin 2015).

There are many more ways in which tourism has degraded the physical environment of islands. It is not uncommon for tourist infrastructure, including air and seaports, hotels, shops, and service facilities, to be built near those natural features that attract tourists to the islands in the first place, including near beaches and coral reefs. This proximity increases the likelihood of damage, even if this takes place in small increments (Harrison 2004). The likelihood of damage increases with mass tourism. Construction of facilities also damages those areas such as mangrove swamps that help protect the environment and threatened species (C. M. Hall

2001). Damaging the local natural environment may also adversely affect the sustainability of fishing-based communities, such as local inshore fisheries (Lowe 2004).

Tourists also tend to consume much more fresh water than the local population, straining supplies on some small islands, especially on low-lying atolls (Dodds and Graci 2010). This additional demand on the water supply is even more problematic where tourism is highly seasonal, such as in Mallorca (Tortella and Tirado 2011) and the Aegean islands (Gikas and Tchobanoglous 2009). This seasonality concentrates the demand for water at peak times and may require more expensive solutions such as the shipment of water in tankers and the construction of water desalination systems to meet that peak capacity, despite being underused the rest of the year (Karagiannis and Soldatos 2007). It also increases the competition for water with local farmers at a time of the year when it is most needed for irrigation (Gössling et al. 2015).

The waste and pollution generated by tourism can also affect the natural environment. Disposal of waste has become a problem in many small tourism-dependent islands, including contaminating surface and groundwater and dumping in sensitive areas (Kapmeier and Gonçalves 2018). One of the most prominent examples of sewage and solid waste disposal challenges is in the Maldives, an archipelago that receives more than one million tourists annually (K. Brown et al. 1997; Kundur and Murthy 2013). Here, the entire coral island lagoon on Thilafushi has been filled in with waste, at a rate of about 330 tons per day, producing an artificial garbage island (Kapmeier and Gonçalves 2018). Although this may have solved the waste management program on the main island of Malé, it has created longer-term environmental and health consequences that may even deter tourists from travelling to the Maldives in the future (Sancha Pastor 2019).

There is one final environmental consequence of tourism that is becoming increasingly relevant to the island tourism sector: the production of greenhouse gases (GHGs) as a result of jet travel, accommodation, and tourist activities at the destinations (Gössling 2010). It has been estimated that tourism was responsible for 8 percent of global carbon emissions in 2013 and that this sector's share of total emissions is growing (Lenzen et al. 2018). Although there is no global estimate of the carbon footprint from tourism specific to islands, it is undoubtedly higher per tourist than on most mainland destinations because of the greater reliance on jet travel. On Iceland, from 2010 to 2015, the average tourist consumed 1.35 tons of CO_2, aviation accounted for 50–82 percent of total emissions, and the total GHG emissions linked to tourism tripled from 600,000 to 1.8 million tons of CO_2 over this five-year period (Sharp et al. 2016). Approximately five million tourists per year travel to the Greek island of Crete

IT IS NOT UNCOMMON FOR TOURIST INFRASTRUCTURE, INCLUDING AIR AND SEAPORTS, HOTELS, SHOPS, AND SERVICE FACILITIES, TO BE BUILT NEAR THOSE NATURAL FEATURES THAT ATTRACT TOURISTS TO THE ISLANDS IN THE FIRST PLACE, INCLUDING NEAR BEACHES AND CORAL REEFS. THIS PROXIMITY INCREASES THE LIKELIHOOD OF DAMAGE, EVEN IF THIS TAKES PLACE IN SMALL INCREMENTS (HARRISON 2004).

in the Aegean Sea, with each trip accounting for 429 kg of CO_2, 81 percent of which is due to transportation (Vourdoubas 2019). So, although islands are victims of the outcomes of climate change, those that rely on international tourism are also linked closely to carbon consumption as a consequence of their economic development goals (K. Nurse et al. 2018). This makes them both complicit and vulnerable to international regulations to reduce carbon emissions and to a growing shift in consumer choice away from destinations that involve greater GHG emissions.

MODELLING THE IMPACTS OF TOURISM ON ISLANDS

With the growth of mass tourism, there have been various attempts to quantify the carrying capacity or the intensity of tourism on small islands. One of the most common methods is the Tourism Penetration Index (McElroy 2003, 2006; McElroy and de Albuquerque 1998, 1999). This rather simple composite measure combines three dimensions: economic (visitor spending per capita), sociocultural (the average daily visitor density/1,000 population), and environmental penetration (number of hotel rooms per square kilometre). Although it has been criticized for being a simplistic representation of the complexity associated with each of these dimensions (Bertram 2006; Torres-Delgado and Saarinen 2014), it has been used extensively because it is easy to interpret, and uses data that is generally available.

Attempts to measure the perception of this presence by host populations have also been made, including by Doxey (1975) with his Irridex scale. Doxey's scale suggests that as tourism increases at sites, attitudes by the local communities evolve from initial euphoria to apathy, then annoyance, and finally antagonism. In reality, the attitudes of local populations to tourism are much more complex and nonlinear (Carmichael 2000).

It is not just the perceptions of residents that have been examined. The Tourism Area (or Destination) Life Cycle (TALC) concept was developed to describe the development of tourist destinations from initial discovery and exploration, to development as the number of tourists increases rapidly, to consolidation and a levelling-off of tourist numbers as the destination matures (Butler 1980). At this stage several outcomes are possible, ranging from rejuvenation to decline. Rejuvenation does not occur automatically. It requires planning and entrepreneurship by the destination stakeholders (Kusumah and Nurazizah 2016). Singapore is given as an example of a tourist destination that went through a decline in the 1990s (Toh, Khan and Koh 2001). The Isle of Man is used as an example of rejuvenation. It evolved from a seaside destination catering to families to one that emphasized high-end cultural, sporting and activity tourism in more rural areas (Baum 1998). The TALC has been described as "one of the most significant contributions to studies of tourism development … because it challenges the notion of tourism studies having a simplistic theoretical base" (C. M. Hall 2006, xv) and as "a valuable framework for the description and analysis of tourism development processes" (Prosser 1995, 318). As a result of their small scale, physical separation from other mainland destinations, and relatively small domestic tourist base, islands have figured prominently in the application of the TALC (Lagiewski 2006), including in the Pacific (Choy 1992; Douglas 1997), Tenerife (Rodríguez et al. 2008), Cyprus (Ioannides 1992), the Caribbean (Weaver 2006; Wilkinson 1987), and Hawai'i (Johnston 2001).

Cruise ship docked in Castries, Saint Lucia, Caribbean Islands. *Source*: **NAPA**

CRUISE SHIP TOURISM AND CONNECTIONS TO ISLANDS

Although consisting of only 2 percent of the travel sector as a whole, cruise ship tourism is one of the fastest-growing segments within tourism (Cruise Lines International Association [CLIA] 2019a). The number of ocean cruise ship passengers increased by more than 60 percent from 2009 to 2018 (28.5 million) with Caribbean destinations accounting for almost one-third of this total (Brida and Zapata-Aguirre 2010; CLIA 2019b). Moreover, since many of the ports of call are small islands, there is a special relationship between cruise ship tourism and island tourism. In fact, cruise ships are now becoming so large and with an explicit emphasis on conspicuous consumption that some of them might be considered mobile artificial islands or theme parks (Wood 2000), and it is not uncommon for passengers to never leave the ship even when docked at ports of call (Rodrigue and Notteboom 2013).

As is the case with all tourism, the actual impacts of cruise ship tourism on local communities vary. Islands have often lobbied for more cruise ship passengers without a clear understanding of the benefits and costs associated with this activity (Wilkinson 1999). On paper, benefits of visits include spending by passengers and crew while at ports of call, purchases of goods and services by the ships, and the taxes charged by port jurisdictions (Brida and Zapata-Aguirre 2008). However, the significance of these benefits is ambiguous (Brida and Zapata-Aguirre 2010). MacNeill and Wozniak (2018) found that in general, despite the optimistic picture painted by the cruise ship industry, local tourism boosters, and even the UN World Tourism Organization, the predicted gains in employment and income failed to materialize, corruption increased, and there were substantial negative environmental impacts. This was especially the case in places that had little regulation, low tax rates, and failed to engage local communities in tourism planning. The intensity of cruise ship tourism on local ports of call can also create a backlash within local communities. Although in some ports, residents have a very positive attitude towards cruise ship development (Del Chiappa and Abbate 2016), it is not uncommon for cruise ship passengers to vastly outnumber local residents, contributing to the stress felt by local residents and the perceived deterioration of residents' quality of life (Jordan and Vogt 2017).

Even before the most recent viral pandemic, cruise ships were prone to the spread of infectious diseases. Outbreaks of the Norwalk virus and other influenza-like illnesses are well documented and cruise ship companies have been forced to respond quickly to prevent contagion (Kak 2007; Minooee and Rickman 1999). The recent COVID-19 coronavirus pandemic has upended all sectors of the tourism economy, but perhaps cruise ship tourism most of all. Stories of cruise ships sailing from port to port desperately seeking a place to dock while passengers and crew get increasingly ill will likely have a longer-term impact on this sector of the industry. Consequently, those islands that had grown to depend on cruise ship tourism for a large portion of their market will also be slower to recover.

MASS AND SUSTAINABLE ISLAND TOURISM

Sometimes referred to as "the smokeless industry," when mass tourism emerged at a large scale following the Second World War, it was generally considered to be a beacon of hope for developing countries (Müller 2000), and it has indeed contributed to much of the economic development of small island developing states as they were emerging from colonial rule (Andriotis 2004; Carlsen and Butler 2011; G. Pratt 2002). Although the phrase mass tourism has already been noted, it is important to consider it more thoroughly in the context of how it is distinguished from sustainable or alternative tourism. Also referred to as Conventional Mass Tourism (or CMT), mass tourism is characterized by large numbers of visitors concentrated in a relatively small space, leading to a high intensity of use. This form of tourism is largely controlled by non-local corporations with relatively low local linkages (Weaver 1993). CMT has been the norm since the 1950s and the travel and tourism industry, through transportation, marketing, accommodations, cruise ships, and services, has become integrated around this model.

By the 1970s, criticism of this model of tourism was growing. Not only was it viewed as damaging the social and ecological fabric of developing countries, with ambiguous economic benefits (Krippendorf 1987; Wilkinson 1989), it was also considered philosophically as just another form of neocolonial dominance, this time by reproducing colonial outposts as "pleasure peripheries" (Turner and Ash 1975) for the developed world. Globally, mass tourism linked to jet travel and greenhouse gas emissions is now considered an important and growing component of disturbances on ecosystems (Gössling et al. 2012). The term "Anthropocene" has entered mainstream discourse to refer to the sustained global impact that human activity is having on the Earth, primarily but not exclusively through impacts on ecosystems. Islands have been used both positively and negatively as examples of this era (DeLoughrey 2019; Pugh 2018; Urry 2010). The failure of the mass tourism model to consider the long-term implications of this activity to the survival of planet Earth in the Anthropocene may be its greatest criticism (Cheer and Lew 2017a; Gren and Huijbens 2015).

Partly as a response to these problems, the concept of sustainable tourism began to emerge as an alternative to this conventional model (Butler 1999; Carlsen and Butler 2011; Hall et al. 2015). Unlike CMT, sustainable or alternative tourism was said to be low-density or dispersed, seek authenticity and meaningful interactions between tourists and members of the host communities, and have stronger local economic linkages (Weaver 1993). Emerging from the broader discussion of sustainability and sustainable development, Middleton and Hawkins

(1998) have defined sustainable tourism as where the cumulative effects of tourist activities (e.g., numbers and types of visitors) at a place, plus the actions of the servicing businesses, can continue into the foreseeable future without damaging the quality of the environment upon which activities are based. Cassidy and Hume (2015, 53) state that it is "tourism which is developed and maintained in such a manner and scale that it remains viable in the long run and does not degrade the environment in which it exists to such an extent that it prohibits the successful development of other activities." Both of these definitions are linked closely to the broader definition of sustainable development posited by the 1987 Brundtland Commission report, named after the former Prime Minister of Norway, and also referred to as the UN World Commission on Environment and Development and *Our Common Future*.

GLOBALLY, MASS TOURISM LINKED TO JET TRAVEL AND GREENHOUSE GAS EMISSIONS IS NOW CONSIDERED AN IMPORTANT AND GROWING COMPONENT OF DISTURBANCES ON ECOSYSTEMS.

The literature on sustainable island tourism is summarized well by Dodds and Graci (2010). Although many of the characteristics and challenges associated with sustainable island tourism are also shared by mainland destinations, the features that are often associated with small islands make them special cases. This includes their small size, higher dependence on tourism, and isolation, but it also includes their more fragile ecosystems and stronger, more close-knit social networks and sense of identity (Kelman 2010a). The resilience that has been associated with islands also surfaces with respect to approaches to tourism (Butler 2017). For example, a number of authors have spoken to the ability of islanders to respond and adapt quickly after natural disasters (Becken et al. 2014; Kelman 2014; Pelling and Uitto 2001) and social crises. As an example of a social crisis, Alberts and Baldacchino (2017) show how the government and the community of Aruba responded quickly and compassionately to the death of a young American female tourist, thereby minimizing the negative impact of this incident on the numbers of visitors. Gurtner (2016) notes a similar resilience on Bali, despite several terrorist attacks. In an edited collection of case studies on tourism and resilience (Cheer and Lew 2017b), many of the chapters consisted of island communities implementing resilient tourism practices. For example, resiliency emerged on the Channel Island of Jersey because of the diversity of tourism themes it adopted, including cultural heritage, the natural environment, and sea and fishing charters (Fleury and Johnson 2017). On the island of Zanzibar, off the east coast of Tanzania, the Maasai nomads were the ones becoming resilient.

Driven off their traditional lands, many of them migrated seasonally to Zanzibar to participate in the tourism sector, in the process becoming adept at new ways of social learning and self-organization (Hooli 2017).

Carlsen and Butler (2011) conclude their book on sustainable island tourism by reminding us that, despite the exaggerated economic importance of tourism to many small islands, tourism is as much about the cultural life of the residents as it is about the governance associated with tourism policies. Many islands are grappling with the implications of either overtourism and undertourism. Positive, sustainable outcomes are more likely where local communities are actively engaged in the decision-making on their own terms (Peterson 2011), where tourism corporations practice social responsibility after listening to communities (E. Hughes and Scheyvens 2016), and where regulations and policies are enforced rather than just treated as rhetoric. We are reminded that an understanding of successful long-term sustainable tourism practices is not just useful for islands. They have been referred to as the harbingers of tourism resilience and new discoveries (Peterson 2011) with innovations that can be adapted to any community where tourism plays a prominent role. As Kelman et al. (2015, 38) state, "Island thinking should not mean insular thinking."

Sustainable Island Tourism and the Coronavirus COVID-19 Pandemic

At the time of writing, the coronavirus COVID-19 pandemic has, at least temporarily, ended almost all international tourism. Many airlines have discontinued their routes and most countries have either closed their borders or required visitors to self-isolate for a period of time that would likely have exceeded the length of their vacations. Although it is too early for the scholarly literature to tell us the impacts of this global event on island tourism, it is likely that places highly dependent on tourism are going to experience the greatest economic impacts. Even when this is no longer a prominent public health issue, an ensuing global recession may adversely affect the demand for international tourism to a greater degree than the 2008–09 recession. Sadly, this economic shock to island economies, and how they may be able to adapt to this shock, may serve as an excellent example of the duality of island vulnerability and resilience. In our highly connected world, almost all places have shown that their public health systems and their economies are vulnerable to external shocks. The loss of local employment in the island resorts and related sectors, bankruptcies, and the loss of government revenue from tourism may make the economy of some islands even more unstable. Greater national debt may force some island governments to enact yet another round of structural adjustment and austerity measures, including cutting back on public services, in order to qualify for international aid. At the same time, most tourism-dependent islands have experienced catastrophes of one sort or another in the past and have adapted to them at the level of the household and the community. Therefore, it will be interesting to see how the quality of life of islanders is ultimately affected by this tragic event.

Key Readings

Baldacchino, Godfrey, ed. 2006b. *Extreme Tourism: Lessons from the World's Cold Water Islands.* London, UK: Routledge.

Baum, Tom. 1997. "The Fascination of Islands: A Tourist Perspective." In *Island Tourism: Trends and Prospects*, edited by Douglas Lockhart and David Drakakis-Smith, 21–35. London, UK: Pinter.

Brida, Juan Gabriel, and Sandra Zapata-Aguirre. 2010. "Cruise Tourism: Economic, Socio-Cultural and Environmental Impacts." *International Journal of Leisure and Tourism Marketing* 1 (3): 205–226.

Butler, Richard W. 1993. "Tourism Development in Small Islands." In *The Development Process in Small Island States*, edited by Douglas Lockhart, Patrick Schembri, and David Smith, 71–91. London, UK: Routledge.

Carlsen, Jack, and Richard Butler, eds. 2011. *Island Tourism: Sustainable Perspectives*. Wallingford, UK: CABI.

Conlin, Michael V., and Tom Baum, eds. 1995. *Island Tourism: Management Principles and Practice*. Chichester, UK: Wiley.

Connell, John. 2011. *Medical Tourism*. Wallingford, UK: CABI.

DeLoughrey, Elizabeth M. 2019. *Allegories of the Anthropocene*. Durham, NC: Duke University Press.

Dodds, Rachel, and Sonya Graci. 2010. *Sustainable Tourism in Island Destinations*. London, UK: Earthscan.

Gillis, John R. 2007. "Island Sojourns." *The Geographical Review* 97 (2): 274–287.

Gössling, Stefan, and Geoffrey Wall. 2007. "Island Tourism." In *A World of Islands: An Island Studies Reader*, edited by Godfrey Baldacchino, 429–453. Charlottetown, PE: Island Studies Press.

MacNeill, Timothy, and David Wozniak. 2018. "The Economic, Social, and Environmental Impacts of Cruise Tourism." *Tourism Management* 66: 387–404.

McElroy, Jerome L. 2006. "Small Island Tourist Economies Across the Life Cycle." *Asia Pacific Viewpoint* 47 (1): 61–77.

McElroy, Jerome L., and Klaus de Albuquerque. 1998. "Tourism Penetration Index in Small Caribbean Islands." *Annals of Tourism Research* 25 (1): 145–168.

Pelling, Mark, and Juha I. Uitto. 2001. "Small Island Developing States: Natural Disaster Vulnerability and Global Change." *Global Environmental Change Part B: Environmental Hazards* 3 (2): 49–62.

Scheyvens, Regina, and Janet Momsen. 2008. "Tourism in Small Island States: From Vulnerability to Strengths." *Journal of Sustainable Tourism* 16 (5): 491–510.

Sharpley, Richard. 2012. "Island Tourism or Tourism on Islands?" *Tourism Recreation Research* 37 (2): 167–172.

Chapter Eleven

Islands in the Age of Sustainability and Sustainable Development

Those identified as imminent climate refugees are being held up like ventriloquists to present a particular (Western) "crisis of nature." (Farbotko and Lazrus 2012, 382)

INTRODUCTION

The concept of island sustainability is multidimensional. At the very least, it embodies and integrates many of the topics covered in earlier chapters, including the physical, economic, and social components of island life. Therefore, it is appropriate that an examination of the relationship among sustainability, sustainable development, and islands should take place at this juncture. Moreover, although it is already challenging to outline the breadth of topics related to island sustainability and sustainable development in a single chapter, it would be incomplete if we did not incorporate the related concepts of island vulnerability and resilience into this broader discussion of sustainability.

DEFINING AND MEASURING ISLAND SUSTAINABILITY, VULNERABILITY, AND RESILIENCE

At its most basic, sustainability means something that is continued or upheld (Kelman and Randall 2017). It is generally identified with the United Nations World Commission on Environment and Development's (WCED) 1987 report, *Our Common Future*, otherwise known as the Brundtland Commission, named after the Chair of the Commission and the former Prime Minister of Norway. However, the term originated fifteen years earlier as part of the UN's Stockholm Conference on the Human Environment and a special issue of research articles titled "A Blueprint for Survival" published in *The Ecologist* (Basiago 1995; E. Goldsmith et al. 1972).

Over the past thirty years, in academia, government policy, and in public debate, sustainability has increasingly been used as an adjective accompanying the process of development (Redclift and Springett 2015). The definition most often used to describe sustainable development comes from *Our Common Future* which describes it as "development that meets the needs of the present without compromising the ability of future generations to meet their own needs" (UNWCED 1987, Chapter I.3.27). There are those who continue to argue passionately that the term "sustainable development" is a contradiction, because development is so often associated with economic growth and increasing consumption within a world of finite resources, while sustainability implies continuity into an indefinite future (Redclift 2005; Spaiser et al. 2017). The term has also been criticized for its vagueness and ambiguity (Mebratu 1998).

In the arena of international policy, sustainable development is embodied within the seventeen sustainable development goals (SDGs): a set of broad goals (and many more subgoals and objectives) that are to be achieved by 2030 (United Nations 2015). As we will see below, island governments were instrumental in shaping the SDGs (Quirk and Hanich 2016) and, perhaps because of their small size and boundedness, islands are themselves seen as ideal settings to implement innovative approaches to achieve the SDGs (Crossley and Sprague 2014). This connection among sustainability, sustainable development, and small islands exists at least in part because islands have been idealized as closed systems or miniature worlds. As Farbotko (2010b, 55) states, "Islands are often imagined to be small, complete worlds knowable in their entirety." The implication of this analogy is that, if we can understand how processes such as sustainability and sustainable development work in these small living laboratories, we may learn how to adapt sustainable practices to other jurisdictions and at larger scales (Deschenes and Chertow 2004).

Originally applied to island archaeology and anthropology (Evans 1973; Rainbird 2007), this conceptualization of islands as closed systems has been useful for biogeography and ecology, fields that often examine natural systems as well as human intervention in those systems at smaller scales (C. M. Hall 2012). It has also been extrapolated to social activities, implying that human behaviour on small islands can be applied elsewhere (Baldacchino 2006c; C. M. Hall 2010). Unfortunately, unlike micro-level natural systems, human systems are messy and rarely (arguably never) evolve on islands without being influenced by external factors. There is also a philosophical critique of the trope of islands as laboratories. Understanding an island as a laboratory suggests certainty and discreteness, two characteristics associated with a positivist philosophical framework (Farbotko 2010b). Although knowledge based on natural, objective phenomena is important, islands cannot be fully understood solely on observation and logic. Islanders are just as likely to exist within a phenomenological world, where individuals have agency and act on the basis of uncertainty, perception, and incomplete information (Hay 2006, 2013; Vannini and Taggart 2013).

It is also difficult to discuss sustainability without also speaking about resilience and vulnerability. While vulnerability may be the degree to which a jurisdiction is exposed to external and internal shocks such as natural disasters, economic collapses, or social unrest, resilience

UNFORTUNATELY, UNLIKE MICRO-LEVEL NATURAL SYSTEMS, HUMAN SYSTEMS ARE MESSY AND RARELY (ARGUABLY NEVER) EVOLVE ON ISLANDS WITHOUT BEING INFLUENCED BY EXTERNAL FACTORS.

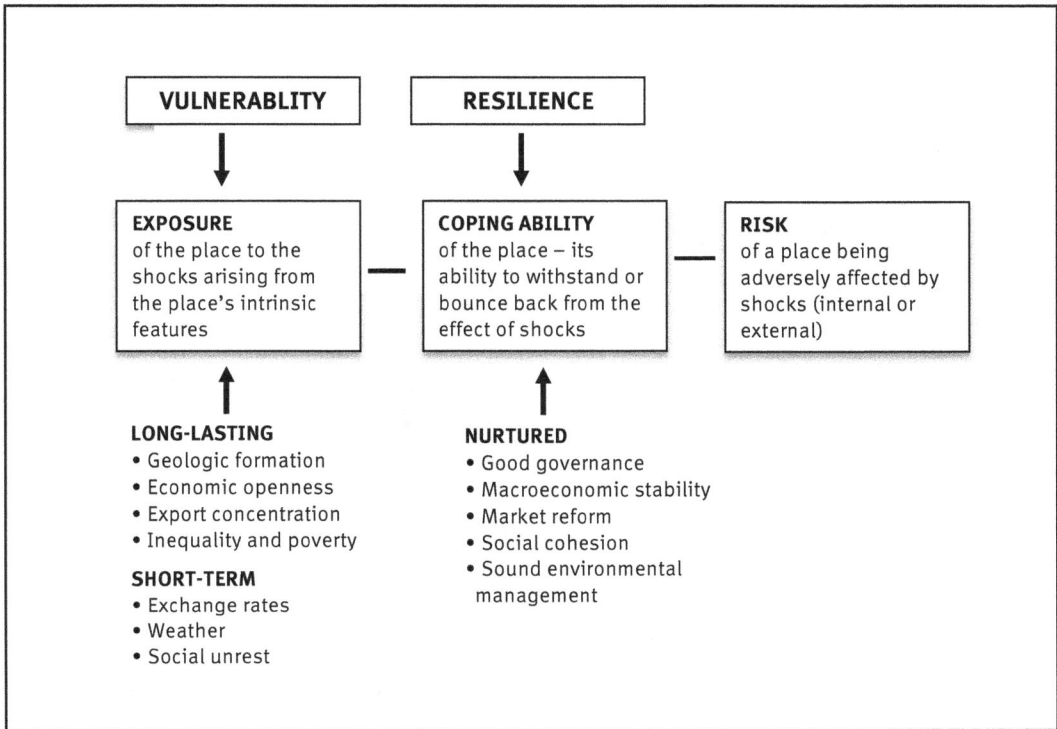

Figure 11.1   Island Vulnerability and Resilience. *Source:* **Adapted from Briguglio and Kinsaga 2004, 49.**

suggests an ability to recover or adapt to those shocks (Angeon and Bates 2015; Briguglio and Kisanga 2004). Figure 11.1 (adapted from Briguglio and Kisanga 2004, 49) shows the interaction between vulnerability and resilience, suggesting that the ultimate risk facing a place is a function of the characteristics that may make it more vulnerable, such as the degree of economic specialization, the exposure to climatic disasters or unstable governance, and the ability to bounce back or evolve so as to withstand future shocks. It also suggests that some of these features may be inherent or permanent. For example, it is difficult to change the geography of a low-lying coral atoll to make it less susceptible to tsunamis, but other characteristics of the island, such as the management and enforcement of effective environmental policies, and effective governance, are more likely to be nourished and evolve over time.

If vulnerability and resilience are not two sides of the same coin, then it must also be true that places can be both vulnerable and resilient, and that the experience of vulnerability can produce resilience (Connell 2013a). For example, immigrants and refugees to Canterbury, New Zealand, and Tohoku, Japan, may have been socially vulnerable because of their socio-economic status in these communities (Uekusa and Matthewman 2017). At the same time, their past experiences with disasters provided them with a resilience or an "earned strength" following earthquakes that the general population was less likely to possess (Uekusa and Matthewman 2017). In the Seychelles, fishermen recognized the vulnerability to their livelihood of marine pollution, overfishing, import dependency, authoritarian governance, and mass tourism, while also becoming more resilient by earning additional informal revenue from

> **THERE IS AN INHERENT CONTRADICTION BETWEEN RESILIENCE—
> PROTECTING GROWTH AND BOUNCING BACK TO A FORMER
> STATE—AND SUSTAINABILITY THAT MAY INVOLVE ENDING
> GROWTH OR CHANGING CONDITIONS FOR A BETTER FUTURE.**

tourism, developing a barter system, and becoming more aware of their own environmental responsibilities (Philpot, Gray, and Stead 2015).

For many, there is significant overlap in the meaning and application of resilience and sustainability (Adger 2003; Edwards 2009). Both share similar goals and applications, particularly as they relate to climate change and human-nature interactions (Lew et al. 2016). However, resilience implies (sustainable) development or perhaps even sustained economic growth while sustainability suggests either conservation of resources or the use of existing resources in perpetuity, even if this means reducing population or economic decline (Kelman and Randall 2017). So there is an inherent contradiction between resilience—protecting growth and bouncing back to a former state—and sustainability that may involve ending growth or changing conditions for a better future (Kelman and Randall 2017). We have already seen how some of the shared island characteristics, such as size, specialization, and isolation, may make islands more vulnerable, but these same characteristics mean that islands may not have experienced the luxury of unbridled growth, having been forced to sustain themselves with the resources at hand.

While it is difficult to distinguish between the concepts of sustainability, vulnerability, and resilience, it is equally challenging to measure them. Although economic and human development indicators may help give us a general overview of the state of a society's well-being, these measures also come with their own flaws. We have already critiqued the pioneering work on island vulnerability indicators by Briguglio and colleagues (Baldacchino and Bertram 2009) while acknowledging the value of this work in raising the global profile of islands in matters related to sustainable development and the consequences of climate change.

The more recent work by Briguglio and his colleagues (see Briguglio et al., 2009) has focused instead on developing measures of resilience, where resilience is measured by the broad categories of macroeconomic stability, market efficiency, good governance, and social development. When his composite measure of resilience was applied to eighty-six world states and territories, he found that the three most resilient jurisdictions, and four of the top ten, were islands (Iceland – 1st; New Zealand – 2nd; Singapore – 3rd; Hong Kong – 8th) (Briguglio 2014). At the other extreme, only one island (Madagascar) was among the ten least resilient jurisdictions. Briguglio's resilience indicators have been criticized for many of the same reasons as his earlier vulnerability indices, including their bias towards a liberal democratic ideology (i.e., higher resilience is correlated with freer trade, a separation of government and judiciary, and a stable democracy) (Bertram 2011). Measuring vulnerability or resilience of islands from a macro lens without also incorporating the agency and local knowledge of human beings may be misleading and inaccurate (Niusulu 2018). Words are, after all, culturally constructed. In some Indigenous languages such as Tongan, there are no equivalent words for vulnerability or resilience, making the application of these concepts in these contexts misleading (Kelman

2018). Nonetheless, it is still a useful starting point to compare islands across various structural characteristics associated with resilience.

Despite the extensive literature that has emerged over the past generation, whether small islands are especially vulnerable to external economic, cultural, and environmental influences or are uniquely resilient to withstand and adapt to these changes is still uncertain. This uncertainty might be best expressed by Pete Hay (2006, 21) who stated that "whether islands are characterised by vulnerability or resilience; whether they are victims of change, economically dependent, and at the mercy of unscrupulous neo-colonial manipulation, or whether they are uniquely resourceful in the face of such threats" is one of the most contested fault lines in island studies.

Sustainability and the History of Islands

Earlier, we discussed the challenges associated with developing an accurate understanding of the historical relationship between Indigenous islanders and their physical environment, especially in the absence of written and oral records. Most of our understanding of human interaction with the physical environment is based on the fossil record, carbon dating, and an interpretation of the archaeology of human settlements. In examining this body of historical research, two competing narratives have emerged. One suggests that islanders placed few demands on the capacity of islands and lived in harmony with nature. This perspective was aptly summarized as "Most continental dwellers regard islands as desirable places to visit, as quaint and backward ... where people live in close harmony with nature and do not desire to harm these environments" (Nunn 2004b, 312). The components of this narrative, based largely on early European observations of Indigenous islanders, assumes that most food was obtained from the immediate surroundings and that cultural and ceremonial practices and taboos were put in place to maintain a balance between human needs and ecosystem supplies (M. Gale 2013). An example might be the cultural taboos in parts of Polynesia and Micronesia regarding harvesting certain crops or fish under certain conditions, ostensibly to maintain these resources for future generations (P. Cohen and Foale 2011; Foale et al. 2011). Another practice might be a highly developed, village-based, land and marine tenure system put in place to avoid a tragedy of the commons–type depletion of resources in the lagoons and reefs, such as existed in Tonga (Malm 2001). Even cultural practices such as induced abortion, cannibalism, and inter-island warfare have all been seen as examples of ways to maintain a long-term balance between island populations and resources (Thomas 2019; Ulijaszek 2006).

This same narrative is also used to explain the ability of islanders to survive natural disasters. Campbell (2009) indicates that traditional disaster risk reduction was built around food security (including food storage, biodiversity, and the production of surpluses), settlement security (such as building villages on elevated sites and using resilient construction techniques), and community cooperation (such as inter-island exchanges and ceremonies and controls over consumption). Growing a more diverse food supply, preserving foods more effectively, and using traditional materials and designs for housing are all specific examples of coping mechanisms that have made islands more resilient to extreme climatic events (Weir, Dovey, and Orcherton 2016).

Despite inter-island rivalry, temporary or permanent relocation was also relatively common during environmental and social crises, including during previous periods of sea-level rise

(Nunn 2007). Inevitably, this narrative concludes that many of these resilient practices have been lost with European contact, at least in part because of colonialism, development, and globalization (Otto 1993). As evidence to support this narrative of a rich and harmonious human-environment interaction with a high carrying capacity, many researchers have noted that island populations prior to European contact were often greater than current populations on small islands. For example, it is estimated that the island of Yap in the Federated States of Micronesia (FSM) at one point supported a population of 40,000 and now has only 7,500 people (Underwood 1969).

A counter-narrative to this idea that Indigenous islanders lived in harmony with nature is one that suggests island societies were in a perpetual state of tension and conflict with their island environments, characterized by periodic collapses of island populations as they overshot their islands' carrying capacities. There are those who have suggested that this situation existed on at least some Pacific islands. For example, Malm (2007) and Kirch (2007) use estimates of population change on the Hawaiian Islands to illustrate the hypothetical dynamics associated with carrying capacity. They argue that if you make the conservative assumption that a

> IN EXAMINING THIS BODY OF HISTORICAL RESEARCH, TWO COMPETING NARRATIVES HAVE EMERGED. ONE SUGGESTS THAT ISLANDERS PLACED FEW DEMANDS ON THE CAPACITY OF ISLANDS AND LIVED IN HARMONY WITH NATURE.

canoe of twenty-five people landed on Hawai'i 2,500 years ago and the subsequent population growth rate was 2 percent annually, the Hawaiian population when Cook landed in 1778 would not be a few hundred thousand as he estimated but instead would be 2.32×10^{13} or 4,000 times the population of the entire planet in the mid-1980s. They conclude that the island population must have gone through significant fluctuations in the millennium prior to first European contact.

The other example that is often cited as proof of disharmony between an island population and its physical environment is that of Rapa Nui/Easter Island. The environmental changes and depopulation that occurred between 1250 and 1650 AD has been suggested by Diamond (2005) and others as one of the clearest examples of "ecocide" (Flenley and Bahn 2007). Diamond's hypothesis was that the Indigenous society's population growth and exploitation of resources exceeded the carrying capacity of the island, resulting in massive deforestation and then depopulation through migration or starvation. Much of this argument may be traced back to a racial and ethnic bias against the Indigenous inhabitants, perpetuated in the popular media by Heyerdahl and others who either believed that the Rapa Nuians first encountered by the Europeans could not have been capable of constructing the 800+ moai statues, or that by doing so they destroyed their own environment (Heyerdahl 1961; Holton 2004).

Considerable criticism of this ecocide argument has emerged. First, there is no evidence that Rapa Nui started with a large population base, and if the original population estimates were much lower, then a significant depopulation could not have occurred (T. Hunt and Lipo 2009; Lipo, DiNapoli, and Hunt 2018). Most of the depopulation that did occur likely took place after European contact as a result of the slave trade and introduced diseases (Malm 2007). Second, there are many other more plausible explanations for the destruction of the palm forest, including the introduction of rats that ate the seeds of the palms (T. Hunt 2006), a prolonged drought, and a Little Ice Age at about AD 1300 that adversely affected the ability to produce food on the island (Nunn and Britton 2001). The argument that the Indigenous society at the time should have known of the consequences of their actions or that they were somehow ecologically ignorant (Bahn and Flenley 1992; Kirch 1984; Ponting 1991) may be yet another ahistorical Anglo-European bias against a non-Caucasian culture (Ingersoll, Ingersoll, and Bove 2017).

A less contested example of historical human-physical environment conflict on islands is that of the Maori on Aotearoa/New Zealand. The Maori introduced invasive species, burned much of the forests to create grasslands, and contributed to species extinction long before Europeans arrived (Baillie and Bayne 2019; Duncan and Blackburn 2004; McGlone 1989;

> A COUNTER-NARRATIVE TO THIS IDEA SUGGESTS ISLAND SOCIETIES WERE IN A PERPETUAL STATE OF TENSION AND CONFLICT WITH THEIR ISLAND ENVIRONMENTS, CHARACTERIZED BY PERIODIC COLLAPSES OF ISLAND POPULATIONS AS THEY OVERSHOT THEIR ISLANDS' CARRYING CAPACITIES.

Nunn 2004; McWethy et al. 2014; Wilmshurst et al. 2008). European contact accelerated the environmental changes taking place on many small islands, including a rapid growth in the number of invasive species introduced to islands, both intentionally (e.g., pigs and goats) and unintentionally (e.g., rats, cats, and wild dogs), and accelerated deforestation at least in part to allow for widespread grazing (Nunn 2004b; Wodzicki 1950)

Regardless of which narrative is more correct, the many human impacts on island flora and fauna, including examples such as the extinction of the flightless moa birds of New Zealand and the deforestation of many southern Pacific islands, suggests that Indigenous peoples "actively manipulated, modified, and, at times, degraded the ecosystems they lived in, producing environmental changes that in turn required ecological adaptation and social adjustments" (Clarke 1990, 235). Nunn (1992) refers to the change in the natural environment on islands in the Pacific as a form of "transported landscape," a term first used by Anderson (1952). This concept suggests that the first islanders brought their cultural concepts with them when they migrated to islands and, in so doing, shaped their new landscapes to fit their cultures.

Globalisation and Sustainability on Islands

The most common perspective on the impacts of globalisation on islands is that it has damaged their social, economic, and ecological futures. As has been the case with many communities, there are numerous examples where island societies have suffered as a result of being exposed to global cultural and economic forces. For example, two of the causes of food insecurity and obesity problems in Pacific island countries are the introduction of high-fat processed food following the Second Worlds War and food regulatory approaches required by the World Trade Organization (WTO) agreements (R. Hughes and Lawrence 2005). However, there are many other examples where the consequences of globalization are more mixed or may have resulted in positive outcomes for islanders. This includes access to formal education, easier and less expensive transportation and communications, and new technologies that have allowed island diasporas to maintain resilient networks (Gough et al. 2010; Hau'ofa 1994). Much of the research appears to show that traditional ecological practices and aspects of modernization interact to allow islands to become more sustainable, including adapting to extreme environmental hazards and climate change. For example, Lauer and colleagues (2013) studied the response to a major tsunami that struck Simbo, one of the Solomon Islands, in 2007. They found that colonialism and globalization had made the island more vulnerable in several ways, including more settlements being built along the coasts, a larger population, and the erosion of social institutions such as customary land sharing and local ecological knowledge. At the same time, villagers who had acquired education and professional knowledge off the island were able to build collaboration and trust across stakeholders at international, national, and local levels and practice effective leadership during a time of crisis, something that may have been beyond the capacity of the traditional leadership system.

In islands across the Philippines, Indonesia, and Timor-Leste, Hiwasaki and others (2014) found that communities could better predict and be better prepared for typhoons, storms, and heavy rainfall events if they combined the local knowledge and practices related to these kinds of events with Western climate modelling. There are many other examples across Oceania where traditional knowledge, customary resource management and capable leadership, and social institutions have combined with positivist scientific observation and modelling to create greater resilience to climate change and to manage biodiversity more effectively (McMillen et al. 2014).

ISLAND SUSTAINABILITY AND CLIMATE CHANGE

Although many places are threatened by the impacts of human-induced climate change, small islands have arguably been affected earlier and more intensively (Betzold 2015). Chapter Twenty-Nine of the Fifth Assessment Report (AR5) of the Intergovernmental Panel on Climate Change (IPCC), a section of the report that focuses specifically on small islands, confirms the observations from earlier IPCC reports that atolls and low-lying coastal areas of islands are at severe risk of flooding, erosion, degrading of their freshwater supplies from wave overwash and saltwater intrusion, and increased coral bleaching and reef degradation (L. Nurse et al. 2014). They also note that the impacts and the adaptations are highly variable across islands given their physical and social diversity.

A large and important literature has emerged in the past decade on the vulnerability, sustainability, and resilience of islands to climate change (Mimura 1999). As a group, Small Island Developing States (SIDS) and the Alliance of Small Island States (AOSIS) have effectively organized on the basis of a shared concern for sustainability, sustainable development, and environmental awareness (Kelman and Randall 2017). Therefore, it is no coincidence that the United Nations recognized SIDS as a formal group at the 1992 Rio de Janeiro Conference on Environment and Development (i.e., the "Earth Summit"), an occasion where member states and non-governmental agencies came together to discuss sustainability. Sustainability and sustainable development have been the most galvanizing causes for small islands since that recognition of shared interests. This includes having a significant influence at the 1994 Barbados Programme of Action on the sustainable development of SIDS, the 2005 Mauritius Strategy, where fifteen islands in the Indian and Pacific Oceans committed to implementing sustainable development, and the 2014 SAMOA Pathway conference, a meeting that emphasized both sustainable development and poverty eradication.

Most analyses of the outcomes of these agreements suggest that, although some progress has been made in some places, a great deal of work still needs to be done (C. M. Hall 2015; United Nations Secretariat 2010). This earlier emphasis by small island states on sustainable development has evolved over the past generation to encompass, and even been eclipsed by,

ALTHOUGH SENSATIONALIZED, THE IMAGES AND STORIES OF ISLANDS DISAPPEARING AS A RESULT OF SEA-LEVEL RISE HAVE PROVEN TO BE POWERFUL SYMBOLS WITH WHICH TO COMMUNICATE TO THE GENERAL PUBLIC THE CONSEQUENCES OF HUMAN IMPACTS ON THE EARTH (C. M. HALL 2010).

the debate on climate change adaptation and mitigation. Island governments and civil society leaders have been instrumental, and arguably very successful, in getting international agreement at various climate change conferences on the consequences of and solutions to human-induced global warming (Betzold 2010; de Águeda Corneloup and Mol 2014; Kelman 2010; Ourbak and Magnan 2018). Although sensationalized, the images and stories of islands disappearing as a result of sea-level rise have proven to be powerful symbols with which to communicate to the general public the consequences of human impacts on the Earth (C. M. Hall 2010). This has played out at a series of global meetings, including the Kyoto Protocol meeting of 1997, the Copenhagen Summit in 2009, and the Paris Climate Change Agreement of 2016.

As a result of being among the earliest and most directly affected by sea-level rise and global warming, islands have also been among the first jurisdictions to test adaptation strategies (Betzold 2015). While recognizing that social and natural characteristics differ, many of these strategies hold promise for other jurisdictions, especially other islands but also coastal mainland jurisdictions. These lessons and strategies are not limited to technological solutions. By being among the first places to contemplate mass migrations, small islands have also forced the world to grapple with issues linked to postcolonialism, social justice, and the moral responsibility that major greenhouse gas emitters have for the future prospects of those they have affected by their actions (Ferdinand 2018).

A CRITIQUE OF SUSTAINABILITY AS APPLIED TO ISLANDS

The consequences and adaptation to climate change, and especially rising sea levels, is now portrayed by most stakeholders as the most important sustainability issue facing islands now and in the future. Although there are very few people who would deny that human-induced climate change will continue to have dire consequences for islanders, there are those who suggest that research has focused too much on the environmental factors that cause vulnerability to the exclusion of the social factors that may either amplify or diminish the damages from climate change (Barnett and Campbell 2010). In the rush to address the outcomes of climate change, other short-term structural problems that affect the quality of life of islanders and are unsustainable may receive less attention. Some have argued that many of the underlying social and political problems on small developing islands, including poverty, income inequality, and tenuous and ineffective governance, existed well before the more obvious alarm bells over the consequences of rising sea levels (Baldacchino and Kelman 2014; Kelman 2013). Implementing long-term strategies to avoid disaster (e.g., building a new sea wall to protect a shoreline, or getting agreement from developed nations to limit their greenhouse gas emissions) without addressing the structural, political, and social issues contributing to environmental degradation and poor land use will do little to solve these problems (Morrison 2017). Despite the urgency to address the challenges associated with climate change on some islands, it may be that the prevailing narrative of climate change is a simplistic outcome to a much more complex problem (Kelman et al. 2015).

Even when climate change adaptation strategies are in place, they will be less effective and more costly if solutions to these underlying social and political challenges are not also addressed. The most significant and effective climate change solutions have something in common with the most effective strategies to achieve sustainable development: they require long-term effective and equitable management and governance, something that is unlikely to occur under the best of circumstances, much less on small islands where it is not uncommon to have human resource shortages and an aversion to the redistribution of power by governments, companies, and international agencies that hold the levers of power (Connell 2018).

IMPLEMENTING LONG-TERM STRATEGIES TO AVOID DISASTER (E.G., BUILDING A NEW SEA WALL TO PROTECT A SHORELINE, OR GETTING AGREEMENT FROM DEVELOPED NATIONS TO LIMIT THEIR GREENHOUSE GAS EMISSIONS) WITHOUT ADDRESSING THE STRUCTURAL, POLITICAL, AND SOCIAL ISSUES CONTRIBUTING TO ENVIRONMENTAL DEGRADATION AND POOR LAND USE WILL DO LITTLE TO SOLVE THESE PROBLEMS (MORRISON 2017).

Figure 11.2   A collapsed church in Port-Au-Prince, Haiti on August 22, 2010.   *Source:* **Arindambanerje**

Haiti's situation may illustrate this problem. Prior to the devastating earthquake that took place on this Caribbean island in 2010 (Figure 11.2), and likely as a function of the availability of international development aid, Haiti had created a comprehensive climate change plan. Unfortunately, since Haiti and the world community did not view an earthquake as an imminent threat, it had not done the same level of planning to prepare for an earthquake (Baldacchino and Kelman 2014; Ratter 2018). The problem with Haiti's preparation for possible extreme events points to a broader issue with small developing islands. Being at the forefront of the climate change and sustainable development movement, many of these places have produced comprehensive climate change adaptation and sustainable development plans. Unfortunately, many of these plans and policies have yet to be implemented (Robinson 2017) or are top-down and shaped by actors such as external aid agencies and the local political elite with little or no input or understanding of the local needs and contexts (Aiafi 2017; Kelman and West 2009). These problems may be exacerbated because of the power of external stakeholders and funders, and the difficulty of finding resources to fund the recommendations. But they may also be a reflection of a weak and inadequate governance system and an imbalance in power relations such that the interests of existing institutions that value short-term gain and self-interest win out over the longer-term sustainability interests of the islands as a whole (Connell 2018).

Winners and Losers

This leads us to the question of identifying the winners and losers in a world in which the outcomes of sea-level rise are becoming more apparent and sustainable development strategies are being implemented. In other words, sustainability for whom (Nunn, 2004b)? With respect to sea-level rise and storm surges, one group of losers might be those who own waterfront

Figure 11.3   Surfers at Pacific Rim National Park, Tofino, BC. *Source:* **Chase Clausen**

properties and will have to bear the costs of adapting their properties while the resale value of these properties decreases (Baldacchino and Kelman, 2014). Ironically, this group may include some of the wealthiest residents who bought in these locations precisely because of the water access and view. The middle class and poor who rent or lease in low-lying areas, and may need to continue living there in order to access jobs and services, are also disadvantaged because of higher rental costs or limited supplies of alternative housing. On the other hand, the short-term winners might be those who own land away from the waterfront that is less susceptible to shoreline erosion or storm damage. They may find the value of their properties will increase relative to properties in other locations. Baldacchino and Kelman (2014) describe another group of winners as the "kinetic elite," a term that has been used elsewhere to describe the relatively wealthy and mobile residents who have the flexibility to move elsewhere with no loss to their quality of life (Baldacchino 2018a).

Island communities that portray themselves as exemplars of sustainability can also emerge as winners or losers. Tofino, British Columbia (Figure 11.3), and Samsø, Denmark, both use renewable energy and/or sustainable development strategies to brand themselves as ecotourism or sustainable tourist destinations (Dodds 2012; Grydehøj and Kelman 2017). Consequently, these communities may find themselves in "eco-island traps," where they are continually competing to be the most green island, and diminishing returns on sustainability investments distracts them from funding more pressing social, environmental, and governance problems.

Very few would argue that small island organizations and SIDS leaders have not been successful in bringing attention to the current and future climate-related sustainability problems of islands. The most extreme examples of this attention have been with respect to the possible disappearance of entire island countries due to rising sea levels. This possibility has been raised especially in relation to atolls where the average island may be only several metres above current sea levels. As reported first in the popular media with respect to the island archipelago of Tuvalu, this has led to the use of the term "climate refugee." The estimates of the number of people who may be displaced within and between countries by changes to the environment range from twenty-five million to one billion (International Organization for Migration 2014). There has also been speculation, partly as a result of examples such as the purchase of land on Fiji by the Kiribati government, that some island governments are preparing for the possibility of a large-scale relocation of their island populations (Hermann and Kempf 2017). Ironically, islands like Tuvalu that are most at risk of sea-level rise and related climate change outcomes may be most useful to the environmental movement as victims when they are completely underwater (Lazrus 2012), a phenomena Farbotko (2010b) calls "wishful sinking." At the same time, the assumption about climate change being a key factor influencing migration must also be questioned.

The focus on sinking islands has largely been on Tuvalu and Carteret Island, a small atoll just off the coast of Bougainville, Papua New Guinea (Connell 2016b). Despite these very public stories, many believe that the actual disappearance of islands is not inevitable and that much of this public discourse is an exaggeration of the likely outcomes (Kench et al. 2015; Nunn 2004; Rankey 2011). In the case of Tuvalu especially, Farbotko and Lazrus (2012) remind us that mobility has been a long standing and accepted component of Tuvaluans and many other Pacific peoples, whether this was in response to natural disasters or other sustainability challenges. For example, although 3,000 of the 14,000 Tuvaluans may have moved to New Zealand and Australia, most of this migration took place before the term climate change was widely used and the motivating factors had less to do with a climate crisis and more to do with historical connections and economic opportunities (Weber 2014). Even without climate change, we would expect migration to increase as a result of growing inequality, poverty, better opportunities elsewhere, and the impacts of globalization (Connell 2013b). As such, migration has always served, and will continue to serve, as a legitimate and voluntary adaptation strategy to evolving physical, economic, social, and political circumstances. Many islanders also eschew the climate refugee label, reminding mainlanders that they have always been adaptive and resilient to changing circumstances (i.e., islanders as resilient, not vulnerable) and prefer to maintain control or agency over their own futures. If islanders are forced to migrate, then they would prefer to view this decision as a form of "migration with dignity" rather than as a capitulation to external forces (McNamara 2015).

Patrick Nunn (2004b) points out that while most of the attention in the climate change discussion has focussed on the consequences of rising sea levels and increased incidence of natural disasters, the more significant impacts on the sustainability of islands over the next century might instead be an increase in the sea temperature and greater variability in precipitation. Even a modest increase in the temperature of the sea is already starting to lead to destruction

MANY ISLANDERS ALSO ESCHEW THE CLIMATE REFUGEE LABEL, REMINDING MAINLANDERS THAT THEY HAVE ALWAYS BEEN ADAPTIVE AND RESILIENT TO CHANGING CIRCUMSTANCES (I.E., ISLANDERS AS RESILIENT, NOT VULNERABLE) AND PREFER TO MAINTAIN CONTROL OR AGENCY OVER THEIR OWN FUTURES.

(bleaching) of coral reefs. This removes one of the natural protective barriers to shorelines, decreases the food security of coastal communities, and eliminates one of the major features that initially attracted tourists and provided foreign revenue (Hoegh-Guldberg 2011; Nunn et al. 1999; Reaser, Pomerance, and Thomas 2000). In one specific example, Uyarra et al. (2005) found that more than 80 percent of tourists to Bonaire would be unwilling to return if coral bleaching affected the quality of this environmental feature. Since the livelihoods of many islanders are significantly dependent on subsistence agriculture, and the supplies of fresh water for irrigation are already limited on many small islands, longer and more frequent periods of drought will have devastating impacts on those in the smaller, outlying rural island communities to sustain themselves with local food (Barnett 2011). One of the challenges is communicating the impacts of the problem. Although rising sea levels and storm surges make for compelling media stories, variability in precipitation and sea surface temperature changes have been more difficult to measure, model, and communicate, especially at the scale of individual islands.

ISLANDS AND THE ANTHROPOCENE

There is growing acceptance that the Holocene has ended and the planet has entered a new epoch where humankind has fundamentally transformed global atmospheric, biological, geochemical, and hydrological systems and the relationship between humans and the Earth (S. Lewis and Maslin 2015; Pugh 2018). Referred to as the Anthropocene, "This is not just an environmental crisis, but a geological revolution of human origin" (Bonneuil and Fressoz 2016, 11). Although the human production of greenhouse gases and their consequences are a critical component of the Anthropocene, it also encompasses the collapse of biodiversity, the modification to the water cycles on continents, and the transformation of the nitrogen and phosphorous cycles. It is estimated that one-third of the Earth's terrestrial biosphere has been changed by humankind, up from 5 percent in 1750, and 84 percent of the ice-free land surface is under direct human influence, including built-up urban areas, croplands and rangelands (Bonneuil and Fressoz 2016; Ellis 2011). So what do islands have to do with this broader philosophical and transformational change? In her recent book, DeLoughrey (2019) argues that this discussion should not be limited to the natural and social sciences, but is also about the experiences of Indigenous peoples, and especially those from island regions in the South that have suffered the impacts of colonialism and continue to be at the forefront of this human transformation of the planet. The island's simultaneous boundedness and permeability have always resulted in exaggerated changes in the habitats of islands. However, the colonialism of the past half-millennium and now climate change are resulting in even more radical changes on islands, including mass extinctions, habitat loss, pollution, and sea-level rise (Rick et al. 2013). Despite criticism over the trope of the island as a living laboratory, small islands do experience intense human-environment interactions. Studying these relationships can tell us much about future changes on islands and other environments in the Anthropocene (Larjosto 2018).

Key Readings

Barnett, Jon, and John Campbell. 2010. *Climate Change and Small Island States: Power, Knowledge, and the South Pacific*. Oxon, UK: Earthscan.

Connell, John. 2013. *Islands at Risk? Environments, Economies and Contemporary Change*. Cheltenham, UK: Edward Elgar.

DeLoughrey, Elizabeth M. 2019. *Allegories of the Anthropocene*. Durham, UK: Duke University.

Farbotko, Carol. 2010. "Wishful Sinking: Disappearing Islands, Climate Refugees and Cosmopolitan Experimentation." *Asia Pacific Viewpoint* 51 (1): 47–60.

Gough, Katherine V., Tim Bayliss-Smith, John Connell, and Ole Mertz. 2010. "Small Island Sustainability in the Pacific: Introduction to the Special Issue." *Singapore Journal of Tropical Geography* 31 (1): 1–9.

Hall, C. Michael. 2012. "Island, Islandness, Vulnerability and Resilience." *Tourism Recreation Research* 37 (2): 177–181.

Kelman, Ilan. 2018. "Islands of Vulnerability and Resilience: Manufactured Stereotypes?" *Area* 50 (1): 1–8.

Lazrus, Heather. 2012. "Sea Change: Island Communities and Climate Change." *Annual Review of Anthropology* 41: 285-301.

Nunn, Patrick D. 2007. *Climate, Environment and Society in the Pacific During the Last Millennium*. Oxford, UK: Elsevier.

Nurse, Leonard et al. 2014. "Small Islands." *In Climate Change 2014: Impacts, Adaptation, and Vulnerability. Part B: Regional Aspects. Contribution of Working Group II to the Fifth Assessment Report of the Intergovernmental Panel on Climate Change*, edited by Vicente R. Barros et. al, 1613-1654. Cambridge, UK: Cambridge University Press.

Ourbak, Timothée, and Alexandre K. Magnan. 2018. "The Paris Agreement and Climate Change Negotiations: Small Islands, Big Players." *Regional Environmental Change* 18 (8): 2201-2207.

Rainbird, Paul. 2007. *The Archaeology of Islands*. Cambridge, UK: Cambridge University.

Ratter, Beate. 2018. *Geography of Small Islands: Outposts of Globalization*. Cham, CH: Springer.

Robinson, Stacey-Ann. 2017. "Climate Change Adaptation Trends in Small Island Developing States." *Mitigation and Adaptation Strategies for Global Change* 22: 669-691.

Lofoten Archipelago, Norway. *Source*: **R7 Photo**

Conclusions and Future Directions in Island Studies

Islands must be lifted from "the sludge of the unadorned nouns" because of the critical inquiries into power that they evoke, provoke, and sustain: once they have become the focus of study, everything becomes part of a sprawling archipelago. One encounters islands everywhere. (Mountz 2015, 636)

The purpose of this book is to provide an introduction to the many dimensions of island studies. In so doing, it also shows the roles that islands have played, and continue to play, in allowing us to better understand this world of islands. In this conclusion, we will address several of the ways in which islands and island studies have made these larger, global contributions. We will also revisit the themes of vulnerability and resilience, isolation and connectedness, and diversity and cohesion. Finally, we will speak to island studies as an interdisciplinary field of enquiry, and how the institutions of island studies are important in framing issues among researchers, governments, students, and the general public.

REVISITING THE THEMES

This book started with the suggestion that there are at least three overarching themes within island studies: vulnerability and resilience, isolation and connectedness, and diversity and cohesion. Although the words "dichotomies" and "contradictions" were used to describe these pairs of themes, it was also noted that to present islands in this way is too simplistic. There is a duality to islands, such that places can be vulnerable *and* resilient, remote *and* connected, unique *and* heterogeneous, all at the same time. It is not unreasonable to consider that the state of vulnerability may also lead to the creation of resilient social and economic coping mechanisms, and islands that may appear to be remote and closed systems to outsiders (i.e., mere specks of land in a vast sea) may be perceived by islanders to be central, open, and connected in the movement of people, products, and ideas. Also, while each island will have its unique properties, some features such as identity, islandness, and the importance of place may be expressed in similar ways across many islands, including a shared connection with the boundary between land and sea, and a desire to protect that environment (Stratford 2008).

> **THERE IS A DUALITY TO ISLANDS, SUCH THAT PLACES CAN BE VULNERABLE *AND* RESILIENT, REMOTE *AND* CONNECTED, UNIQUE *AND* HETEROGENEOUS, ALL AT THE SAME TIME.**

Beyond their vulnerabilities, islands should be seen as "places of hope and innovation" (Kueffer and Kinney 2017, 320).

Research in island studies has become more relational, such that "islands form part of complex networks of relations, assemblages, and flows" (Chandler and Pugh 2018, 65). This perspective has been used to describe many different island environments, including encounters by French scientists on the sub-Antarctic Kerguelen Islands (Prince 2018); development that may endanger the sense of islandness on Lieyu, a small island on the edge of Taiwan's Kinmen archipelago (Lee, Huang, and Grydehøj 2017); and islandness across the network of connections linking the many islands in the Venetian lagoon (Grydehøj and Casagrande 2020). This trend has been accompanied by the use of the term archipelago, metaphorically and objectively, to conceptualize our understanding of the world as a sea of islands, island chains, artificial islands, and mainlands filled with freshwater islands (Pugh 2013; Stratford et al. 2011). This relational approach brings a new vitality to research and an appreciation for the role that island studies has played and will continue to play in addressing broader issues of the role of small islands in the age of the Anthropocene (Pugh 2018).

WHAT HAVE ISLANDS AND ISLAND STUDIES DONE FOR THE WORLD?

Defining an Island

Prior to reading this book, you may have assumed that defining an island was straightforward: it was simply a piece of land surrounded by water. However, we are now aware that this is too simplistic. Islands exist in multiple ways that are not fully understood solely by their physical geography. They are defined by their social, economic, and political roles and the identities and values of islanders. They, and the islandness associated with them, are constructs of the sea (Baldacchino 2015b; Hay 2013). Their existence allows us to critique broader concepts of space and place, scale, remoteness and isolation, boundedness and connection, and transitional spaces or edges. Islands have been used to critique absolute versus relative remoteness, what it means to be at an edge or to be ill-connected (Bocco 2016).

Islands have also led us to critique the idea of places as closed systems or living laboratories, especially when humans are part of those systems. However, for those outside island studies, these tropes can offer a way to better understand the relationships between connectedness and separation, between dependence and independence, and between vulnerability and resilience. Islands lead us to question the social and economic relationships between places. The fact that the isolation of islands may simultaneously lead to stagnation and despair as well as resilience and dynamism holds lessons that are relevant for the future of our planet (Hay 2013). Conceptualizations of different kinds of "islands," including mountain peaks, forest fragments, and cultural communities, allow us to better understand complex issues and what really takes place on the edges of spaces and places (Diamond 1975; Eriksen 1993; Pardini, Nichols, and Püttker 2017).

Geopolitical Pawns and Powers

At first glance, islands are classic examples of geopolitical pawns. Historically, they have been used and abused by larger, more powerful states and corporations, leading to resource depletion, environmental degradation, slavery, and genocide. In part because they are often thought to be out of sight and therefore out of mind, islands have been harmed by political superpowers precisely because they are on the edges of international law and conventions, and outside the consciousness of mainland residents. They have been seen as politically invisible (Mountz 2011). The most extreme expression of this power imbalance may have been the use of island territories by the United States, France, and Britain as test sites for nuclear devices in the "empty spaces" of Oceania (e.g., the Marshall Islands, Mororua, Fangataufa, Malden, and Christmas Islands). The consequences of these tests on Indigenous peoples and their islands continues to this day, more than sixty years after the detonations (dé Ishtar 2003; Maclellan 2005). These and other activities have perpetuated a racist legacy built on a social hierarchy that started with the first European contact (Maclellan 2005, 2019).

There are many more examples where, because of unequal power relations, islanders have been coerced into accepting activities that would otherwise be politically unpopular if carried out closer to home (Vine 2009). This includes the detention of asylum seekers on Manus Island, Christmas Island, and Nauru by Australia as part of the "Pacific Solution," an enforcement archipelago strategy of detention with traumatic repercussions that continues to reverberate through families and communities to this day (Mountz 2017). It is noteworthy that Christmas Island has recently been reopened by Australia as a quarantine site for Australian citizens being evacuated from China who may have been exposed to the 2020 novel coronavirus.

Islands have a long history as friction points between states with competing geopolitical interests. While they may once have been perceived as politically irrelevant, they have become more important, often for reasons related to security and military strategy (Herr 2006). Tensions between China and other regional states over uninhabited rocks in the South and East China Seas, and the conversion of some of these pieces of land into artificial islands by China to serve as military bases, is only one of many examples where islands are at the centre of the theory and practice of international relations (Baldacchino 2016). Although disputes such as this often focus on the conflicts, many territorial disagreements involving islands have been resolved peacefully, providing lessons to other islands and coastal jurisdictions on how they may be able to solve their own conflicts (Baldacchino 2017).

All this suggests that islands are not much more than pawns in a chess match played by more powerful states. However, to portray this as a purely one-sided game is too simplistic. For example, ratification of the United Nations Convention on the Law of the Sea in 1994 gave island states and territories the rights and responsibilities to manage marine and seabed resources over vast regions. So archipelagos that may have consisted of only a small land area now have both an obligation and an opportunity to manage the resources on and under the sea over areas larger than most mainland countries. The fact that many islanders prefer to think of themselves as members of "large ocean states" rather than "small island states" is more

> THE FACT THAT MANY ISLANDERS PREFER TO THINK OF
> THEMSELVES AS MEMBERS OF "LARGE OCEAN STATES"
> RATHER THAN "SMALL ISLAND STATES" IS MORE THAN JUST
> A LINGUISTIC CHOICE; IT REFLECTS HOW THEY FEEL ABOUT
> THEMSELVES AND THEIR RELATIONSHIPS WITH THE REST
> OF THE WORLD.

than just a linguistic choice; it reflects how they feel about themselves and their relationships with the rest of the world. In addition to new opportunities, these changes to international law have fundamentally altered the relationships between island states and transnational corporations, and created new economic and environmental uncertainties into these relationships in the process, for example by signing fishing licenses and agreements to allow seabed mining of minerals and rare metals (Sammler 2016; Valencia 1997). It has also changed the way that subnational island jurisdictions and small island states are perceived by these companies and global states. For example, the difference between defining the Spratly Islands in the South China Sea as either rocks or islands may have very different repercussions on territorial ownership and how marine resources are developed (Gjetnes 2001).

Islands have shown a degree of political nimbleness in negotiating the terms of aid packages that one might not expect given their scale. In other words, their political capital or capacity cannot be measured solely by counting the number of employees dedicated to international relations. Their success may be because islands are continuously connected to the rest of the world: culturally, economically, and politically. While mainland states and institutions engage in external relations as one of a set of government functions, external relations on small islands are a part of most government functions and the everyday lives of residents. It is no wonder then that so many subnational island jurisdictions (SNIJ) are excellent examples of the practice of asymmetrical power relations and have developed some of the most innovative autonomy arrangements in the world (Hepburn 2012). We have already seen that maintaining a semiautonomous political relationship as a SNIJ while in a constant state of negotiations with a metropole is often preferred over the uncertainty of political independence. This situation has even been referred to as a new form of "Islandian" sovereignty (Korson 2018; Prinsen and Blaise 2017; Prinsen, Lafoy, and Migozzi 2017).

Harbingers of the Earth's Climate Crisis?

Much has been written about the plight of small islands as a result of global warming, including sea-level rise and the growing number and intensity of extreme weather events. The way of life of many of the 600 million islanders has already been compromised by climate change, suggesting that they are now harbingers of the future that other coastal communities will likely face (Hirsch 2015). As early as the 1990s, the degradation of coral reefs as a result of increased sea temperatures was evident (Hoegh-Guldberg 1999). Some have even speculated that a consequence of continued global warming will be the end of tourism as we know it (P. Burns and Bibbings 2009). We have seen the impact of groupings of small island states such as AOSIS in shaping international climate change policy with, for example, the "1.5 to Stay Alive" campaign, referring to the need to keep global average temperatures from increasing by more than 1.5 degrees Celsius from preindustrial levels. International policy has also been influenced by the impassioned leadership from small islands including from the former President of the Maldives, Mohamed Nasheed, the former Foreign Minister of the Marshall Islands, Tony deBrum, and Fijian Prime Minister Frank Bainimarama.

Islands have experienced natural and human external threats in the past and they have incorporated coping and adaptation strategies, especially at the local level, to meet these challenges (Kelman 2016). One of the most compelling ways in which adaptation to climate change and disasters is transferable to other places is when locally based traditional knowledge is integrated with science before it is used in policy, education, and action (Hiwasaki, Luna, and Shaw 2014; McMillen et al. 2014). Some of the key lessons on how to reduce risk from disasters and climate change relate to islands' models of governance and innovative leadership, their interconnected support structures, alliance-building, mobility strategies, and their ability to incorporate local and traditional knowledge into solutions (Finucane and Keener 2015; Kelman and Khan 2013). The implication that you can learn from local contexts may make islands the poster children for climate change adaptation strategies as well as the importance of space, spatial relations and power structures (Petzold and Ratter 2019). Despite this potential, we should be cautious about adopting strategies that may have been successful in one place but would be less appropriate in places facing different social and economic conditions (Tompkins and Hurlston 2005).

Beyond the ecological impacts, this connection among small islands, climate change, and sea-level rise also has moral, justice, and international legal implications. This is particularly relevant where the magnitude of the problem exceeds the resilience of the people and threatens the survival of their home and state. In these situations, the possibility of abandoning the land becomes real (Ferreira 2018). In these situations, Stoutenburg (2015) and others ask: 1) who determines whether and when a state has disappeared? 2) if land no longer exists or becomes uninhabitable, what rights does the state have to the marine territories where the land once existed? 3) is the former state still a member of international organizations like the United Nations? and, 4) what legal and financial responsibilities do the corporations—those most responsible for the production of greenhouse gases—have to fix the problems they caused (Byravan and Rajan 2010)? International law has not yet evolved to handle these questions of environmental justice, at least partly because it has emerged as a product of colonialism and the exploitation of resources, not their preservation (Storr 2016). However, the situation of some low-lying islands is prompting these broader legal discussions. This issue is not solely legal in nature. It is also a question of morality and social justice. What moral rights do small island nations have to relocate when their livelihoods and existence are threatened by the outcomes of climate change (Risse 2009)?

WHAT MORAL RIGHTS DO SMALL ISLAND NATIONS HAVE TO RELOCATE WHEN THEIR LIVELIHOODS AND EXISTENCE ARE THREATENED BY THE OUTCOMES OF CLIMATE CHANGE (RISSE 2009)?

Evolution and Biodiversity

Islands have played a fundamental role in shaping our understanding of ecology and evolution (Vitousek, Adsersen, and Loope 1995). Even in their early writings, Darwin, Wallace, and others paid little attention to the island as a special context. However, with the analysis of adaptive radiation on the Galápagos Islands (Darwin) and the Malay Archipelago (Wallace), this changed. One of the most enduring and prominent contributions of islands has been in the field of biogeography and the dynamics of closed systems. Although the seminal work by MacArthur and Wilson (1967) was initially specific to islands, it has led to the development of more general models in evolutionary biology and biogeography (Whittaker et al. 2017). Many of the richest biodiversity hotspots, with one-fifth of the world's plant and vertebrate species, are found on islands and a combination of sea-level rise and agricultural expansion is threatening these areas (Courchamp et al. 2014; Habel et al. 2019). In the last thirty to forty years we have seen a reduction of about half of the coral reef cover in the Atlantic, Indo-Pacific, and Great Barrier reef systems (Birkeland 2015). Not only is there an intrinsic loss when the endemic species in these areas become extinct, there is also long-term damage to food security and human health (Perrings and Gadgil 2003). Less certain is the loss of marine and terrestrial organisms that are often used in traditional medicines and could be used in future pharmaceuticals (Arrieta, Arnaud-Haond, and Duarte 2010; B. Hunt and Vincent 2006). The biologically and culturally connected nature of islands means that loss in one area has ripple effects throughout the network of islands (Kueffer and Kinney 2017).

Relationships Between Indigenous and Non-Indigenous Peoples

Encounters between islanders and outsiders have been problematic from the start. Although there are exceptions, Europeans viewed islands and islanders first in the Caribbean and then in the Pacific as inconsequential or marginal, and the Indigenous islanders as ignorant. Jolly (2007) suggests that foreigners and Indigenous islanders had vastly different visions of regions. While foreigners visualized the space around them functionally and represented it in cartography, Indigenous peoples conceptualized space in terms of "sharing a deep genealogy of cultural and historical connection—the great ocean of Hau'ofa's vision" (Jolly 2007, 530).

Several examples throughout this book have demonstrated the nature of these Eurocentric attitudes. They may have started in adventure novels and atlases where islanders were portrayed as savage, dimwitted, or sly to readers in Anglo-European society. Intended or otherwise, these portrayals of islands and islanders in adventure novels such as *Treasure Island* instilled in young males imperialist values, acceptance of a rigid class structure, and fueled a desire to "get rich quick" that was emerging alongside the rise of mercantilism and the industrial revolution (Mathison 2016). Eurocentrism continued with the longstanding but

THIS STRONG SOCIAL CAPITAL IS NOT JUST USEFUL TO RESIST EXTERNAL THREATS; IT ALSO SHOWS THE MUTUAL AND REINFORCING LINKS BETWEEN CULTURE AND ECONOMIC DEVELOPMENT, AT LEAST PARTLY MEDIATED BY STABLE GOVERNANCE (BALDACCHINO 2005A).

now discredited belief that islanders were not intelligent enough to have intentionally settled Oceania. It continues to be reflected in research that implies islanders on places such as Rapa Nui/Easter Island committed ecocide because they failed to understand the carrying capacity of their islands. Although it is changing, we still see islanders portrayed in simplistic ways in television shows and movies.

Social Systems and Social Networks

Many island communities are known to have a high degree of social cohesion. The rich social networks and abundance of social capital that assist in allowing islanders to cope with external shocks, including from resort-based tourism, immigration, and extreme weather events, which are not always present in mainland communities. On islands, residents need to prepare and recover from disaster; they cannot easily flee from impending crises. This strong social capital is not just useful to resist external threats; it also shows the mutual and reinforcing links between culture and economic development, at least partly mediated by stable governance (Baldacchino 2005a). Combining agency and place creates governance models in small island communities that can provide lessons to many other small and medium-sized jurisdictions (Baldacchino, Greenwood, and Felt 2009).

Economic and Political Entrepreneurship

Small island jurisdictions experience economic challenges not faced by other places. Not only do they have small domestic markets and finite resources, but they may also face high transportation costs, questionable governance practices, and the threat of natural disasters. Sometimes these vulnerabilities lead to epic failures, such as the poor use of royalties from Nauru's mining sector, and accusations of money laundering and other financial shenanigans. At the same time, relative to other small mainland jurisdictions, the overall quality of life of the residents of many small islands appears to be quite high. Many islands have managed to harness their assets well, have found niches for their products and services, are adept at negotiating favourable terms with external actors, and generally have populations that seem satisfied, even as suggested by measures such as the Happy Planet Index (Abdallah et al. 2009). Tourism and financial services are two of many sectors where small islands have excelled (Baldacchino 2015a). Ironically, the small scale, remoteness, and specialization that is so often associated with vulnerability may be the very features that have allowed them to succeed (Baldacchino 2006e). This resourcefulness of jurisdiction, combined with "the triple accidents of size, geography, and sovereignty" (Baldacchino and Bertram 2009, 153) has also led to economic success stories and lessons in political economy and entrepreneurship that may be applicable to other small states and territories (Baldacchino and Milne 2000).

Mobility and Migration

Islands have an intense relationship with mobility and migration (R. King 2009). This relationship is often historical, multidimensional, and integrated into the cultures and economies of islands. This is perhaps best illustrated by islands in Oceania, where Indigenous peoples have a history of discovery, as well as intra- and inter-archipelagic movement that has existed for thousands of years (Jolly 2007). Although it may not constitute a large share of overall global migration, this human circulation is an important feature of the twenty-first century, leading to distinct transnational mobility patterns and linkages and novel ways of defining home (Baldacchino 2018a). It has been suggested that the experiences islands are facing is creating a new politics of mobility (Baldacchino 2018a). In some jurisdictions, such as Prince Edward Island, Malta, and Tasmania, policies exist to recruit and retain economic migrants or high-net-worth individuals as a critical element of a broader regional economic development strategy. In its more extreme forms, and to greater or lesser success, islands have engaged in passport sales programs (van Fossen 2018). Border policing policies have also been evolving rapidly in order to deal with the waves of asylum seekers arriving at intermediate points on islands such as Lampedusa, Lesvos, and Cyprus in the Mediterranean, islands that are on paths to final destinations for the persecuted and less fortunate (R. King 2009; Mainwaring 2014).

The other impact of temporary or permanent migration has been the role of remittances and the cultural and political impacts of the island diaspora. Most states receive remittances and have established relationships with their current and former citizens living abroad. However, because remittances constitute such a large share of the revenues of some islands, this private capital has even greater significance for economic development and poverty reduction, while at the same time leading to population loss and skills drain (Connell 2015). As a microcosm of migration, islands are shaping family and community relationships and island cultures (Connell and Conway 2008), providing lessons for other jurisdictions.

Social Determinants of Health and Obesity

Malnutrition is a severe problem in many developing countries, but it has led to successful strategies to combat the resulting health outcomes. In the same way, overnutrition or obesity has become a health challenge on many small islands, particularly those in Oceania (S. Gillespie and Haddad 2001). Not only are there direct health consequences to this problem but these non-communicable diseases strain government health budgets and contribute to lower productivity (Anderson 2013). The fact that lifestyle changes have produced these serious health outcomes after only thirty years allows epidemiologists an opportunity to better understand the relationships among nutrition, lifestyle changes, diet, and other social, economic, and environmental changes (Zimmet et al. 1990). Strategies to address the obesity problems, including interventions in food stores and mass media communications (Gittelsohn et al. 2007), and the replacement of imported foods with locally grown foods through education campaigns (Englberger et al. 2011), also have broader application to other non-island communities.

DESPITE THE HISTORICAL ENDURANCE OF THE TOURISM IMAGINARY OF ISLAND PARADISES, AND THE GROWTH OF THE CRUISE SHIP INDUSTRY AND ITS STRONG RELATIONSHIP WITH ISLANDS, THERE ARE SIGNS THAT THE CLOSE RELATIONSHIP BETWEEN TOURISM AND MARKETING ISLANDS AS A BRAND IS UNDER THREAT.

Tourism Marketing, Branding, and Islands as Paradises

There has been growing interest in destination marketing, place branding, and particularly the ways in which tourist destinations are portrayed to consumers (Baldacchino and Khamis 2018). We have already seen that the tourism marketing sector has long portrayed islands as utopian paradises and as escapes from a mundane life, linking islands and holidays as one of the most durable tropes in Western civilization (Berg and Edelheim 2012). Islanders are stereotypically seen as inviting, exotic, and erotic (Brislin 2003). Baldacchino and Khamis (2018, 372) suggest that "islands are now, often unwittingly, the objects of what may be the most lavish, global, and consistent branding exercise in human history." Despite the historical endurance of the tourism imaginary of island paradises, and the growth of the cruise ship industry and its strong relationship with islands, there are signs that the close relationship between tourism and marketing islands as a brand is under threat. In some cases, overtourism is straining the environmental and cultural capacities of small islands. In other cases, the authentic experiences desired by the twenty-first-century tourist have disappeared or been distorted as a result of mass tourism. Finally, there may be increasing resistance on the part of the environmentally conscious, responsible tourist to contribute to the production of greenhouse gases directly related to air travel to distant island destinations (Fiorello and Bo 2012; Gössling et al. 2012)

THE INSTITUTIONS OF ISLAND STUDIES

There are many island studies scholars who have suggested that this interdisciplinary field is emerging and growing (Grydehøj 2017; Stratford 2008). In fact, it seems as though island studies has been emerging for the past twenty-five years (Randall 2020). This emergence has taken place across a range of institutions, including through supra-national organisations such as the Alliance of Small Island States (AOSIS), as well as island-focused research centres, with an emphasis on specific issues such as biodiversity, renewable energy, sustainability, or development. A growing number of these centres are affiliated with universities or colleges and therefore have a post-graduate educational component to their research and public engagement mandates. The two best examples of this kind of institution are the University of Prince Edward Island's (Canada) Institute of Island Studies and the University of Malta's Islands and Small States Institute, both of which recently celebrated thirty years of operation. There are indications that the number of these island-centric academic initiatives is increasing, including proposals from China's Hainan Island, the University of the Highlands and Islands in Scotland, and the University of Groningen in the Netherlands. At other universities, including the University of the Ryukyus' Research Institute for Islands and Sustainability (RIIS) on Okinawa, Japan, the scope of their island studies research activities is expanding.

Because there are so few degree programs in island studies, most academics who practice island studies have been trained in cognate disciplines such as geography, economics, biology, sociology, or environmental studies. They come together in international conferences through their professional associations, such as the International Small Islands Studies Association (ISISA) and the Small Island Cultures Research Initiative (SICRI). They also publish their work in dedicated, open-access, peer-reviewed journals such as *Island Studies Journal* and *SHIMA*. The interdisciplinary, place-based nature of island studies means it has the potential to better understand broad issues and challenges from multiple perspectives. However, by not being represented as part of the academic administrative structures at many university institutions, it also means island studies leads a tenuous existence.

Other fields of enquiry that began as interdisciplinary organizations, such as Environmental Studies and Women's or Gender Studies, emerged within the disciplinary mainstream as a result of existential crises that could not be addressed to the satisfaction of students and society by the existing institutional structures. So the crisis that precipitated Environmental Studies may have been the impacts of humankind on the planet's environment, and the emergence of Women's Studies may have occurred as a result of the subjugation and absence of women from intellectual enquiry combined with the rise of the feminist movement. So what is the existential crisis that might result in a heightened interest in island studies? The answer may point to a multi-pronged crisis: a combination of the rising geopolitical importance of islands, their sustainable development challenges, and the recognition of the critical importance of islands as real and metaphorical symbols of the human-nature conflict in the age of the Anthropocene.

Figure and Image Credits

Chapter One

Figure 1.1 Chris Allan/Shutterstock.com
Figure 1.2 Natalia Bratslavsky/Shutterstock.com
Figure 1.3 Wikimedia Commons/Rigobert Bonne, public domain
Figure 1.4 Wikimedia Commons, public domain
Figure 1.5 Wikimedia Commons/Jackopoid, CC BY-SA 3.0:
https://creativecommons.org/licenses/by-sa/3.0/deed/en
Figure 1.6 Andy Strangeway, Wikimedia Commons, CC BY-SA 3.0:
https://creativecommons.org/licenses/by-sa/2.0/deed.en
Figure 1.7 Wikimedia Commons, public domain
Figure 1.8 Wikimedia Commons, public domain
Figure 1.9 Rainer Lesniewski/Shutterstock.com
Figure 1.10 Wikimedia Commons/Jerzystrzelecki, CC BY 3.0:
https://creativecommons.org/licenses/by/3.0/deed.en
Additional images: Santa Cruz del Islote by lulejt/Shutterstock.com,
Aerial view high above The Palm Island, Dubai by Mo Azizi

Chapter Two

Figure 2.1 Wikimedia Commons, public domain
Figure 2.2 Wikimedia Commons, public domain
Figure 2.3 Wikimedia Commons, public domain
Figure 2.4 Wikimedia Commons, public domain
Figure 2.5 NASA/public domain
Figure 2.6 Wikimedia Commons/fearlessRich, CC BY 2.0:
https://creativecommons.org/licenses/by/2.0/deed/en
Figure 2.7 USGS, National Park Service
Figure 2.8 WRONAart/Shutterstock.com
Figure 2.9 Wikimedia Commons/F.W., CC BY-SA 3.0 DE:
https://creativecommons.org/licenses/by-sa/3.0/de/deed.en
Figure 2.10 Wikimedia Commons, public domain

Figure 2.11 Wikimedia Commons/Maximilian Dörrbecker, CC BY-SA 3.0:
https://creativecommons.org/licenses/by-sa/3.0/deed.en
Additional images: A lithograph of the island of Rakata, Wikimedia Commons, public domain,
Satellite image of New Orleans and Mississippi delta, TommoT/Shutterstock.com,
Hawaii Kauai Waialeale sign, Luc Kohnen/Shutterstock.com,
Male Bee Hummingbird, Melinda Fawver/Shutterstock.com

Chapter Three

Figure 3.1 Wikimedia Commons, public domain
Figure 3.2 Wikimedia Commons/Dylan Kereluk, CC BY 2.0:
https://creativecommons.org/licenses/by/2.0/deed.en
Figure 3.3 Wikimedia Commons/Sebastian Münster, public domain
Figure 3.4 Wikimedia Commons/Chris Light, CC BY-SA 4.0:
https://creativecommons.org/licenses/by-sa/4.0/deed.en
Figure 3.5 Flicker.com/Dennis Sylvester Hurd, public domain
Figure 3.6 Wikimedia Commons/Paul Gauguin, public domain
Figure 3.7 Wikimedia Commons, public domain

Chapter Four

Fig 4.1 Wikimedia Commons, public domain
Fig 4.2 dustin77a/Shutterstock.com
Additional images: Traditional Maori wood carved canoes at Waitangi, New Zealand.
Patricia Hofmeester/Shutterstock.com,
Queen Lili'uokalani statue, Theodore Trimmer/Shutterstock.com

Chapter Five

Figure 5.1 Wikimedia Commons/Edward Lynch, public domain
Figure 5.2 Wikimedia Commons, public domain
Figure 5.3 Wikimedia Commons/Kwami, GNU Free Documentation:
https://commons.wikimedia.org/wiki/Commons:GNU_Free_Documentation_License,_
version_1.2
Figure 5.4 Wikipedia Commons, public domain
https://commons.wikimedia.org/wiki/File:Vanuatu_map.png
Figure 5.5 Russ Heinl/Shutterstock.com
Additional images: West Point lighthouse, PEI, Verena Joy/Shutterstock.com,
Cook Islander farmer holds a watermelon, ChameleonsEye/Shutterstock.com

Chapter Six

Figure 6.1 Wikimedia Commons/CIA World Factbook, public domain
Figure 6.2 Wikimedia Commons/OCHA, CC BY 3.0:
https://creativecommons.org/licenses/by/3.0/deed.en
Figure 6.3 Wikimedia Commons/TUBS, CC BY-SA 3.0:
https://creativecommons.org/licenses/by-sa/3.0/deed.en

Figure 6.4 Wikimedia Commons/Osiris, CC BY-SA 3.0:
https://creativecommons.org/licenses/by-sa/3.0/deed.en
Figure 6.5 Wikimedia Commons/Stasyan117, CC A-SA 4.0:
https://creativecommons.org/licenses/by-sa/4.0/deed.en
Figure 6.6 Wikimedia Commons/JOSH tw, CC BY-SA 3.0:
https://creativecommons.org/licenses/by-sa/3.0/deed.en
Additional image: Banana plantation, Brian Snelson/Flicker.com,
CC BY 2.0 License: https://creativecommons.org/licenses/by/2.0/

Chapter Seven

Figure 7.1 Wikimedia Commons/CNX OpenStax, CC BY 4.0,
https://creativecommons.org/licenses/by/4.0/deed.en
Figure 7.2 Wikimedia Commons/Ben Moore, CC BY-SA 3.0:
https://creativecommons.org/licenses/by-sa/3.0/
Figure 7.3 Wikimedia Commons/OCHA, CC BY 3.0:
https://creativecommons.org/licenses/by/3.0/deed.en
Figure 7.4 Alejandro Carnicero/Shutterstock
Figure 7.5 Author, derived from World Bank Open Data, CC BY-4.0:
https://datacatalog.worldbank.org/public-licenses#cc-by
Figure 7.6 Statistics Iceland
Figure 7.7 Wikimedia Commons/Jamaicajoe, CC BY-SA 3.0:
https://creativecommons.org/licenses/by-sa/3.0/deed.en
Figure 7.8 CIA World Factbook, public domain
Figure 7.9 Wikimedia Commons/Quizimodo, CC BY-SA 3.0:
https://creativecommons.org/licenses/by-sa/3.0/deed.en

Chapter Eight

Figure 8.1 Wikimedia Commons/John Snow, public domain.
Figure 8.2 Author, adapted from Cliff, Andrew, and Peter Haggett. 1995.
"The Epidemiological Significance of Islands." Health & Place 1 (4): 199-209.
Figure 8.3 Wikimedia Commons, public domain
https://commons.wikimedia.org/wiki/File:World_location_map.svg
Figure 8.4 Author, adapted from Anja Heilmann
https://www.researchgate.net/figure/The-main-determinants-of-health-35_fig2_314500772)
Figure 8.5 Michael Mozart, Flicker.com, CC BY 2.0:
https://creativecommons.org/licenses/by/2.0/
Figure 8.6 Wikipedia/Human Development Index, CC BY-SA 3.0:
https://creativecommons.org/licenses/by-sa/3.0/

Chapter Nine

Figure 9.1 Wikimedia Commons/TUBS, CC BY-SA 3.0:
https://creativecommons.org/licenses/by-sa/3.0/
Figure 9.2 Wikimedia Commons/ARM Image Library, public domain
https://commons.wikimedia.org/wiki/File:Nauru_satellite.jpg
Figure 9.3 Wikimedia Commons/Alinor, CC BY-SA 3.0:
https://creativecommons.org/licenses/by-sa/3.0/
Figure 9.4 Wikimedia Commons, Ian Macky, public domain

Chapter Ten

Figure 10.1 Author (From statistics found at KNOEMA Data Atlas, Accessed January 7, 2019)
Figure 10.2 Source: Godfrey Baldacchino and Geoff Bertram.
From Baldacchino and Bertram. 2009. Permission has been granted by Dr.
Baldacchino to use this figure.
Figure 10.3 Compiled by the author with assistance from Sarah Davison. From various sources,
including the World Bank, Knoema, and WTTC Annual Reports by country.
Additional images: Colourful signs on Flores, Barbara Bednarz/Shutterstock.com
Cruise ship docked in Castries, Saint Lucia, Caribbean Islands, NAPA/Shutterstock.com
Airplane by motive56/Shutterstock.com

Chapter Eleven

Figure 11.1 Author, adapted from Briguglio and Kinsaga 2004, 49.
Figure 11.2 Arindambanerje/Shutterstock.com
Figure 11.3 Chase Clausen/Shutterstock.com
Additional images: Mabul Island, Malaysia, Rick Carey/Shutterstock.com

Conclusion

Image: Lofoten Archipelago, Norway, R7 Photo/Shutterstock.com

References

Abarcar, Paolo, and Caroline Theoharides. 2017. "The International Migration of Healthcare Professionals and the Supply of Educated Individuals Left Behind." Paper presented at NEUDC Annual Conference 2017, Tufts University, Medford, MA, November 4-5, 2017. https://appam.confex.com/data/extendedabstract/appam/2018/Paper_27374_extendedabstract_1634_0.pdf.

Abdallah, Saamah, Sam Thompson, Juliet Michaelson, Nic Marks, and Nicola Steuer. 2009. The Happy Planet Index 2.0: Why Good Lives Don't Have to Cost the Earth. London, UK: The New Economic Foundation. http://www.happyplanetindex.org/learn/download-report.html.

Abe, Tetsuto, Shun'ichi Makino, and Isamu Okochi. 2008. "Why Have Endemic Pollinators Declined on the Ogasawara Islands?" *Biodiversity and Conservation* 17: 1465-1473.

Ackrén, Maria. 2014. "Greenlandic Paradiplomatic Relations." In *Security and Sovereignty in the North Atlantic*, edited by Lassi Hainenen, 42-61. London, UK: Palgrave Macmillan.

Adamo, Antonino. 2018. "A Cursed and Fragmented Island: History and Conflict Analysis in Bougainville, Papua New Guinea." *Small Wars & Insurgencies* 29 (1): 164-186.

Adger, W. Neil. 2003. "Building Resilience to Promote Sustainability." *IHDP Update* 2: 1-3.

Adersen, Henning. 1995. "Research on Islands: Classic, Recent, and Prospective Approaches." In *Islands: Biological Diversity and Ecosystem*, edited by Peter Vitousek, Lloyd L. Loope, and Henning Adersen, 7-21. New York: Springer.

Agence Française du Développement. 2015. "Annual Report 2014." https://issuu.com/objectif-developpement/docs/afd-annual-report-2014.

Aguiló, Eugeni, Joaquín Alegre, and Maria Sard. 2005. "The Persistence of the 'Sun and Sand' Tourism Model." *Tourism Management* 26 (2): 219-231.

Aiafi, Potoae Roberts. 2017. "The Nature of Public Policy Processes in the Pacific Islands." *Asia and the Pacific Policy Studies* 4 (3): 451-466.

Alberts, Arjen, and Godfrey Baldacchino. 2017. "Resilience and Tourism in Islands: Insights from the Caribbean." In *Tourism and Resilience*, edited by Richard Butler, 150-162. Wallingford, UK: CABI.

Alexeyeff, Kalissa. 2004. "Sea Breeze: Globalization and Cook Islands Popular Music." *The Asia Pacific Journal of Anthropology* 5 (2): 145-158.

Allen, Oliver. 1980. *The Pacific Navigators*. Alexandria, VI: Time-Life.

Amoamo, Maria. 2011. "Remoteness and Myth Making: Tourism Development on Pitcairn Island." *Tourism Planning & Development* 8 (1): 1-19.

Amoamo, Maria. 2013. "Development on the Periphery: A Case Study of the Sub-National Island Jurisdiction of Pitcairn Island." *Asia Pacific Viewpoint* 54 (1): 91-108.

Amoamo, Maria. 2017. "Resilience and Tourism in Remote Locations: Pitcairn Island." In *Tourism and Resilience*, edited by Richard Butler, 163-180. Wallingford, UK: CABI.

Anckar, Carsten. 2008. "Size, Islandness, and Democracy: A Global Comparison." *International Political Science Review* 29 (4): 433-459.

Anckar, Dag. 2010. "Small is Democratic: But Who is Small?" *Arts and Social Sciences Journal* 1: ASSJ2.

Anderson, Atholl. 1997. "Prehistoric Polynesian Impact on the New Zealand Environment: Te Whenua Hou." In *Historical Ecology in the Pacific Islands,* edited by Patrick Vinton Kirch and Terry L. Hunt, 147-165. New Haven, CT: Yale University Press.

Anderson, Edgar. 1952. *Plants, Man, and Life.* Berkeley: University of California.

Anderson, Ian. 2013. "The Economic Costs of Noncommunicable Diseases in the Pacific Islands: A Rapid Stocktake of the Situation in Samoa, Tonga, and Vanuatu." Health, Nutrition, and Population (HNP) Discussion Paper. Washington, DC: The World Bank Group. http://documents.worldbank.org/curated/en/291471468063255184/pdf/865220WP0Econo0Box385176B000PUBLIC0.pdf.

Andriotis, Konstantinos. 2004. "Problems of Island Tourism Development: The Greek Insular Regions." In *Coastal Mass Tourism: Diversification and Sustainable Development in Southern Europe,* edited by Bill Bramwell, 114-132. Clevedon, UK: Channel View.

Androus, Zachary, and Neyooxet Greymorning. 2016. "Critiquing the SNIJ Hypothesis with Corsica and Hawai'i." *Island Studies Journal* 11 (2): 447-464.

Angeon, Valérie, and Samuel Bates. 2015. "Reviewing Composite Vulnerability and Resilience Indexes: A Sustainable Approach and Application." *World Development* 72 (C): 140-162.

Appleyard, Reginald. 1989. "Migration and Development: Myths and Realities." *International Migration Review* 23 (3): 486-499.

Archer, Brian, Chris Cooper, and Lisa Ruhanen. 2005. "The Positive and Negative Impacts of Tourism." In *Global Tourism*, 3rd ed., edited by William Theobald, 79-102. Burlington, MA: Elsevier.

Arias, Pedro, Cora Dankers, Pascal Liu, and Paul Pilkauskas. 2003. *The World Banana Economy: 1985-2002.* Rome: Food and Agriculture Organization of the United Nations.

Armstrong, Harvey W., and Robert A. Read. 2003. "Small States and Small Island States: Implications of Size, Location and Isolation for Prosperity." In *On the Edge of the Global Economy: Implications of Economic Geography for Small and Medium-Sized Economies at Peripheral Locations,* edited by Jacques Poot, 191-223. Cheltenham, UK: Edward Elgar.

Arrieta, Jesús M., Sophie Arnaud-Haond, and Carlos M. Duarte. 2010. "What Lies Underneath: Conserving the Oceans' Genetic Resources." *Proceedings of the National Academy of Sciences* 107 (43): 18318-18324.

Ash, Jillian, and Jillian Campbell. 2016. "Climate Change and Migration: The Case of the Pacific Islands and Australia." *Journal of Pacific Studies* 36 (1): 53-72.

Asis, Maruja. 2006. "The Philippines' Culture of Migration." *Migration Information Source: The Online Journal of the Migration Policy Institute.* http://www.migrationpolicy.org/article/philippines-culture-migration.

Australian Agency for International Development. 2006. *Pacific 2020 – Challenges and Opportunities for Growth.* Canberra: Australian Agency for International Development.

Ayres, Ron. 2002. "Cultural Tourism in Small-Island States: Contradictions and Ambiguities." In *Island Tourism and Sustainable Development: Caribbean, Pacific, and Mediterranean Experiences*, edited by Yorghos Apostolopoulos and Dennis Cole, 145-160. Westport, CT: Praeger.

Bagolin, Izete, and Flavio Comim. 2008. "Human Development Index (HDI) and its Family of Indexes: An Evolving Critical Review." *Revista de Economia* 34 (2): 7-28.

Bahn, Paul G., and John Flenley. 1992. *Easter Island, Earth Island.* London, UK: Thames and Hudson.

Baillie, Brenda R., and Karen M. Bayne. 2019. "The Historical Use of Fire as a Land Management Tool in New Zealand and the Challenges for its Continued Use." *Landscape Ecology* 34: 2229-2244.

Baines, Graham. 2014. "The Political Economy of Transitioning to a Green Economy in Nauru." In *Transitioning to a Green Economy: Political Economy of Approaches in Small States,* edited by Nadine Smith, Anna Halton, and Janet Strachan, 158-177. London, UK: Commonwealth Secretariat.

Baldacchino, Godfrey. 2004. "The Coming of Age of Island Studies." *Tijdschrift voor Economische en Sociale Geografie* 95 (3): 272-283.

Baldacchino, Godfrey. 2005a. "The Contribution of 'Social Capital' to Economic Growth: Lessons from Island Jurisdictions." *The Round Table* 94 (378): 31-46.

Baldacchino, Godfrey. 2005b. "Editorial: Islands – Objects of Representation." *Geografiska Annaler: Series B, Human Geography* 87 (4): 247-251.

Baldacchino, Godfrey, ed. 2006a. *Extreme Tourism: Lessons from the World's Cold Water Islands*. London, UK: Routledge.

Baldacchino, Godfrey. 2006b. "Innovative Development Strategies from Non-Sovereign Island Jurisdictions: A Global Review of Economic Policy and Governance Practices." *World Development* 34 (5): 852-867.

Baldacchino, Godfrey. 2006c. "Islands, Island Studies, Island Studies Journal." *Island Studies Journal* 1 (1): 3-18.

Baldacchino, Godfrey. 2006d. "Managing the Hinterland Beyond: Two Ideal-Type Strategies of Economic Development for Small Island Territories." *Asia Pacific Viewpoint* 47 (1): 45-60.

Baldacchino, Godfrey. 2006e. "Small Islands Versus Big Cities: Lessons in the Political Economy of Regional Development from the World's Small Islands." *The Journal of Technology Transfer* 31 (1): 91-100.

Baldacchino, Godfrey. 2006f. "Warm Versus Cold Water Island Tourism." *Island Studies Journal* 1 (2): 183-200.

Baldacchino, Godfrey. 2007. "Introducing a World of Islands." In *A World of Islands*, edited by Godfrey Baldacchino, 1-29. Charlottetown, PE: Island Studies Press.

Baldacchino, Godfrey. 2008. "Studying Islands: On Whose Terms? Some Epistemological and Methodological Challenges to the Pursuit of Island Studies." *Island Studies Journal* 3 (1): 37-56.

Baldacchino, Godfrey. 2010a. "The Island Lure." *International Journal of Entrepreneurship and Small Business* 9 (4): 373-377.

Baldacchino, Godfrey. 2010b. *Island Enclaves: Offshoring Strategies, Creative Governance, and Subnational Island Jurisdictions*. Montréal: McGill-Queen's University Press.

Baldacchino, Godfrey. 2011. "Introduction: Singing, Islands, and Island Songs." In *Island Songs: A Global Repertoire*, edited by Godfrey Baldacchino, xix-xli. Plymouth, UK: Scarecrow Press.

Baldacchino, Godfrey. 2012a. "Islands and Despots." *Commonwealth & Comparative Politics* 50 (1): 103-120.

Baldacchino, Godfrey. 2012b. "The Lure of the Island: A Spatial Analysis of Power Relations." *Journal of Marine and Island Cultures* 1 (2): 55-62.

Baldacchino, Godfrey. 2013. "Only Ten: Islands as Uncomfortable Fragmented Polities." In *The Political Economy of Divided Islands,* edited by Godfrey Baldacchino, 1-17. Basingstoke, UK: Palgrave Macmillan.

Baldacchino, Godfrey, ed. 2015a. *Entrepreneurship in Small Island States and Territories*. London, UK: Routledge.

Baldacchino, Godfrey. 2015b. "Smallness and Islandness: Wither the Twain Shall Meet?" In *Insularity: Small Worlds in Cultural and Linguistic Perspectives*, edited by Ralf Heimrath and Arndt Kremer, 31-44. Würzburg, DE: Königshausen and Neumann.

Baldacchino, Godfrey. 2016. "Diaoyu Dao, Diaoyutai, or Senkaku? Creative Solutions to a Festering Dispute in the East China Sea from an 'Island Studies' Perspective." *Asia Pacific Viewpoint* 57 (1): 16-26.

Baldacchino, Godfrey. 2017. *Seven Protocols to Festering Island Disputes: 'Win-Win' Solutions for the Diaoyu/Senkaku Islands*. London, UK: Routledge.

Baldacchino, Godfrey. 2018a. "Connectivity, Mobility and Island Life: Parallel Narratives from Malta and Lesvos." *Symposia Melitensia* 14: 7-17.

Baldacchino, Godfrey, and Geoffrey Bertram. 2009. "The Beak of the Finch: Insights into the Economic Development of Small Economies." *The Round Table: Commonwealth Journal of International Affairs* 98 (401): 141-160.

Baldacchino, Godfrey, and Robert Greenwood, eds. 1998. *Competing Strategies of Socio-Economic Development for Small Islands*. Charlottetown, PE: Island Studies Press.

Baldacchino, Godfrey, and Ilan Kelman. 2014. "Critiquing the Pursuit of Island Sustainability: Blue and Green, With Hardly a Colour in Between." *Shima: The International Journal of Research into Island Cultures* 8 (2): 1-21.

Baldacchino, Godfrey, and Susie Khamis. 2018b. "Preface." In *The Routledge International Handbook of Island Studies: A World of Islands*, edited by Godfrey Baldacchino, xix-xxxv. London, UK: Routledge.

Baldacchino, Godfrey, and David Milne, eds. 2000. *Lessons from the Political Economy of Small Islands: The Resourcefulness of Jurisdiction*. New York: St. Martin's and Macmillan Press.

Baldacchino, Godfrey, and David Milne. 2006. "Exploring Sub-national Island Jurisdictions: An Editorial Introduction." *The Round Table* 95 (386): 487-502.

Baldacchino, Godfrey, and David Milne, eds. 2009. *The Case for Non-Sovereignty: Lessons From Sub-National Island Jurisdictions*. London: Routledge.

Baldacchino, Godfrey, Robert Greenwood, and Lawrence Felt. 2009. *Remote Control: Governance Lessons For and From Small, Insular, and Remote Regions*. St. John's, NL: Institute of Social and Economic Research, Memorial University of Newfoundland.

Baldwin, Douglas. 1993. "L.M. Montgomery's Anne of Green Gables: The Japanese Connection." *Journal of Canadian Studies* 28 (3): 123-133.

Banks, Glenn. 2008. "Understanding 'Resource' Conflicts in Papua New Guinea." *Asia Pacific Viewpoint* 49 (1): 23-34.

Bansal, Harvir, and Horst A. Eiselt. 2004. "Exploratory Research of Tourist Motivations and Planning." *Tourism Management* 25 (3): 387-396.

Banton, Arnold H. 1951. "A Genetic Study of Mediterranean Anaemia in Cyprus." *American Journal of Human Genetics* 3 (1): 47-64.

Barfelz, Abby L. 2011. "The Little Island That Could: How Reforming Cultural Preservation Policies Can Save Easter Island and the World's Heritage." *Michigan State University College of Law International Law Review* 20 (1): 149-177.

Bargh, Maria. 2014. "A Blue Economy for Aotearoa New Zealand?" *Environment, Development, and Sustainability* 16 (3): 459-470.

Barnes, Geraldine. 1995. "Vinland the Good: Paradise Lost." *Parergon* 12 (2): 75-96.

Barnett, Jon. 2011. "Dangerous Climate Change in the Pacific Islands: Food Production and Food Security." *Regional Environmental Change* 11 (1): 229-237.

Barnett, Jon, and W. Neil Adger. 2003. "Climate Change and Atoll Countries." *Climatic Change* 61 (3): 321-337.

Barnett, Jon, and John Campbell. 2010. *Climate Change and Small Island States: Power, Knowledge, and the South Pacific*. Oxon, UK: Earthscan.

Barratt, Peter. 2003. *Bahama Saga: The Epic Story of the Bahama Islands*. Bloomington, Indiana: 1st Books.

Barrett, James, Roelf Beukens, Ian Simpson, Patrick Ashmore, Sandra Poaps, and Jacqui Huntley. 2000. "What Was the Viking Age and When Did It Happen? A View from Orkney." *Norwegian Archaeological Review* 33 (1): 1-39.

Bartlett, Mark S. 1957. "Measles Periodicity and Community Size." *Journal of the Royal Statistical Society - Series A* 120 (1): 48-70.

Bartmann, Barry. 2006. "In or Out: Sub-National Island Jurisdictions and the Antechamber of Para-Diplomacy." *The Round Table* 95 (386): 541-559.

Basiago, Andrew. 1995. "Methods of Defining 'Sustainability'." *Sustainable Development* 3 (3): 109-119.

Bass, Stephen, and Barry Dalal-Clayton. 1995. *Small Island States and Sustainable Development: Strategic Issues and Experience.* London, UK: International Institute for Environment and Development.

Baum, Tom. 1997. "The Fascination of Islands: A Tourist Perspective." In *Island Tourism: Trends and Prospects*, edited by Douglas Lockhart and David Drakakis-Smith, 21-35. London, UK: Pinter.

Baum, Tom. 1998. "Taking the Exit Route: Extending the Tourism Area Life Cycle Model." *Current Issues in Tourism* 1 (2): 167-175.

Bayliss-Smith, Tim. 1974. "Constraints on Population Growth: The Case of the Polynesian Outlier Atolls in the Precontact Period." *Human Ecology* 2 (4): 259-295.

Bayliss-Smith, Tim, Richard Bedford, Harold Brookfield, and Marc Latham. 1988. *Islands, Islanders and the World: The Colonial and Post-Colonial Experience of Eastern Fiji.* Cambridge, UK: Cambridge University Press.

Beattie, Andrew J., Mark Hay, Bill Magnusson, Rocky de Nys, James Smeathers, and Julian F. V. Vincent. 2011. "Ecology and Bioprospecting." *Austral Ecology* 36 (3): 341-356.

Becken, Susanne, Roché Mahon, Hamish G. Rennie, and Aishath Shakeela. 2014. "The Tourism Disaster Vulnerability Framework: An Application to Tourism in Small Island Destinations." *Natural Hazards* 71 (1): 955-972.

Beckford, George L. 1969. "The Economics of Agricultural Resource Use and Development in Plantation Economies." *Social and Economic Studies* 18 (4): 321-347.

Bedford, Richard, and Graeme Hugo. 2008. "International Migration in a Sea of Islands: Challenges and Opportunities for Insular Pacific Spaces." Population Studies Centre Discussion Papers 69. Hamilton, NZ: Population Studies Centre, University of Waikato.

Belich, James. 1986. *The Victorian Interpretation of Racial Conflict: The Maori, the British, and the New Zealand Wars.* Kingston, Ontario: McGill-Queen's University Press.

Bellwood, Peter. 1978. *Man's Conquest of the Pacific – The Prehistory of Southeast Asia and Oceania.* Auckland: Collins.

Benitez, Silvia. 2001. "Visitor Use Fees and Concession Systems in Protected Areas: Galápagos National Park Case Study." Ecotourism Program Technical Report Series #3. Arlington, VA: The Nature Conservancy.

Bennett, Andrew F. 1998. *Linkages in the Landscape: The Role of Corridors and Connectivity in Wildlife Conservation.* Gland, CH: International Union for Conservation of Nature (IUCN).

Berg, Ina, and Johan Edelheim. 2012. "The Attraction of Islands: Travellers and Tourists in the Cyclades (Greece) in the Twentieth and Twenty-First Centuries." *Journal of Tourism and Cultural Change* 10 (1): 84-98.

Bernal, Richard, and Henry Lowe. 2019. *Medical and Wellness Tourism in Jamaica.* Kingston, JAM: Ian Randle.

Berry, Andrew. 2007. "Evolution on Islands." In *A World of Islands*, edited by Godfrey Baldacchino, 143-174. Charlottetown, PE: Island Studies Press.

Berry, Andrew J. and Rosemary Gillespie. 2018. "Evolution." In *The Routledge International Handbook of Island Studies*, 72-100. London, UK: Routledge.

Bertram, Geoffrey. 2004. "The MIRAB Model in the 21st Century." Paper presented at *Changing Islands - Changing Worlds, Islands of the World, International Small Islands Studies Association International Conference VIII, Taiwan, November 1-7, 2004.*

Bertram, Geoffrey. 2006. "Introduction: The MIRAB Model in the Twenty-First Century." *Asia Pacific Viewpoint* 47 (1): 1-13.

Bertram, Geoffrey. 2011. "Profiling Vulnerability and Resilience: A Manual for Small States." *Island Studies Journal* 6 (1): 99-102.

Bertram, Geoffrey. 2013. "Pacific Island Economies." In *The Pacific Islands: Environment and Society,* rev. ed., edited by Moshe Rapaport, 325-340. Honolulu: University of Hawai'i Press.

Bertram, Geoffrey, and Bernard Poirine. 2018. "Economics and Development." In *The Routledge International Handbook of Island Studies: A World of Islands,* edited by Godfrey Baldacchino, 202-246. London, UK: Routledge.

Bertram, Geoffrey, and Raymond F. Watters. 1985. "The MIRAB Economy in South Pacific Microstates." *Pacific Viewpoint* 26 (3): 497-519.

Bertram, Geoffrey, and Raymond F. Watters. 1986. "The MIRAB Process: Earlier Analyses in Context." *Pacific Viewpoint* 27 (1): 47-59.

Besnier, Niko. 2011. *On the Edge of the Global: Modern Anxieties in a Pacific Island Nation.* Stanford, CA: Stanford University Press.

Betzold, Carola. 2010. "'Borrowing' Power to Influence International Negotiations: AOSIS in the Climate Change Regime, 1990–1997." *Politics* 30 (3): 131-148.

Betzold, Carola. 2015. "Adapting to Climate Change in Small Island Developing States." *Climatic Change* 133 (3): 481-489.

Biglaiser, Glen, and Karl R. DeRouen. 2007. "Following the Flag: Troop Deployment and US Foreign Direct Investment." *International Studies Quarterly* 51 (4): 835-854.

Birkeland, Charles. 2015. "Coral Reefs in the Anthropocene." In *Coral Reefs in the Anthropocene,* edited by Charles Birkeland, 1-15. Dordrecht, NLD: Springer.

Blake, Raymond B. 2015. *Lions or Jellyfish: Newfoundland-Ottawa Relations Since 1957.* Toronto: University of Toronto Press.

Bocco, Gerardo. 2016. "Remoteness and Remote Places" A Geographic Perspective." *Geoforum* 77: 178-181.

Boeckx, Cedric. 2012. *Syntactic Islands.* Cambridge, UK: Cambridge University Press.

Boekstein, Mark. 2014. "From Illness to Wellness: Has Thermal Spring Health Tourism Reached a New Turning Point?" *African Journal of Hospitality, Tourism, and Leisure* 3 (2): 1-11.

Bolles, Lynn A. 1992. "Sand, Sea, and the Forbidden." *Transforming Anthropology* 3 (1): 30-34.

Bonneuil, Christophe, and Jean-Baptiste Fressoz. 2016. *The Shock of the Anthropocene: The Earth, History, and Us.* London, UK: Verso.

Boorstin, Daniel. 1983. *The Discoverers: A History of Man's Search to Know His World and Himself.* New York: Random House.

Bouchard, Christian, and William Crumplin. 2010. "Neglected No Longer: The Indian Ocean at the Forefront of World Geopolitics and Global Geostrategy." *Journal of the Indian Ocean Region* 6 (1): 26-51.

Bramwell, David. 2010. "Island Hot Spots: The Challenge of Climate Change." In *Beyond Cladistics: The Branching of a Paradigm,* edited by David Williams and Sandra Knapp, 91-100. Berkeley: University of California Press.

Bramwell, David. 2011. "Introduction: Islands and Plants." In *The Biology of Island Floras,* edited by David Bramwell and Juli Caujapé-Castells, 1-10. Cambridge, UK: Cambridge University Press.

Braudel, Fernand. 1972. *The Mediterranean and the Mediterranean World in the Age of Philip II: Volume 1.* Berkeley: University of California Press.

Bravo, Karen E. 2011. "Challenges to Caribbean Economic Sovereignty in a Globalizing World." *Michigan State International Law Review* 20 (1): 33-56.

Breathnach, Proinnsias. 1998. "Exploring the 'Celtic Tiger' Phenomenon: Causes and Consequences of Ireland's Economic Miracle." *European Urban and Regional Studies* 5 (4): 305-316.

Brewis, Alexandra A., and Stephen T. McGarvey. 2000. "Body Image, Body Size, and Samoan Ecological and Individual Modernization." *Ecology of Food and Nutrition* 39 (2): 105-120.

Brida, Juan Gabriel, and Sandra Zapata-Aguirre. 2008. "The Impacts of the Cruise Industry on Tourism Destinations." Paper presented at *Congress on Sustainable Tourism as a Factor of Local Development, Monza, Italy, November 7-9, 2008.*

Brida, Juan Gabriel, and Sandra Zapata-Aguirre. 2010. "Cruise Tourism: Economic, Socio-Cultural, and Environmental Impacts." *International Journal of Leisure and Tourism Marketing* 1 (3): 205-226.

Briguglio, Lino. 1995. "Small Island Developing States and Their Economic Vulnerabilities." *World Development* 23 (9): 1615-1632.

Briguglio, Lino. 2014. "A Vulnerability and Resilience Framework for Small States." In *Building the Resilience of Small States: A Revised Framework*, edited by Denny Bynoe-Lewis, 1-10. London, UK: Commonwealth Secretariat.

Briguglio, Lino, and Eliawony J. Kisanga, eds. 2004. *Economic Vulnerability and Resilience of Small States.* Msida, MT: Islands and Small States Institute, L-Università ta' Malta.

Briguglio, Lino, Brian Archer, Jafar Jafari, and Geoffrey Wall, eds. 1996. *Sustainable Tourism in Islands and Small States: Issues and Policies Volume I.* London, UK: Cassell.

Briguglio, Lino, Richard Butler, David Harrison, and Walter Leal Filho, eds. 1996. *Sustainable Tourism in Islands and Small States: Case Studies Volume II.* London, UK: Cassell.

Briguglio, Lino, Gordon Cordina, Nadia Farrugia, and Stephanie Vella. 2009. "Economic Vulnerability and Resilience: Concepts and Measurements." *Oxford Development Studies* 37 (3): 229-247.

Brinklow, Laurie. 2012. *Here for the Music.* Charlottetown, PE: Acorn Press.

Brinklow, Laurie. 2013. "Stepping-stones to the Edge: Artistic Expressions of Islandness in an Ocean of Islands." *Island Studies Journal* 8 (1): 39-54.

Brinklow, Laurie. 2015. "Artists and the Articulation of Islandness, Sense of Place, and Story in Newfoundland and Tasmania." PhD diss., University of Tasmania.

Brinklow, Laurie, Frank Ledwell, and Jane Ledwell, eds. 2000. *Message in a Bottle: The Literature of Small Islands.* Charlottetown, PE: Island Studies Press.

Brislin, Tom. 2003. "Exotics, Erotics, and Coconuts: Stereotypes of Pacific Islanders." In *Images That Injure: Pictorial Stereotypes in the Media*, 2nd ed., edited by Paul Martin Lester and Susan Dente Ross, 103-112. Westport, CT: Praeger.

Brown, Katrina, R. Kerry Turner, Hala Hameed, and Ian Bateman. 1997. "Environmental Carrying Capacity and Tourism Development in the Maldives and Nepal." *Environmental Conservation* 24 (4): 316-325.

Brown, Lawrence A., and Rickie L. Sanders. 1981. "Toward a Development Paradigm of Migration, with Particular Reference to Third World Settings." In *Migration Decision-Making: Multidisciplinary Approaches to Microlevel Studies in Developed and Developing Countries*, edited by Robert W. Gardner and Gordon F. DeJong, 148-185. New York: Pergamon.

Brown, Richard, John Connell, and Eliana V. Jimenez-Soto. 2014. "Migrants' Remittances, Poverty, and Social Protection in the South Pacific: Fiji and Tonga." *Population, Space and Place* 20 (5): 434-454.

Burney, David A. 2009. "Climate Change." In *Encyclopedia of Islands*, edited by Rosemary Gillespie and David Clague, 169-171. Berkeley: University of California Press.

Burney, David A., and Timothy F. Flannery. 2005. "Fifty Millennia of Catastrophic Extinctions After Human Contact." *Trends in Ecology and Evolution* 20 (7): 395-401.

Burns, Peter, and Lyn Bibbings. 2009. "The End of Tourism? Climate Change and Societal Challenges." *Twenty-First Century Society* 4 (1): 31-51.

Burns, William. 2000. *The Possible Impacts of Climate Change on Pacific Island State Ecosystems.* Occasional Paper for the Pacific Institute for Studies in Development, Environment, and Security. Oakland, CA: Pacific Institute. https://www.sprep.org/att/IRC/eCOPIES/Pacific_Region/331.pdf.

Butler, Richard W. 1980. "The Concept of a Tourist Area Cycle of Evolution: Implications for Management of Resources." *Canadian Geographer/Le Géographe Canadien* 24 (1): 5-12.

Butler, Richard W. 1993. "Tourism Development in Small Islands." In *The Development Process in Small Island States*, edited by Douglas G. Lockhart, David W. Drakakis-Smith, and Patrick J. Schembri, 71-91. London, UK: Routledge.

Butler, Richard W. 1999. "Sustainable Tourism: A State-of-the-Art Review." *Tourism Geographies* 1 (1): 7-25.

Butler, Richard W. 2006. "Epilogue: Contrasting Coldwater and Warmwater Island Tourism Destinations." In *Extreme Tourism: Lessons from the World's Cold Water Islands,* edited by Godfrey Baldacchino, 247-257. Oxford, UK: Elsevier.

Butler, Richard W. 2012. "Islandness: It's All in the Mind." *Tourism Recreation Research* 37 (2): 173-176.

Butler, Richard W., ed. 2017. *Tourism and Resilience.* Wallingford, UK: CABI.

Byravan, Sujatha, and Sudhir Chella Rajan. 2010. "The Ethical Implications of Sea-Level Rise Due to Climate Change." *Ethics & International Affairs* 24 (3): 239-260.

Byron, Margaret. 2000. "Return Migration to the Eastern Caribbean: Comparative Experiences and Policy Implications." *Social and Economic Studies* 49 (4): 155-188.

Cabanda, Exequiel. 2017. "Higher Education, Migration and Policy Design of the Philippine Nursing Act of 2002." *Higher Education Policy* 30 (4): 555-575.

Cambridge Dictionary. n.d. "Island." http://dictionary.cambridge.org/dictionary/british/island.

Cameron, Catherine M., and John B. Gatewood. 2008. "Beyond Sun, Sand and Sea: The Emergent Tourism Programme in the Turks and Caicos Islands." *Journal of Heritage Tourism* 3 (1): 55-73.

Campbell, Ian. 1987. *A History of the Pacific Islands.* Christchurch, NZ: University of Canterbury Press.

Campbell, John. 2009. "Islandness: Vulnerability and Resilience in Oceania." *Shima: The International Journal of Research into Island Cultures* 3 (1): 85-97.

Campling, Liam. 2006. "A Critical Political Economy of the Small Island Developing States Concept: South-South Cooperation for Island Citizens?" *Journal of Developing Studies* 22 (3): 235-285.

Cao-Lormeau, Van-Mai. 2016. "Tropical Islands as New Hubs for Emerging Arboviruses." *Emerging Infectious Diseases* 22 (5): 913-915.

Cao-Lormeau, Van-Mai, and Didier Musso. 2014. "Emerging Arboviruses in the Pacific." *The Lancet* 384 (9954): 1571-1572.

Carlquist, Sherwin. 1965. *Island Life: A Natural History of the Islands of the World.* Garden City, NY: Natural History Press.

Carlquist, Sherwin. 1974. *Island Biology.* New York: Columbia University Press.

Carlsen, Jack, and Richard Butler, eds. 2011. *Island Tourism: Sustainable Perspectives.* Wallingford, UK: CABI.

Carmichael, Barbara A. 2000. "A Matrix Model for Resident Attitudes and Behaviours in a Rapidly Changing Tourist Area." *Tourism Management* 21 (6): 601-611.

Carpenter, Kevin. 1984. *Desert Isles and Pirate Islands.* Frankfurt: P. Lang.

Carr, Liam M., and Daniel Y. Liu. 2016. "Measuring Stakeholder Perspectives on Environmental and Community Stability in a Tourism-Dependent Economy." *International Journal of Tourism Research* 18 (6): 620-632.

Case, Ted. 1978. "A General Explanation for Insular Body Size Trends in Terrestrial Vertebrates." *Ecology* 59 (1): 1-18.

Cassels, Susan. 2006. "Overweight in the Pacific: Links Between Foreign Dependence, Global Food Trade, and Obesity in the Federated States of Micronesia." *Globalization and Health* 2 (1): 10.

Cassidy, Frances, and Margee Hume. 2015. "Advancing the Global Perspective of Tourism by Examining Core and Peripheral Destinations." In *Handbook of Research on Global Hospitality and Tourism Management*, edited by Angelo A. Camillo, 37-53. Hershey, PA: IGI Global.

Cave, Jenny, Keith G. Brown, and Godfrey Baldacchino. 2012. "Come Visit, but Don't Overstay: Critiquing a Welcoming Society." *International Journal of Culture, Tourism and Hospitality Research* 6 (2): 145-153.

Censky, Ellen, Karim Hodge, and Judy Dudley. 1998. "Over-Water Dispersal of Lizards Due to Hurricanes." *Nature* 395 (6702): 556.

Centers for Disease Control and Prevention [CDC]. 2016. "Epidemiology." https://www.cdc.gov/careerpaths/k12teacherroadmap/epidemiology.html.

Central Intelligence Agency [CIA]. 2017a. "Country Comparison: Population Growth Rate." The World Factbook. Accessed November 9, 2019. https://www.cia.gov/library/publications/the-world-factbook/rankorder/2002rank.html.

Central Intelligence Agency [CIA]. 2017b. "GDP – Per Capita (PPP)." USA Central Intelligence Agency, World Factbook. Accessed November 16, 2019. https://www.cia.gov/library/Publications/the-world-factbook/rankorder/2004rank.html.

Central Intelligence Agency [CIA]. 2019. "Central America: Cayman Islands." The World Factbook. Accessed November 3, 2019. https://www.cia.gov/library/publications/the-world-factbook/geos/cj.html.

Chan, Nicholas. 2018. "'Large Ocean States': Sovereignty, Small Islands, and Marine Protected Areas in Global Ocean Governance." *Global Governance: A Review of Multilateralism and International Organizations* 24 (4): 537-555.

Chand, Satish, and Ron Duncan. 1997. "Resolving Property Issues as a Precondition for Growth: Access to Land in the Pacific Islands." In *The Governance of Common Property in the Pacific Region*, edited by Peter Larmour, 33-46. Canberra: Australian National University Press.

Chandler, David, and Jonathan Pugh. 2018. "Islands of Relationality and Resilience: The Shifting Stakes of the Anthropocene." *Area* 52: 65-72.

Chappell, David. 1999. "The Postcontact Period." In *The Pacific Islands: Environment & Society*, edited by Moshe Rapaport, 134-143. Honolulu: The Bess Press.

Cheer, Joseph M., and Alan A. Lew. 2017. "Sustainable Tourism Development: Towards Resilience in Tourism." *Interaction* 45 (1): 10-15.

Cheer, Joseph M., and Alan A. Lew, eds. 2017. *Tourism, Resilience, and Sustainability: Adapting to Social, Political and Economic Change*. Oxon, UK: Routledge.

Cheng, Margaret Harris. 2010. "Asia-Pacific Faces Diabetes Challenge." *The Lancet* 375 (9733): 2207-2210.

Chieftains' Live Over Ireland: Water from the Well. 2000. DVD. Directed by Maurice Linnane. London, UK: Eagle Rock Entertainment.

Choy, Dexter J. L. 1992. "Life Cycle Models for Pacific Island Destinations." *Journal of Travel Research* 30 (3): 26-31.

Church, John A., Peter U. Clark, Anny Cazenave, Jonathan M. Gregory, Svetlana Jevrejeva, Anders Levermann, Mark A. Merrifield, Glenn A. Milne, R. Steven Nerem, Patrick D. Nunn, et al. 2013. "Sea Level Change." In *Climate Change 2013: The Physical Science Basis. Contribution of Working Group I to the Fifth Assessment Report of the Intergovernmental Panel on Climate Change*, edited by Thomas F. Stocker, Dahe Quin, Gian-Kasper Plattner, Melinda M.B. Tignor, Simon K. Allen,

Judith Boschung, Alexaner Nauels, Yu Xia, Vincent Bex, and Pauline M. Midgley, 1137-1216. Cambridge, UK: Cambridge University Press.

Citron, Kenneth M., and Jack Pepys. 1964. "An Investigation of Asthma Among the Tristan da Cunha Islanders." *British Journal of Diseases of the Chest* 58 (3): 119-123.

Claridge, Elin. 2009. "Wallace, Alfred Russel." In *Encyclopedia of Islands*, edited by Rosemary G. Gillespie and David A. Clague, 962-967. Berkeley: University of California Press.

Clark, Eric. 2009. "Island Development." In *International Encyclopaedia of Human Geography*, edited by Rob Kitchin and Nigel Thrift, 607-610. Oxford, UK: Elsevier.

Clark, Eric. 2013. "Financialization, Sustainability, and the Right to the Island: A Critique of Acronym Models of Island Development." *Journal of Marine and Island Cultures* 2 (2): 128-136.

Clark, Eric, Karin Johnson, Emma Lundholm, and Gunnar Malmberg. 2007. "Island Gentrification and Space Wars." In *A World of Islands: An Island Studies Reader*, edited by Godfrey Baldacchino, 481-510. Charlottetown, PE: Island Studies Press.

Clarke, Harold D., William Mishler, and Paul Whiteley. 1990. "Recapturing the Falklands: Models of Conservative Popularity, 1979–83." *British Journal of Political Science* 20 (1): 63-81.

Clarke, William C. 1990. "Learning from the Past: Traditional Knowledge and Sustainable Development." *The Contemporary Pacific* 2 (1): 233-253.

Clegg, Peter, and Peter Gold. 2011. "The UK Overseas Territories: A Decade of Progress and Prosperity?" *Commonwealth and Comparative Politics* 49 (1): 115-135.

Clegg, Peter, Justin Daniel, Emilio Pantojas-Garcia, and Wouter Veenendaal. 2017. "The Global Financial Crisis and its Aftermath: Economic and Political Recalibration in the Non-Sovereign Caribbean." *Canadian Journal of Latin American and Caribbean Studies/Revue Canadienne des Études Latino-Américaines et Caraïbes* 42 (1): 84-104.

Cliff, Andrew, and Peter Haggett. 1985. *The Spread of Measles in Fiji and the Pacific: Spatial Components in the Transmission of Epidemic Waves Through Island Communities*. Canberra: Australian National University Press.

Cliff, Andrew, and Peter Haggett. 1995. "The Epidemiological Significance of Islands." *Health & Place* 1 (4): 199-209.

Cliff, Andrew, and Peter Haggett. 2004. "Time, Travel, and Infection." *British Medical Bulletin* 69 (1): 87-99.

Cliff, Andrew, Peter Haggett, and Matthew Smallman-Raynor. 2000. *Island Epidemics*. Oxford, UK: Oxford University Press.

Cliff, Andrew, Peter Haggett, John Keith Ord, and G. R. Versey. 1981. *Spatial Diffusion: An Historical Geography of Epidemics in an Island Community*. Cambridge, UK: Cambridge University Press.

Cobb, Sharon C. 2001. "Globalization in a Small Island Context: Creating and Marketing Competitive Advantage for Offshore Financial Services." *Geografiska Annaler: Series B, Human Geography* 83 (4): 161-174.

Cobham, Alex, and Steven J. Klees. 2016. "Global Taxation: Financing Education and Other Sustainable Development Goals. A Report to the International Commission on Financing Global Education Opportunity." Paper prepared for the International Commission on Financing Global Education Opportunity. New York: The Education Commission. http://www.taxjustice.net/wp-content/uploads/2016/11/Global-Taxation-Financing-Education.pdf.

Cocola-Gant, Agustin. 2018. "Tourism Gentrification." In *Handbook of Gentrification Studies*, edited by Loretta Lees with Martin Phillips, 281-293. Cheltenham, UK: Edward Elgar.

Coffin, Millard. 2009. "Granitic Islands." In *Encyclopedia of Islands*, edited by Rosemary G. Gillespie and David A. Clague, 380-382. Berkeley: University of California Press.

Cohen, Anthony. 1985. *The Symbolic Construction of Community*. New York: Routledge.

Cohen, Philippa, and Simon Foale. 2011. "Fishing Taboos: Securing Pacific Fisheries for the Future." *SPC Traditional Marine Resource Management and Knowledge Information Bulletin* 28: 3-13.

Cole, Stroma. 2007. "Beyond Authenticity and Commodification." *Annals of Tourism Research* 34 (4): 943-960.

Conkling, Philip. 2007. "On Islanders and Islandness." *Geographical Review* 97 (2): 191-201.

Conlin, Michael V., and Tom Baum, eds. 1995. *Island Tourism: Management Principles and Practice.* Chichester, UK: Wiley.

Connell, John. 1991. "Island Microstates: The Mirage of Development." *The Contemporary Pacific* 3 (2): 251-287.

Connell, John. 2003a. "An Ocean of Discontent? Contemporary Migration and Deprivation in the South Pacific." In *Migration in the Asia Pacific: Population, Settlement, and Citizenship Issues,* edited by Robyn Iredale, Charles Hawksley and Stephen Castles, 55-77. Cheltenham, UK: Edward Elgar.

Connell, John. 2003b. "Island Dreaming: The Contemplation of Polynesian Paradise." *Journal of Historical Geography* 29 (4): 554-581.

Connell, John. 2006. "Nauru: The First Failed Pacific State?" *The Round Table: Commonwealth Journal of International Affairs* 95 (363): 47-63.

Connell, John. 2007. "Islands, Idylls, and the Detours of Development." *Singapore Journal of Tropical Geography* 28 (2): 116-135.

Connell, John. 2008. "Niue: Embracing a Culture of Migration." *Journal of Ethnic and Migration Studies* 34 (6): 1021-1040.

Connell, John. 2011a. "Epilogue: Memories and Island Music." In *Island Songs: A Global Repertoire,* edited by Godfrey Baldacchino, 261-280. Plymouth, UK: Scarecrow Press.

Connell, John. 2011b. *Medical Tourism.* Wallingford, UK: CABI.

Connell, John. 2013a. *Islands at Risk? Environments, Economies and Contemporary Change.* Cheltenham, UK: Edward Elgar.

Connell, John. 2013b. "Medical Tourism in the Caribbean Islands: A Cure for Economies in Crisis?" *Island Studies Journal* 8 (1): 115-130.

Connell, John. 2013c. "Soothing Breezes? Island Perspectives on Climate Change and Migration." *Australian Geographer* 44 (4): 465-480.

Connell, John. 2015. "The Pacific Diaspora." In *Migration and Development: Perspectives from Small States,* edited by Wonderful Hope Khonje, 244-264. London, UK: Commonwealth Secretariat.

Connell, John. 2016a. "Greenland and the Pacific Islands: An Improbable Conjunction of Development Trajectories." *Island Studies Journal* 11 (2): 465-484.

Connell, John. 2016b. "Last Days in the Carteret Islands? Climate Change, Livelihoods, and Migration on Coral Atolls." *Asia Pacific Viewpoint* 57 (1): 3-15.

Connell, John. 2018. "Islands: Balancing Development and Sustainability?" *Environmental Conservation* 45 (2): 111-124.

Connell, John. 2019. "Another Pause for Independence? The 2018 New Caledonia Referendum." *The Round Table* 108 (3): 1-18.

Connell, John, and Richard Brown. 2005. *Remittances in the Pacific: An Overview.* Manila: Asian Development Bank. http://www.adb.org/sites/default/files/publication/28799/remittances-pacific.pdf.

Connell, John, and Dennis Conway. 2000. "Migration and Remittances in Island Microstates: A Comparative Perspective on the South Pacific and the Caribbean." *International Journal of Urban and Regional Research* 24 (1): 52-78.

Connell, John, and Chris Gibson. 2004. "World Music: Deterritorializing Place and Identity." *Progress in Human Geography* 28 (3): 342-361.

Connell, John, and Carmen Voigt-Graf. 2006. "Towards Autonomy? Gendered Migration in Pacific Island Countries." In *Migration Happens: Reasons, Effects, and Opportunities of Migration in the South Pacific*, edited by Katarina Ferro and Margot Wallner, 43-62. Wien, DE: LIT Verlag.

Connolly, Lesley. 2013. "The Troubled Road to Peace: Reflections on the Complexities of Resolving the Political Impasse in Madagascar." *African Centre for the Constructive Resolution of Disputes, Policy & Practice Brief* 21. https://www.accord.org.za/publication/troubled-road-peace.

Conroy, John D. 2012. "A Guide to Subsistence Affluence." *SSRN Electronic Journal.* https://dx.doi.org/10.2139/ssrn.2059357.

Constantakopoulou, Christy. 2007. *The Dance of the Islands: Insularity, Networks, the Athenian Empire and the Aegean World.* Oxford, UK: Oxford University Press.

Conway, Dennis, and Robert B. Potter. 2007. "Caribbean Transnational Return Migrants as Agents of Change." *Geography Compass* 1 (1): 25-45.

Cook, David, Nína Saviolidis, Brynhildur Davíðsdóttir, Lára Jóhannsdóttir, and Snjólfur Ólafsson. 2019. "Synergies and Trade-Offs in the Sustainable Development Goals: The Implications of the Icelandic Tourism Sector." *Sustainability* 11 (15): 1-23.

Corbett, Jack. 2015. *Being Political: Leadership and Democracy in the Pacific Islands.* Honolulu: University of Hawai'i Press.

Corbett, Jack, and Wouter Veenendaal. 2016. "Westminster in Small States: Comparing the Caribbean and Pacific Experience." *Contemporary Politics* 22 (4): 432-449.

Corbett, Jack, and Wouter Veenendaal. 2018. *Democracy in Small States: Persisting Against All Odds.* Oxford, UK: Oxford University Press.

Courchamp, Franck, Benjamin D. Hoffmann, James C. Russell, Camille Leclerc, and Céline Bellard. 2014. "Climate Change, Sea-Level Rise, and Conservation: Keeping Island Biodiversity Afloat." *Trends in Ecology & Evolution* 29 (3): 127-130.

Cox, Alan. 2008. *Plate Tectonics: How it Works.* Palo Alto, CA: Blackwell.

Cox, Jeremy. 2009. "The Money Pit: An Analysis of Nauru's Phosphate Mining Policy." *Pacific Economic Bulletin* 24 (1): 174-186.

Craig, Robin K. 2006. "Are Marine National Monuments Better Than National Marine Sanctuaries? U.S. Ocean Policy, Marine Protected Areas, and the Northwest Hawaiian Islands." *Sustainable Development Law & Policy* 7 (1): 27-31.

Crawford, Barbara E. 1987. *Scandinavian Scotland.* Vol. 2 of *Scotland in the Early Middle Ages.* Leicester, UK: Leicester University Press.

Crick, Malcolm. 1989. "Representations of International Tourism in the Social Sciences: Sun, Sex, Sights, Savings, and Servility." *Annual Review of Anthropology* 18 (1): 307-344.

Croes, Robertico, Seung Hyun Lee, and Eric D. Olson. 2013. "Authenticity in Tourism in Small Island Destinations: A Local Perspective." *Journal of Tourism and Cultural Change* 11 (1-2): 1-20.

Crosby, Alfred. 1986. *Ecological Imperialism: The Biological Expansion of Europe, A.D. 900-1900.* Cambridge, UK: Cambridge University Press.

Crossley, Michael, and Terra Sprague. 2014. "Education for Sustainable Development: Implications for Small Island Developing States (SIDS)." *International Journal of Educational Development* 35 (2): 86-95.

Cruise Lines International Association [CLIA]. 2019a. "Cruise Lines International Association (CLIA) Reveals Growth in Global and North American Passenger Numbers and Insights." Press release, April 9, 2019. https://cruising.org/news-and-research/press-room/2019/april/clia-reveals-growth.

Cruise Lines International Association [CLIA]. 2019b. "State of the Cruise Industry Outlook 2020." https://cruising.org/-/media/research-updates/research/state-of-the-cruise-industry.pdf.

D'Arcy, Paul. 1998. "No Empty Ocean: Trade and Interaction Across the Pacific Ocean to the Middle of the Eighteenth Century." In *Studies in the Economic History of the Pacific Rim*, edited by Sally Miller, A.J.H. Latham, and Dennis O. Flynn, 21-44. London, UK: Routledge.

Dagnini, Jérémie. 2010. "The Importance of Reggae Music in the Worldwide Cultural Universe." *Études Caribéennes* 16. https://doi.org/10.4000/etudescaribeennes.4740.

Dalrymple, Kate, Ian Williamson, and Jude Wallace. 2003. "Cadastral Systems Within Australia." *Australia Surveyor* 48 (1): 37-49.

Dalton, Brian J. 1966. "A New Look at the Maori Wars of the Sixties." *Australian Historical Studies* 12 (46): 230-247.

Damgaard, Jannick, Thomas Elkjaer, and Niels Johannesen. 2018. "Piercing the Veil of Tax Havens." *International Monetary Fund: Finance & Development Quarterly* 55 (2): 50-53.

Danson, Mike, and Kathryn Burnett. 2014. "Enterprise and Entrepreneurship on Islands." In *Exploring Rural Enterprise: New Perspectives on Research, Policy & Practice*, edited by Colette Henry and Gerard McElwee, 151-174. Bingley, UK: Emerald Group.

Darwent, David F. 1969. "Growth Poles and Growth Centers in Regional Planning: A Review." *Environment and Planning A*, 1 (1): 5-32.

Davidson, Basil. 1992. "Columbus: The Bones and Blood of Racism." *Race and Class* 33 (3): 17-25.

Davis, James, Jessica Busch, Zoe Hammatt, Rachel Novotny, Rosanne Harrigan, Andrew Grandinetti, and David Easa. 2004. "The Relationship Between Ethnicity and Obesity in Asian and Pacific Islander Populations: A Literature Review." *Ethnicity & Disease* 14 (1): 111-118.

Dawson, Jackie, and Harvey Lemelin. 2015. "Last Chance Tourism: A Race to be Last?" In *The Practice of Sustainable Tourism*, edited by Michael Hughes, David Weaver, and Christof Pfoor, 155-167. London, UK: Routledge.

Day, A. Grove. 1987. *Mad About Islands: Novelists of a Vanished Pacific*. Honolulu: Mutual Publishing.

Daye, Marcella. 2005. "Mediating Tourism: An Analysis of the Caribbean Holiday Experience in the UK National Press." In *The Media and the Tourist Imagination: Converging Cultures*, edited by David Crouch, Rhona Jackson, and Felix Thompson, 14-26. Oxon, UK: Routledge.

de Águeda Corneloup, Inés, and Arthur P. J. Mol. 2014. "Small Island Developing States and International Climate Change Negotiations: The Power of Moral 'Leadership'." *International Environmental Agreements: Politics, Law and Economics* 14 (3): 281-297.

dé Ishtar, Zohl. 2003. "Poisoned Lives, Contaminated Lands: Marshall Islanders Are Paying a High Price for the United States Nuclear Arsenal." *Seattle Journal for Social Justice* 2 (1): 287-307.

Defoe, Daniel. (1719) 1883. *The Life and Strange Surprising Adventures of Robinson Crusoe of York, Mariner, as Related by Himself*. Reprint, London, UK: W. Taylor.

Del Chiappa, Giacomo, and Tindara Abbate. 2016. "Island Cruise Tourism Development: A Resident's Perspective in the Context of Italy." *Current Issues in Tourism* 19 (13): 1372-1385.

DeLoughrey, Elizabeth M. 2007. *Routes and Roots: Navigating Caribbean and Pacific Island Literatures*. Honolulu: University of Hawai'i Press.

DeLoughrey, Elizabeth M. 2013. "The Myth of Isolates: Ecosystem Ecologies in the Nuclear Pacific." *Cultural Geographies* 20 (2): 167-184.

DeLoughrey, Elizabeth M. 2019. *Allegories of the Anthropocene*. Durham, NC: Duke University Press.

Denevan, William, ed. 1992. *The Native Population of the Americas in 1492*, 2nd ed. Madison, WI: University of Wisconsin Press.

Dening, Greg. 2007. "Sea People of the West." *The Geographical Review* 97 (2): 288-301.

Depraetere, Christian, and Arthur Dahl. 2007. "Island Locations and Classifications." In *A World of Islands*, edited by Godfrey Baldacchino, 57-106. Charlottetown, PE: Island Studies Press.

Depraetere, Christian, and Arthur Dahl. 2018. "Locations and Classifications." In *The Routledge International Handbook of Island Studies*, 21-51. London, UK: Routledge.

Déry, Patrick. and Bart Anderson. 2007. "Peak Phosphorus." *Energy Bulletin.* http://www.greb.ca/GREB/Publications_files/Peakphosphorus.pdf.

Deschenes, P. J., and Marian Chertow. 2004. "An Island Approach to Industrial Ecology: Towards Sustainability in the Island Context." *Journal of Environmental Planning and Management* 47 (2): 201-217.

Diamond, Jared. 1975. "The Island Dilemma: Lessons of Modern Biogeographic Studies for the Design of Natural Reserves." *Biological Conservation* 7 (2): 129-146.

Diamond, Jared. 2005. *Collapse: How Societies Choose to Fail or Succeed.* New York: Viking.

Diaz, Vicente M. 2016. "Don't Swallow (or Be Swallowed By) Disney's 'Culturally Authenticated Moana.'" *Indian Country Today.* Washington, DC: Indian Country Media Network. https://newsmaven.io/indiancountrytoday/archive/don-t-swallow-or-be-swallowed-by-disney-s-culturally-authenticated-moana-9NFXz7ZqJEa9h-I3120lrQ.

Dictionary.com. 2020a. "Island." http//www.dictionary.com/browse/island.

Dictionary.com. 2020b. "Insular." https://www.dictionary.com/browse/insular.

Dillinger, Jessica. 2018. "The Most Obese Countries in the World." World Atlas. http://www.worldatlas.com/articles/29-most-obese-countries-in-the-world.html.

Dodds, Klaus. 2013. "Consolidate! Britain, the Falkland Islands and Wider the South Atlantic/Antarctic." *Global Discourse* 3 (1): 166-172.

Dodds, Rachel. 2012. "Sustainable Tourism: A Hope or a Necessity? The Case of Tofino, British Columbia, Canada." *Journal of Sustainable Development* 5 (5): 54-64.

Dodds, Rachel, and Richard Butler. 2019. "The Phenomena of Overtourism: A Review." *International Journal of Tourism Cities* 5 (4): 519-528.

Dodds, Rachel, and Sonya Graci. 2010. *Sustainable Tourism in Island Destinations.* London, UK: Earthscan.

Donne, John. 1959. *Devotions Upon Emergent Occasions: Together with Death's Duel.* Ann Arbor, MI: University of Michigan Press.

Donne, John. 1988. *No Man is an Island.* London, UK: Souvenir.

Doty, Alvah H. 1897. "Quarantine Methods." *The North American Review* 165 (489): 201-212.

Douglas, Ngaire. 1997. "Applying the Life Cycle Model to Melanesia." *Annals of Tourism Research* 24 (1): 1-22.

Doxey, George V. 1975. "A Causation Theory of Visitor-Resident Irritants: Methodology and Research Inferences." In *Travel and Tourism Research Association (TTRA) Sixth Annual Conference Proceedings,* 195-98.

Dubin, Marc. 2009. *The Rough Guide to Cyprus.* London, UK: Penguin.

Duffy, David. 2010. "Changing Seabird Management in Hawai'i: From Exploitation Through Management to Restoration." *Waterbirds* 33 (2): 193-207.

Duncan, Richard P., and Tim M. Blackburn. 2004. "Extinction and Endemism in the New Zealand Avifauna." *Global Ecology and Biogeography* 13 (6): 509-517.

Dunn, Leslie. 2011. "The Impact of Political Dependence on Small Island Jurisdictions." *World Development* 39 (12): 2132-2146.

Edwards, Charlie. 2009. *Resilient Nation.* London, UK: Demos.

Eggenberger, Katrin. 2018. "When is Blacklisting Effective? Stigma, Sanctions, and Legitimacy: The Reputational and Financial Costs of Being Blacklisted." *Review of International Political Economy* 25 (4): 483-504.

Ehrlich, Paul. 1971. *The Population Bomb.* New York: Ballantine.

Elahi, Khandakar Qudrat-I. 2009. "UNDP on Good Governance." *International Journal of Social Economics* 36 (12): 1167-1180.

Ellis, Erle C. 2011. "Anthropogenic Transformation of the Terrestrial Biosphere." *Philosophical Transactions of the Royal Society A* 369 (1938): 1010-1035.

Emmers, Ralf. 2009. *Geopolitics and Maritime Territorial Disputes in East Asia.* London, UK: Routledge.

Encyclopedia of Missions. 1913. "Missionary View of Natives." Mount Holyoke Historical Atlas. https://www.mtholyoke.edu/courses/rschwart/hatlas/mhc_widerworld/hawaii/westernview.html.

Engilis, Andrew Jr., and Maura Naughton. 2004. "U.S. Pacific Islands Regional Shorebird Conservation Plan." Portland, OR: U.S. Department of the Interior, Fish and Wildlife Service. https://www.shorebirdplan.org/wp-content/uploads/2013/01/USPI1.pdf.

Englberger, Lois, Adelino Lorens, Moses E. Pretrick, Bill Raynor, Jim Currie, Allison Corsi, Laura Kaufer, Rupesh I. Naik, Robert Spegal, and Harriet V. Kuhnlein. 2011. "Approaches and Lessons Learned for Promoting Dietary Improvement in Pohnpei, Micronesia." In *Combating Micronutrient Deficiencies: Food-Based Approaches,* edited by Brian Thompson and Leslie Amoroso, 224-253. Wallingford, UK: CABI and Food and Agricultural Organization of the United Nations.

Entwistle, Michael, and Michael J. Oliver. 2015. "Jersey, A Small Island International Finance Center: Adapting to Survive." In *Entrepreneurship on Small Island States and Territories*, edited by Godfrey Baldacchino, 251-267. New York: Routledge.

Eriksen, Thomas Hylland. 1993. "In Which Sense Do Cultural Islands Exist?" *Social Anthropology-Cambridge* 1 (1B): 133-147.

Erlandson, Jon, and Scott Fitzpatrick. 2006. "Oceans, Islands, and Coasts: Current Perspectives on the Role of the Sea in Human Prehistory." *Journal of Island & Coastal Archaeology* 1: 5-32.

Esteves, Paulo, Camila dos Santos, Citlali Ayala, Alexandra Teixeira, and Camila Amorim Jardim. 2019. "South-South Cooperation and the 2030 Agenda: BAPA+40 and Beyond." *Development Cooperation Review* 1 (10-12): 17-22.

Evans, John D. 1973. "Islands as Laboratories for the Study of Culture Process." In *The Explanation of Culture Change: Models in Prehistory*, edited by Colin Renfrew, 517-520. London, UK: Duckworth.

Fagence, Michael. 1999. "Tourism as a Feasible Option for Sustainable Development in Small Island Developing States (SIDS): Nauru as a Case Study." *Pacific Tourism Review* 3 (2): 133-142.

Falkland, Anthony, ed. 1991. *Hydrology and Water Resources of Small Islands: A Practical Guide - A Contribution to the International Hydrological Programme, IHP-III, Project 4.6.* Paris: UNESCO. http://unesdoc.unesco.org/images/0009/000904/090426eo.pdf.

Farbotko, Carol. 2005. "Tuvalu and Climate Change: Constructions of Environmental Displacement in the 'Sydney Morning Herald.'" *Geografiska Annaler Series B: Human Geography* 87 (4): 279-293.

Farbotko, Carol. 2010a. "'The Global Warming Clock is Ticking So See These Places While You Can': Voyeuristic Tourism and Model Environmental Citizens on Tuvalu's Disappearing Islands." *Singapore Journal of Tropical Geography* 31 (2): 224-238.

Farbotko, Carol. 2010b. "Wishful Sinking: Disappearing Islands, Climate Refugees, and Cosmopolitan Experimentation." *Asia Pacific Viewpoint* 51 (1): 47-60.

Farbotko, Carol, and Heather Lazrus. 2012. "The First Climate Refugees? Contesting Global Narratives of Climate Change in Tuvalu." *Global Environmental Change* 22 (2): 382-390.

Farnsworth, Kevin, and Gary J. Fooks. 2015. "Corporate Taxation, Corporate Power, and Corporate Harm." *The Howard Journal* 54 (1): 25-41.

Feld, Steven. 1996. "Waterfalls of Place: An Acoustemology of Place Re-sounding in Bosavi, Papua New Guinea." In *Senses of Place*, edited by Steven Feld and Keith Basso, 91-135. Santa Fe, NM: School of American Research Press.

Ferdinand, Malcom. 2018. "Subnational Climate Justice for the French Outre-mer: Postcolonial Politics and Geography of an Epistemic Shift." *Island Studies Journal* 13 (1): 119-134.

Ferdinand, Malcom, Gert Oostindie, and Wouter Veenendaal. 2019. "A Global Comparison of Non-Sovereign Island Territories: The Search for 'True Equality.'" *Island Studies Journal,* e-pub ahead of print. https://www.islandstudies.ca/sites/default/files/ISJFerdinandetalComparisonNonSovereign-IslandTerritories.pdf.

Ferguson, James. 2008. *A Traveller's History of the Caribbean,* 2nd ed. Northampton, MA: Interlink.

Fernández-Palacios, Jose Maria. 2009. "Island Biogeography, Theory of." In *Encyclopedia of Islands,* edited by Rosemary G. Gillespie and David A. Clague, 486-490. Berkeley: University of California Press.

Ferreira, Vera. 2018. "Climate-Induced Migrations: Legal Challenges." In *Intergenerational Responsibility in the 21st Century,* edited by Julia M. Puaschunder, 107-122. Wilmington, Delaware: Vernon.

Fichtner, Jan. 2016. "The Anatomy of the Cayman Islands Offshore Financial Center: Anglo-America, Japan, and the Role of Hedge Funds." *Review of International Political Economy* 23 (6): 1-30.

Finucane, Melissa L., and Victoria W. Keener. 2015. "Understanding the Climate-Sensitive Decisions and Information Needs of Island Communities." *Journal of the Indian Ocean Region* 11 (1): 110-120.

Fiorello, Amélie, and Damien Bo. 2012. "Community-Based Ecotourism to Meet the New Tourist's Expectations: An Exploratory Study." *Journal of Hospitality Marketing & Management* 21 (7): 758-778.

Firat, Aytekin F. 1995. "Consumer Culture, or Culture Consumed?" In *Marketing in a Multicultural World: Ethnicity, Nationalism, and Cultural Identity,* edited by Janeen A. Costa and Gary J. Bamossy, 105-125. Thousand Oaks, CA: Sage.

Firdous, Naila, Stephen Gibbons, and Bernadette Modell. 2011. "Falling Prevalence of Beta-Thalassaemia and Eradication of Malaria in the Maldives." *Journal of Community Genetics* 2 (3): 173-189.

Fischer, Steven. 2002. *A History of the Pacific Islands.* London, UK: Palgrave Macmillan.

Fischer, Steven. 2005. *Island at the End of the World: The Turbulent History of Easter Island.* London, UK: Reaktion.

Fischer, Steven. 2012. *Islands: From Atlantis to Zanzibar.* London, UK: Reaktion.

Fisher, Denise. 2011. "France in the South Pacific: An Australian Perspective." In *French History and Civilization: Papers from the George Rudé Seminar,* edited by Briony Neilson and Robert Aldrich, 237-254. https://h-france.net/rude/wp-content/uploads/2017/08/FisherVol4.pdf.

Fisher, Denise. 2012. "France: 'In,' 'Of,' or 'From' the South Pacific Region?" *Journal de la Société des Océanistes* 135 (2): 185-200.

Fisher, Graham, Joey Britto, and Emel Thomas. 2015. "British Overseas Territories: Gibraltar, the Caribbean Territories, the Overseas Territory of St. Helena, Ascension, and Tristan da Cuhna." In *Education in the United Kingdom,* edited by Colin Brock, 341-369. London, UK: Bloomsbury.

Fisk, E. K. 1982. "Subsistence Affluence and Development Policy." *Regional Development Dialogue* Special Issue: 1-12.

Fitzpatrick, Scott. 2007. "Archaeology's Contribution to Island Studies." *Island Studies Journal* 2 (1): 77-100.

Flenley, John, and Paul Bahn. 2007. "Conflicting Views of Easter Island." *Rapa Nui Journal: Journal of the Easter Island Foundation* 21 (1): 11-13.

Fletcher, Lisa. 2011. "'…Some Distance To Go': A Critical Survey of Island Studies." *New Literatures Review* 47-48: 17-34.

Fleury, Christian, and Henry Johnson. 2017. "Tourism and Resilience on Jersey: Culture, Environment, and Sea." In *Tourism, Resilience, and Sustainability: Adapting to Social, Political, and Economic Change,* edited by Joseph M. Cheer and Alan A. Lew, 85-102. Oxon, UK: Routledge.

Flint, Elizabeth. 2009. "Midway." In *Encyclopedia of Islands,* edited by Rosemary G. Gillespie and David A. Clague, 631-633. Berkeley: University of California Press.

Foale, Simon, Philippa Cohen, Stephanie Januchowski-Hartley, Amelia Wenger, and Martha Macintyre. 2011. "Tenure and Taboos: Origins and Implications for Fisheries in the Pacific." *Fish and Fisheries* 12 (4): 357-369.

Foster, J. Bristol. 1964. "Evolution of Mammals on Islands." *Nature* 202: 234-235.

Found, William C. 2004. "Historic Sites, Material Culture, and Tourism in the Caribbean Islands." In *Tourism in the Caribbean: Trends, Development, Prospects*, edited by David Timothy Duval, 152-167. London, UK: Routledge.

Fowles, John. 1965. *The Magus*. New York: Little, Brown and Co.

Frankel, Jeffrey. 2016. "Mauritius: African Success Story." In *African Successes, Volume IV: Sustainable Growth*, edited by Sebastian Edwards, Simon Johnson, and David Weil, 295-342. Chicago: University of Chicago Press.

Freeman, Derek. 1999. *The Fateful Hoaxing of Margaret Mead: A Historical Analysis of Her Samoan Research*. Boulder, CO: Westview Press.

Frendo, Henry. 1993. "The Legacy of Colonialism: The Experience of Malta and Cyprus." In *The Development Process in Small Island States*, edited by Douglas G. Lockhart, David W. Drakakis-Smith, and Patrick J. Schembri, 151-160. London, UK: Routledge.

Fridriksson, Sturla. 2009. "Surtsey." In *Encyclopedia of Islands*, edited by Rosemary G. Gillespie and David A. Clague, 883-888. Berkeley: University of California Press.

Fuchs, Christian. 2014. *OccupyMedia!: The Occupy Movement and Social Media in Crisis Capitalism*. Winchester, UK: Zero Books.

Fusco, Leah. 2007. "Offshore Oil: An Overview of Development in Newfoundland and Labrador." Occasional Paper of the 'Oil, Power, and Dependency: Global and Local Realities of the Offshore Oil Industry in Newfoundland and Labrador' Project, directed by Peter Sinclair and Sean Cadigan. St. John's, NL: Memorial University of Newfoundland. http://www.ucs.mun.ca/~oilpower/documents/NL%20oil%207-25-1.pdf.

Gaffin, Stuart R. 1997. *High Water Blues: Impacts of Sea Level Rise on Selected Coasts and Islands*. New York: Environmental Defense Fund.

Gaggero, Jorge, and Magdalena Belén Rua. 2015. "The Role of Global Banks: Financial Asset Management by 'Private Banking'." Centro de Economía y Finanzas para el Desarrollo de la Argentina (CEFID-AR). https://www.taxjustice.net/wp-content/uploads/2013/04/CEFID-AR-role-of-global-banks-2015.pdf.

Gailey, R. Alan. 1959. "Settlement and Population in the Aran Islands." *Irish Geography* 4: 65-78.

Galanello, Renzo, and Raffaella Origa. 2010. "Beta-Thalassemia." *Orphanet Journal of Rare Diseases* 5 (11).

Gale, Maradel K. 2013. "Self-Sufficiency and Sustainability." University of Oregon. Accessed January 16, 2020. https://darkwing.uoregon.edu/~mspp/sustainability.htm.

Gale, Stephen J. 2019. "Lies and Misdemeanours: Nauru, Phosphate and Global Geopolitics." *The Extractive Industries and Society* 6 (3): 737-746.

Gani, Azmat. 2009a. "Governance and Foreign Aid in Pacific Island Countries." *Journal of International Development: The Journal of the Development Studies Association* 21 (1): 112-125.

Gani, Azmat. 2009b. "Some Aspects of Communicable and Non-Communicable Diseases in Pacific Island Countries." *Social Indicators Research* 91 (2): 171-187.

Gardiner, Kevin. 1994. "The Irish Economy: A Celtic Tiger." In *Ireland: Challenging for Promotion – Morgan Stanley Euroletter* August 31: 9-21.

Gascon, Claude, Thomas M. Brooks, Topiltzin Contreras-MacBeath, Nicolas Heard, William Konstant, John Lamoreux, Frederic Launay, Michael Maunder, Russell A. Mittermeier, Sanjay Molur, et al. 2015. "The Importance and Benefits of Species." *Current Biology* 25 (10): R431-R438.

Gaspar, David Barry. 1991. "Antigua Slaves and Their Struggle to Survive." In *Seeds of Change: A Quincentennial Commemoration,* edited by Herman J. Viola and Carolyn Margolis, 130-138. London, UK: Smithsonian.

Gatewood, John B., and Catherine M. Cameron. 2009. "Belonger Perceptions of Tourism and its Impacts in the Turks and Caicos Islands." Final report to the Turks and Caicos Islands Ministry of Tourism. https://www.lehigh.edu/~jbg1/Perceptions-of-Tourism.pdf.

Gikas, Petros, and George Tchobanoglous. 2009. "Sustainable Use of Water in the Aegean Islands." *Journal of Environmental Management* 90 (8): 2601-2611.

Gil-Alana, Luis A., and Edward H. Huijbens. 2018. "Tourism in Iceland: Persistence and Seasonality." *Annals of Tourism Research* 68: 20-29.

Gillespie, Rosemary G. 2007. "Oceanic Islands: Models of Diversity." In *Encyclopedia of Biodiversity,* edited by Simon A. Levin, 1-13. Amsterdam: Elsevier.

Gillespie, Rosemary G., and David A. Clague, eds. 2009. *Encyclopedia of Islands.* Berkeley: University of California Press.

Gillespie, Stuart, and Lawrence Haddad. 2001. *Attacking the Double Burden of Malnutrition in Asia and the Pacific.* Manila: Asian Development Bank.

Gillis, John. 2003. "Islands in the Making of an Atlantic Oceania, 1400-1800." Paper presented at *Seascapes, Littoral Cultures, and Transpacific Exchanges Conference, Library of Congress, Washington, DC, February 12-15, 2003.* http://webdoc.sub.gwdg.de/ebook/p/2005/history_cooperative/www.historycooperative.org/proceedings/seascapes/gillis.html.

Gillis, John. 2007. "Island Sojourns." *The Geographical Review* 97 (2): 274-287.

Gillis, John, and David Lowenthal. 2007. "Introduction." *The Geographical Review* 97 (2): iii-iv.

Girvan, Norman. 2014. "Extractive Imperialism in Holistic Perspective." In *Extractive Imperialism in the Americas,* edited by James Petras and Henry Veltmeyer, 49-61. Leiden, NLD: Brill.

Gjetnes, Marius. 2001. "The Spratlys: Are They Rocks or Islands?" *Ocean Development & International Law* 32 (2): 191-204.

Golding, William. 1954. *Lord of the Flies.* London, UK: Faber and Faber.

Goldsmith, Edward, Robert Allen, Michael Allaby, John Davull, and Sam Lawrence. 1972. *A Blueprint for Survival.* Boston, MA: Houghton Mifflin.

Goldsmith, Rosie. October 14, 2013. "Iceland: Where One in 10 People Will Publish a Book." *BBC News.* Accessed March 9, 2020. http://www.bbc.com/news/magazine-24399599.

González-Salzberg, Damián, and Loveday Hodson. 2019. "A Policy of Quiet Disregard: The Chagos Islands, Islanders, and International Law." *Journal of Brazilian Human Rights* 19: 107-124.

Goodwin, Harold. 2015. "Tourism, Good Intentions, and the Road to Hell: Ecotourism and Volunteering." *Brown Journal of World Affairs* 22 (1): 37-50.

Gosling, Anna, Hallie Buckley, Elizabeth Matisoo-Smith, and Tony Merriman. 2015. "Pacific Populations, Metabolic Disease, and 'Just-So Stories': A Critique of the 'Thrifty Genotype' Hypothesis in Oceania." *Annals of Human Genetics* 79 (6): 470-480.

Gössling, Stefan. 2010. *Carbon Management in Tourism: Mitigating the Impacts on Climate Change.* London, UK: Routledge.

Gössling, Stefan, and Geoffrey Wall. 2007. "Island Tourism." In *A World of Islands: An Island Studies Reader,* edited by Godfrey Baldacchino, 429-453. Charlottetown, PE: Island Studies Press.

Gössling, Stefan, C. Michael Hall, and Daniel Scott. 2015. *Tourism and Water.* Bristol, UK: Channel View.

Gössling, Stefan, Daniel Scott, C. Michael Hall, Jean-Paul Ceron, and Ghislain Dubois. 2012. "Consumer Behaviour and Demand Response of Tourists to Climate Change." *Annals of Tourism Research* 39 (1): 36-58.

Gough, Katherine V., Tim Bayliss-Smith, John Connell, and Ole Mertz. 2010. "Small Island Sustainability in the Pacific: Introduction to the Special Issue." *Singapore Journal of Tropical Geography* 31 (1): 1-9.

Graham, Katherine, and Evelyn Peters. 2002. "Aboriginal Communities and Urban Sustainability." Discussion Paper F27. Family Network. Ottawa: Canadian Policy Research Networks Inc. http://www.urbancenter.utoronto.ca/pdfs/elibrary/CPRNUrban.pdf.

Green, Charles. 2001. *Manufacturing Powerlessness in the Black Diaspora: Inner-City Youth and the New Global Frontier.* Walnut Creek, CA: Altamira.

Greenhill, Basil, and Ann Giffard. 1967. *West Countrymen in Prince Edward's Isle.* Toronto: University of Toronto Press.

Gren, Martin, and Edward H. Huijbens. 2015. *Tourism and the Anthropocene.* Oxon, UK: Routledge.

Grigg, Russell. 2010. "Darwin and the Feugians." *Creation*, pre-publication version. https://creation.com/darwin-and-the-fuegians.

Grydehøj, Adam. 2011. "Making the Most of Smallness: Economic Policy in Microstates and Sub-National Island Jurisdictions." *Space and Polity* 15 (3): 183-196.

Grydehøj, Adam. 2014. "Guest Editorial Introduction: Understanding Island Cities." *Island Studies Journal* 9 (2): 183-190.

Grydehøj, Adam. 2015. "Island City Formation and Urban Island Studies." *Area* 47 (4): 429-435.

Grydehøj, Adam. 2017. "A Future of Island Studies." *Island Studies Journal* 12 (1): 3-16.

Grydehøj, Adam, and Marco Casagrande. 2020. "Islands of Connectivity: Archipelago Relationality and Transport Infrastructure in Venice Lagoon." *Area* 52 (1): 56-64.

Grydehøj, Adam, and Ilan Kelman. 2017. "The Eco-Island Trap: Climate Change Mitigation and Conspicuous Sustainability." *Area* 49 (1): 106-113.

Grydehøj, Adam, Xavier Barceló Pinya, Gordon Cooke, Naciye Doratli, Ahmed Elewa, Ilan Kelman, Jonathan Pugh, Lea Schick, and R. Swaminathan. 2015. "Returning From the Horizon: Introducing Urban Island Studies." *Urban Island Studies* 1 (1): 1-19.

Guilcher, Andre. 1988. *Coral Reef Geomorphology.* Chichester, UK: John Wiley & Sons.

Gunstra, Diane. 1985. "The Island Pattern." *Children's Literature Association Quarterly* 10 (2): 55-57.

Gurtner, Yetta. 2016. "Returning to Paradise: Investigating Issues of Tourism Crisis and Disaster Recovery on the Island of Bali." *Journal of Hospitality and Tourism Management* 28 (1): 11-19.

Gylfason, Thorvaldur, and Gylfi Zoega. 2018. "The Dutch Disease in Reverse: Iceland's Natural Experiment." In *Getting Globalization Right: Sustainability and Inclusive Growth in a Post Brexit Age*, edited by Luigi Paganetto, 13-36. Cham, CH: Springer.

Habel, Jan C., Livia Rasche, Uwe A. Schneider, Jan O. Engler, Erwin Schmid, Dennis Rödder, Sebastian T. Meyer, Natalie Trapp, Ruth Sos del Diego, Hilde Eggermont, et al. 2019. "Final Countdown for Biodiversity Hotspots." *Conservation Letters* 12 (6): e12668.

Haggett, Peter. 1994. "Geographical Aspects of the Emergence of Infectious Diseases." *Geografiska Annaler: Series B, Human Geography* 76 (2): 91-104.

Haggett, Peter. 2000. *The Geographical Structure of Epidemics.* Oxford, UK: Oxford University Press.

Hale, Brack W. 2018. "Mapping Potential Environmental Impacts from Tourists Using Data from Social Media: A Case Study in the Westfjords of Iceland." *Environmental Management* 62 (3): 446-457.

Hall, Adrian, and Allen Fraser. 2004. "Shetland Landscapes." Accessed March 5, 2020. http://www.landforms.eu/shetland.

Hall, C. Michael. 2001. "Trends in Ocean and Coastal Tourism: The End of the Last Frontier?" *Ocean & Coastal Management* 44 (9-10): 601-618.

Hall, C. Michael. 2006. "Introduction." In *The Tourism Area Life Cycle, Volume 2: Conceptual and Theoretical Issues*, edited by Richard W. Butler, xv-xix. Clevedon, UK: Channel View.

Hall, C. Michael. 2010. "Island Destinations: A Natural Laboratory for Tourism – Introduction." *Asia Pacific Journal of Tourism Research* 15 (3): 245-249.

Hall, C. Michael. 2012. "Island, Islandness, Vulnerability, and Resilience." *Tourism Recreation Research* 37 (2): 177-181.

Hall, C. Michael. 2015. "Islands of Sustainability or Analogues of the Challenge of Sustainable Development?" In *Routledge International Handbook of Sustainable Development*, edited by Michael Redclift and Delyse Springett, 55-73. London, UK: Routledge.

Hall, C. Michael, Stefan Gössling, and Daniel Scott. 2015. "Introduction." In *The Routledge Handbook of Tourism and Sustainability*, edited by C. Michael Hall, Stefan Gössling, and Daniel Scott, 1-12. Oxon, UK: Routledge.

Hamacher, Duane W., and Carla Bento Guedes. 2017. "How Far They'll Go: 'Moana' Shows the Power of Polynesian Celestial Navigation." The Conversation. https://theconversation.com/how-far-theyll-go-moana-shows-the-power-of-polynesian-celestial-navigation-72375.

Hamilton, Warren. 1988. "Plate Tectonics and Island Arcs." *Geological Society of America Bulletin* 100 (10): 1503-1527.

Hampton, Mark P. 1994. "Treasure Islands or Fool's Gold: Can and Should Small Island Economies Copy Jersey?" *World Development* 22 (2): 237-250.

Hampton, Mark P. 1996. "Sixties Child? The Emergence of Jersey as an Offshore Finance Centre 1955–71." *Accounting, Business & Financial History* 6 (1): 51-71.

Hampton, Mark P., and John Christensen. 2007. "Competing Industries in Islands: A New Tourism Approach." *Annals of Tourism Research* 34 (4): 998-1020.

Hampton, Mark P., and Michael Levi. 1999. "Fast Spinning into Oblivion? Recent Developments in Money-Laundering Policies and Offshore Finance Centres." *Third World Quarterly* 20 (3): 645-656.

Happy Planet Index. 2016. "Explore the Data." Happy Planet Index. Accessed November 10, 2019. http://www.happyplanetindex.org/countries.

Hardin, Jessica, Amy K. McLennan, and Alexandra Brewis. 2018. "Body Size, Body Norms, and Some Unintended Consequences of Obesity Intervention in the Pacific Islands." *Annals of Human Biology* 45 (3): 285-294.

Harling Stalker, Lynda, and John G. Phyne. 2014. "The Social Impact of Out-Migration: A Case Study from Rural and Small Town Nova Scotia, Canada." *Journal of Rural and Community Development* 9 (3): 203-226.

Harrison, David. 2001. "Islands, Image, and Tourism." *Tourism Recreation Research* 26 (3): 9-14.

Harrison, David. 2004. "Tourism in Pacific Islands." *The Journal of Pacific Studies* 26 (1): 1-28.

Hauʻofa, Epeli. 1993. "Our Sea of Islands." In *A New Oceania: Rediscovering Our Sea of Islands*, edited by Eric Waddell, Vijay Naidu, and Epeli Hauʻofa, 2-16. Suva, Fiji: University of the South Pacific.

Hauʻofa, Epeli. 1994. "Our Sea of Islands." *The Contemporary Pacific* 6 (1): 148-161.

Hauʻofa, Epeli. 1998. "The Ocean in Us." *The Contemporary Pacific* 10 (2): 391-410.

Hauser, Mike. 2016. "Island (Haiku)." Hello Poetry. https://hellopoetry.com/poem/1573605/island-haiku.

Hawley, Nicola L., and Stephen T. McGarvey. 2015. "Obesity and Diabetes in Pacific Islanders: The Current Burden and the Need for Urgent Action." *Current Diabetes Reports* 15 (5): 29.

Hay, Pete. 2002. "A Tale of Two Islands." *Island* 89 (Autumn): 12-25.

Hay, Pete. 2003a. "The Poetics of Island Place: Articulating Particularity." *Local Environment* 8 (5): 553-558.

Hay, Pete. 2003b. "That Islanders Speak, and Others Hear…" *Interdisciplinary Studies in Literature and Environment* 10 (2): 203-205.

Hay, Pete. 2006. "A Phenomenology of Islands." *Island Studies Journal* 1 (1): 19-42.

Hay, Pete. 2013. "What the Sea Portends: A Reconsideration of Contested Island Tropes." *Island Studies Journal* 8 (2): 209-232.

Hayfield, Erika Anne, and Mariah Schug. 2019. "'It's Like They Have a Cognitive Map of Relations': Feeling Strange in a Small Island Community." *Journal of Intercultural Studies* 40 (4): 383-398.

Healy, Robert G. 1994. "Tourist Merchandise as a Means of Generating Local Benefits from Ecotourism." *Journal of Sustainable Tourism* 2 (3): 137-151.

Helgadóttir, Guðrún, Anna Vilborg Einarsdóttir, Georgette Leah Burns, Guðrún Þóra Gunnarsdóttir, and Jóhanna María Elena Matthíasdóttir. 2019. "Social Sustainability of Tourism in Iceland: A Qualitative Inquiry." *Scandinavian Journal of Hospitality and Tourism*: 1-18.

Henry, James S. 2012. "The Price of Offshore Revisited." *Tax Justice Network* 22: 57-168. http://taxjustice.nonprofitsoapbox.com/storage/documents/The_Price_of_Offshore_Revisited_-_22-07-2012.pdf.

Heo, Uk, and Min Ye. 2017. "U.S. Military Deployment and Host-Nation Economic Growth." *Armed Forces & Society* 45 (2): 1-34.

Hepburn, Eve. 2012. "Recrafting Sovereignty: Lessons from Small Island Autonomies?" In *Political Autonomy and Divided Societies: Imagining Democratic Alternatives in Complex Settings*, edited by Alain-G. Gagnon and Michael Keating, 118-133. London, UK: Palgrave Macmillan.

Herman, Doug. 2016. "How the Story of 'Moana' and Maui Holds Up Against Cultural Truths." Smithsonian Magazine. https://www.smithsonianmag.com/smithsonian-institution/how-story-moana-and-maui-holds-against-cultural-truths-180961258.

Hermann, Elfriede, and Wolfgang Kempf. 2017. "Climate Change and the Imagining of Migration: Emerging Discourses on Kiribati's Land Purchase in Fiji." *The Contemporary Pacific* 29 (2): 231-263.

Herr, Richard. 2006. "The Geopolitics of Pacific Islands' Regionalism: From Strategic Denial to the Pacific Plan." *Fijian Studies: A Journal of Contemporary Fiji* 4 (2): 111-125.

Herr, Richard, and Donald Potter. 2006. "Nauru in the Arc of Instability: Too Many Degrees of Freedom?" In *Australia's Arc of Instability: The Political and Cultural Dynamics of Regional Security*, edited by Dennis Rumley, Vivian Louis Forbes, and Christopher Griffin, 199-214. Dordrecht, NLD: Springer.

Heyerdahl, Thor. 1950. *The Kon Tiki Expedition: By Raft Across the South Seas.* London, UK: Allen & Unwin.

Heyerdahl, Thor. 1961. *Reports of the Norwegian Archaeological Expedition to Easter Island and the East Pacific, Volume 1: The Archaeology of Easter Island.* London, UK: Allen & Unwin.

Hill, Roland Paul. 1995. "Blackfellas and Whitefellas: Aboriginal Land Rights, the Mabo Decision, and the Meaning of Land." *Human Rights Quarterly* 17 (2): 303-322

Hines, James R. Jr. 2010. "Treasure Islands." *Journal of Economic Perspectives* 24 (4): 103-126.

Hirano, Asao, Leonard T. Kurland, Robert S. Krooth, and Simmons Lessell. 1961. "Parkinsonism-Dementia Complex, an Endemic Disease on the Island of Guam." *Brain*, 84 (4): 642-661.

Hirsch, Eric. 2015. "'It Won't Be Any Good to Have Democracy If We Don't Have a Country': Climate Change and the Politics of Synecdoche in the Maldives." *Global Environmental Change* 35: 190-198.

Hiwasaki, Lisa, Emmanuel Luna, and Rajib Shaw. 2014. "Process for Integrating Local and Indigenous Knowledge with Science for Hydro-Meteorological Disaster Risk Reduction and Climate Change Adaptation in Coastal and Small Island Communities." *International Journal of Disaster Risk Reduction* 10 (A): 15-27.

Hodge, Allison M., Gary K. Dowse, Paul Z. Zimmet, and Veronica R. Collins. 1995. "Prevalence and Secular Trends in Obesity in Pacific and Indian Ocean Island Populations." *Obesity Research* 3 (S2): 77s-88s.

Hodgetts, Darrin, Neil Drew, Christopher Sonn, Ottilie Stolte, Linda Waimarie Nikora, and Cate Curtis. 2010. *Social Psychology and Everyday Life*. London, UK: Macmillan International Higher Education.

Hoegh-Guldberg, Ove. 1999. "Climate Change, Coral Bleaching, and the Future of the World's Coral Reefs." *Marine and Freshwater Research* 50 (8): 839-866.

Hoegh-Guldberg, Ove. 2011. "Coral Reef Ecosystems and Anthropogenic Climate Change." *Regional Environmental Change* 11 (1): 215-227.

Hogenstijn, Maarten, and Daniel van Middelkoop. 2005. "Saint Helena: Citizenship and Spatial Identities on a Remote Island." *Tijdschrift voor Economische en Sociale Geografie* 96 (1): 96-104

Holm, Bill. 2000. *Eccentric Islands: Travels Real and Imaginary*. Minneapolis: Milkweed.

Holton, Graham E. L. 2004. "Heyerdahl's Kon Tiki Theory and the Denial of the Indigenous Past." *Anthropological Forum* 14 (2):163-181.

Hooli, Lauri Johannes. 2017. "From Warrior to Beach-Boy: Resilience of Maasai in Zanzibar Tourism Business." In *Tourism, Resilience and Sustainability: Adapting to Social, Political, and Economic Change*, edited by Joseph M. Cheer and Alan A. Lew, 103-116. Oxon, UK: Routledge.

Hopley, David. 1994. "Worlds Apart." In *Islands*, edited by Robert Stevenson and Frank Talbot, 14-23. New York: Rodale.

Hopley, David, ed. 2011. *Encyclopedia of Modern Coral Reefs: Structure, Form, and Process*. Dordrecht, NLD: Springer.

Hovgaard, Gestur. 2002. *Coping Strategies and Regional Policies – Social Capital in the Nordic Peripheries: Country Report Faroe Islands*. Stockholm: Nordic Centre for Spatial Development.

Howard, Michael. 1991. *Mining, Politics, and Development in the South Pacific*. Boulder, CO: Westview.

Howe, Kerry. 2000. *Nature, Culture, and History: The 'Knowing' of Oceania*. Honolulu: University of Hawai'i Press.

Howe, Kerry. 2003. *The Quest for Origins: Who First Discovered and Settled the Pacific Islands?* Honolulu: University of Hawai'i Press.

Howe, Kerry, Robert Kiste, and Brij Lal, eds. 1994. *Tides of History: The Pacific Islands in the Early Twentieth Century*. Honolulu: University of Hawaii Press.

Hudson, Ray. 2006. "Regions and Place: Music, Identity, and Place." *Progress in Human Geography* 30 (5): 626-634.

Huebner, Anna. 2011. "Tourism and the (Un)Expected: A Research Note." *Pacific News* 36: 25-28.

Hughes, Emma, and Regina Scheyvens. 2016. "Corporate Social Responsibility in Tourism Post-2015: A 'Development First' Approach." *Tourism Geographies* 18 (5): 469-482.

Hughes, Helen. 1964. "The Political Economy of Nauru." *Economic Record* 40 (92): 508-534.

Hughes, Robert G., and Mark Lawrence. 2005. "Globalisation, Food, and Health in Pacific Island Countries." *Asia Pacific Journal of Clinical Nutrition* 14 (4): 298-305.

Hughes, Robert G., and Geoffrey Marks. 2009. "Against the Tide of Change: Diet and Health in the Pacific Islands." *Journal of the American Dietetic Association* 109 (10): 1701-1703.

Huijbens, Edward H. 2011. "Nation-Branding: A Critical Evaluation. Assessing the Image Building of Iceland." In *Iceland and Images of the North,* edited by Sumarlidi R. Isleifsson with Daniel Chartier, 553-582. Québec City: Presses de l'Université du Québec.

Hunt, Bob, and Amanda C. J. Vincent. 2006. "Scale and Sustainability of Marine Bioprospecting for Pharmaceuticals." *AMBIO: A Journal of the Human Environment* 35 (2): 57-64.

Hunt, Terry L. 2006. "Rethinking the Fall of Easter Island." *American Scientist* 94: 412-419.

Hunt, Terry L. 2007. "Rethinking Easter Island's Ecological Catastrophe." *Journal of Archaeological Science* 34: 485-502.

Hunt, Terry L., and Carl P. Lipo. 2009. "Revisiting Rapa Nui (Easter Island) 'Ecocide'." *Pacific Science* 63 (4): 601-616.

Icelandic Tourist Board. 2019. "Numbers of Foreign Visitors." Accessed December 26, 2019. https://www.ferdamalastofa.is/en/recearch-and-statistics/numbers-of-foreign-visitors.

Imada, Adria L. 2004. "Hawaiians on Tour: Hula Circuits Through the American Empire." *American Quarterly* 56 (1): 111-149.

Ingersoll, Kathleen B., Daniel W. Ingersoll, and Andrew Bove. 2017. "Healing a Culture's Reputation: Challenging the Cultural Labeling and Libeling of the Rapanui." In *Cultural and Environmental Change on Rapa Nui*, edited by Sonia Haoa Cardinali, Kathleen B. Ingersoll, Daniel W. Ingersoll Jr., and Christopher M. Stevenson, 188-202. London, UK: Routledge.

Intergovernmental Panel on Climate Change. 2000. "IPCC Special Report Emissions Scenarios: Summary for Policymakers." United Nations Environment Programme and World Meteorological Organization. https://www.ipcc.ch/report/emissions-scenarios.

International Organization for Migration. 2014. "IOM Outlook on Migration, Environment and Climate Change." Geneva, CH: United Nations. https://publications.iom.int/system/files/pdf/mecc_outlook.pdf.

Ioannides, Dimitri. 1992. "Tourism Development Agents: The Cypriot Resort Cycle." *Annals of Tourism Research* 19 (4): 711-731.

Irwin, Geoffrey. 1989. "Against, Across, and Down the Wind: A Case for the Systematic Exploration of the Remote Pacific Islands." *Journal of the Polynesian Society* 98 (2): 167-206.

Iso-Ahola, Seppo. 1982. "Toward a Social Psychological Theory of Tourism Motivation: A Rejoinder." *Annals of Tourism Research* 9 (2): 256-262.

Jacobsen, Marc. 2019. "Greenland's Arctic Advantage: Articulations, Acts, and Appearances of Sovereignty Games." *Cooperation and Conflict* (Online First). https://doi.org/10.1177/0010836719882476.

Jahan, Selim. 2015. "Human Development Report 2015: Work for Human Development." New York: United Nations Development Program. http://hdr.undp.org/sites/default/files/2015_human_development_report.pdf.

Jędrusik, Maciej. 2011. "Island Studies. Island Geography. But What is an Island?" *Miscellanea Geographica – Regional Studies on Development* 15: 201-212.

Jóhannesson, Gunnar Thór. 2015. "A Fish Called Tourism: Emergent Realities of Tourism Policy in Iceland." In *Tourism Encounters and Controversies: Ontological Politics of Tourism Development*, edited by Gunnar Thór Jóhannesson, Carina Ren, and René van der Duim, 181-200. Surrey, UK: Ashgate.

Jóhannesson, Gunnar Thór, Edward Hákon Huijbens, and Richard Sharpley. 2010. "Icelandic Tourism: Past Directions – Future Challenges." *Tourism Geographies* 12 (2): 278-301.

Johnson, Kate, Sandy Kerr, and Jonathan Side. 2013. "Marine Renewables and Coastal Communities: Experiences from the Offshore Oil Industry in the 1970s and Their Relevance to Marine Renewables in the 2010s." *Marine Policy* 38: 491-499.

Johnston, Charles Samuel. 2001. "Shoring the Foundations of the Destination Life Cycle Model, Part 2: A Case Study of Kona, Hawai'i Island." *Tourism Geographies* 3 (2): 135-164.

Johnston, Rory, Krystyna Adams, Lisa Bishop, Valorie A. Crooks, and Jeremy Snyder. 2015. "'Best Care on Home Ground' versus 'Elitist Healthcare': Concerns and Competing Expectations for Medical Tourism Development in Barbados." *International Journal for Equity in Health* 14 (15): 1-12.

Jolly, Margaret. 1996. "Desire, Difference, and Disease: Sexual and Venereal Exchanges on Cook's Voyages in the Pacific." In *Exchanges: Cross-Cultural Encounters in Australia and the Pacific*, edited by Ross Gibson, 187-217. Sydney, AUS: Museum of Sydney and Historical Houses Trust of New South Wales.

Jolly, Margaret. 2007. "Imagining Oceania: Indigenous and Foreign Representations of a Sea of Islands." *The Contemporary Pacific* 19 (2): 508-545.

Jones, Owen A., and Robert Endean. 1977. *Biology and Geology of Coral Reefs Volume 4: Geology 2.* London, UK: Academic Press.

Jones, Peter, David Hillier, and Daphne Comfort. 2016. "The Environmental, Social, and Economic Impacts of Cruising and Corporate Sustainability Strategies." *Athens Journal of Tourism* 3 (4): 273-286.

Jordan, Evan J., and Christine A. Vogt. 2017. "Residents' Perceptions of Stress Related to Cruise Tourism Development." *Tourism Planning & Development* 14 (4): 527-547.

Josling, Timothy, and Timothy Taylor, eds. 2003. *Banana Wars: The Anatomy of a Trade Dispute.* Cambridge, MA: CABI.

Júlíusdóttir, Magnfríður, Unnur Dís Skaptadóttir, and Anna Karlsdóttir. 2013. "Gendered Migration in Turbulent Times in Iceland." *Norsk Geografisk Tidsskrift/Norwegian Journal of Geography* 67 (5): 266-275.

Jumeau, Ronny. 2013. "Small Island Developing States, Large Ocean States." Expert Group Meeting on Oceans, Seas and Sustainable Development: Implementation and Follow-Up to Rio+20. https://sustainabledevelopment.un.org/content/documents/1772Ambassador%20Jumeau_EGM%20Oceans%20FINAL.pdf

Kaeppler, Adrienne. 1978. "Polynesian Music, Captain Cook, and the Romantic Movement in Europe." *Music Educators Journal* 65 (3): 54-60.

Kahn, Miriam. 1996. "Your Place and Mine: Sharing Emotional Landscapes in Wamira, Papua New Guinea." *Senses of Place*, edited by Steven Feld and Keith Basso, 167-196. Santa Fe, NM: School of American Research Press.

Kahn, Miriam. 2000. "Tahiti Intertwined: Ancestral Land, Tourist Postcard, and Nuclear Test Site." *American Anthropologist* 102 (1): 7-26.

Kak, Vivek. 2007. "Infections in confined spaces: cruise ships, military barracks, and college dormitories." *Infectious disease clinics of North America* 21 (3): 773-784.

Kakazu, Hiroshi. 1994. *Sustainable Development of Small Island Economies.* Boulder, CO: Westview.

Källgård, Anders. 2005. "Fact Sheet: The Islands of Sweden." *Geografiska Annaler: Series B, Human Geography* 87 (4): 295-298.

Kapmeier, Florian, and Paulo Gonçalves. 2018. "Wasted Paradise? Policies for Small Island States to Manage Tourism-Driven Growth While Controlling Waste Generation: The Case of the Maldives." *System Dynamics Review* 34 (1-2): 172-221.

Karagiannis, Ioannis C., and Petros G. Soldatos. 2007. "Current Status of Water Desalination in the Aegean Islands." *Desalination* 203 (1-3): 56-61.

Karlsdóttir, Unnur B. 2013. "Nature Worth Seeing! The Tourist Gaze as a Factor in Shaping Views on Nature in Iceland." *Tourist Studies* 13 (2): 139-155.

Kattan, Gustavo H., and Humberto Alvarez- López. 1996. "Preservation and Management of Biodiversity in Fragmented Landscapes in the Colombian Andes." In *Forest Patches in Tropical Landscapes*, edited by John Schelhas and Russell S. Greenberg, 3-18. Washington, DC: Island Press.

Kaufmann, Daniel, Aart Kraay, and Massimo Mastruzzi. 2003. "Governance Matters III: Governance Indicators for 1996, 1998, 2000, and 2002." *The World Bank Economic Review* 18 (2): 253-287.

Kelman, Ilan. 2010a. "Foreword." In *Sustainable Tourism in Island Destinations*, edited by Rachel Dodds and Sonya Graci, xiii-xv. London, UK: Earthscan.

Kelman, Ilan. 2010b. "Hearing Local Voices From Small Island Developing States For Climate Change." *Local Environment* 15 (7): 605-619.

Kelman, Ilan. 2013. "No Change from Climate Change: Vulnerability and Small Island Developing States." *The Geographical Journal* 180 (2): 120-129.

Kelman, Ilan. 2014. "Climate Change and Other Catastrophes: Lessons from Island Vulnerability and Resilience." *Moving Worlds: A Journal for Transcultural Writings* 14 (2): 127-140.

Kelman, Ilan. 2015. "Difficult Decisions: Migration from Small Island Developing States Under Climate Change." *Earth's Future* 3 (4): 133-142.

Kelman, Ilan. 2016. "Small Island Developing States and Climate Change Adaptation." In *Climate Adaptation Governance in Cities and Regions: Theoretical Fundamental and Practical Evidence*, edited by Jörg G.F. Knieling, 355-369. Chichester, UK: Wiley Blackwell.

Kelman, Ilan. 2018. "Islands of Vulnerability and Resilience: Manufactured Stereotypes?" *Area* 50 (1): 1-8.

Kelman, Ilan, and Shabana Khan. 2013. "Progressive Climate Change and Disasters: Island Perspectives." *Natural Hazards* 69 (1): 1131-1136.

Kelman, Ilan, and James E. Randall. 2017. "Island Resilience and Sustainability." In *The Routledge International Handbook of Island Studies*, edited by Godfrey Baldacchino, 353-367. London, UK: Routledge.

Kelman, Ilan, and Jennifer West. 2009. "Climate Change and Small Island Developing States: A Critical Review." *Ecological and Environmental Anthropology* 5 (1): 1-16.

Kelman, Ilan, Tom R. Burns, and Nora Machado des Johansson. 2015. "Islander Innovation: A Research and Action Agenda on Local Responses to Global Issues." *Journal of Marine and Island Cultures* 4 (1): 34-41.

Kelman, Ilan, Robert Stojanov, Shabana Khan, Oscar Alvarez Gila, Barbora Duží, and Dmytro Vikhrov. 2015. "Viewpoint Paper - Islander Mobilities: Any Change from Climate Change?" *International Journal of Global Warming* 8 (4): 584-602.

Kench, Paul S., D. Thompson, Murray R. Ford, Hiroki Ogawa, and Roger F. McLean. 2015. "Coral Islands Defy Sea-Level Rise Over the Past Century: Records from a Central Pacific Atoll." *Geology* 43 (6): 515-518.

Kendall, David. 2009. "Doomed Island: Nauru's Short-Sightedness and Resulting Decline Are an Urgent Warning to the Rest of the Planet." *Alternatives Journal* 35 (1): 34-38.

Kennedy, Liam, Paul S. Ell, E. M. Crawford, and L. A. Clarkson. 1999. *Mapping the Great Irish Famine*. London, UK: Four Courts.

Keown, Michelle. 2007. *Pacific Islands Writing: The Postcolonial Literatures of Aotearoa/New Zealand and Oceania*. Oxford, UK: Oxford University Press.

Kerr, Sandy. 2005. "What is Small Island Sustainable Development About?" *Ocean & Coastal Management* 48: 503-524.

Kerr, Sandy, Kate Johnson, and Stephanie Weir. 2017. "Understanding Community Benefit Payments from Renewable Energy Development." *Energy Policy* 105 (June): 202-211.

Kier, Gerold, Holger Kreft, Tien Ming Lee, Walter Jetz, Pierre L. Ibisch, Christopher Nowicki, Jens Mutke, and Wilhelm Barthlott. 2009. "A Global Assessment of Endemism and Species Richness Across Island and Mainland Regions." *Proceedings of the National Academy of Sciences* 106 (23): 9322-9327.

Kille, Mary. 2011. *Proving Flight*. Hobart: Forty Degrees South.

King, Brian. 1997. *Creating Island Resorts*. New York: Routledge.

King, Russell. 1993. "The Geographical Fascination of Islands." In *The Development Process in Small Island States*, edited by Douglas G. Lockhart, David W. Drakakis-Smith, and Patrick J. Schembri, 13-37. London, UK: Routledge.

King, Russell. 2009. "Geography, Islands, and Migration in an Era of Global Mobility." *Island Studies Journal* 4 (1): 53-84.

King, Russell, and John Connell, eds. 1999. *Small Worlds, Global Lives*. London, UK: Pinter.

Kirch, Patrick. 1984. *The Evolution of the Polynesian Chiefdoms*. Cambridge, UK: Cambridge University Press.

Kirch, Patrick. 2007. "Hawaii as a Model System for Human Ecodynamics." *American Anthropologist* 109 (1): 8-26.

Kirch, Patrick, and Jean-Louis Rallu, eds. 2007. *Growth and Collapse of Pacific Island Societies: Archaeological and Demographic Perspectives.* Honolulu: University of Hawai'i Press.

Kirtsoglou, Elisabeth, and Dimitrios Theodossopoulos. 2004. "'They are Taking Our Culture Away': Tourism and Culture Commodification in the Garifuna Community of Roatan." *Critique of Anthropology* 24 (2): 135-157.

Kjellgren, Eric. 2001. *Splendid Isolation: Art of Easter Island.* New York: Metropolitan Museum of Art.

Kjellgren, Eric. 2007. *Oceania: Art of the Pacific Islands in the Metropolitan Museum of Art.* New York: Metropolitan Museum of Art.

Klein, Ross. 2018. "Dreams and Realities: A Critical Look at the Cruise Ship Industry." In *Tourists and Tourism: A Reader,* 3rd ed., edited by Sharon Gmelch and Adam Kaul, 247-258. Long Grove, IL: Waveland.

Koechlin, Valerie, and Gianmarco Leon. 2007. "International Remittances and Income Inequality: An Empirical Investigation." *Journal of Economic Policy Reform* 10 (2): 123-141.

Korson, Cadey. 2018. "(Re)Balancing Inequality Through Citizenship, Voter Eligibility, and Islandian Sovereignty in Kanaky/New Caledonia." *Geopolitics*, published online. https://doi.org/10.1080/146 50045.2018.1543270.

Kostas, Triantis. 2011. "Symposium Summary: Island Biogeography." *Frontiers of Biogeography* 3 (1): 21-22.

Kouremenos, Anna, and Laura Dierksmeier. 2019. "Teaching Insularity: Archaeological and Historical Perspectives." *Academia.com,* self-published online. https://www.academia.edu/40722094/_Forthcoming_Teaching_Insularity_Archaeological_and_Historical_Perspectives.

Kovacevic, Milorad. 2010. "Review of HDI Critiques and Potential Improvements." United Nations Development Program, Human Development Research Paper 2010/33.

Krippendorf, Jost. 1987. "Ecological Approach to Tourism Marketing." *Tourism Management* 8 (2): 174-176.

Ksano, Kazuhiko. 2009. "Ephemeral Islands, Geology." In *Encyclopedia of Islands*, edited by Rosemary G. Gillespie and David A. Clague, 259-260. Berkeley: University of California Press.

Kueffer, Christoph and Kealohanuiopuna Kinney. 2017. "What is the Importance of Islands to Environmental Conservation?" *Environmental Conservation* 44 (4): 311-322.

Kundur, Suresh Kumar, and Krishna Murthy. 2013. "Environmental Impacts of Tourism and Management in Maldives." *International Journal of Environmental Sciences* 2 (1): 44-50.

Kusumah, Galih, and Ghoitsa Rohmah Nurazizah. 2016. "Tourism Destination Development Model: A Revisit to Butler's Area Life Cycle." In *Heritage, Culture and Society: Research Agenda and Best Practices in the Hospitality and Tourism Industry,* edited by Salleh Mohd Radzi, Mohd Hafiz Mohd Hanafiah, Norzuwana Sumarjan, and Zurinawati Mohi, 31-36. Boca Raton, FL: CRC Press.

Lagiewski, Richard M. 2006. "The Application of the TALC Model: A Literature Survey." In *The Tourism Area Life Cycle, Volume 1: Applications and Modification,* edited by Richard W. Butler, 27-50. Trowbridge, UK: Cromwell.

Lam, Michelle. 1998. "Consideration of Customary Marine Tenure System in the Establishment of Marine Protected Areas in the South Pacific." *Ocean & Coastal Management* 39 (1-2): 97-104.

Larjosto, Vilja. 2018. "Islands of the Anthropocene [Special Section]." *Area*: 1-9. https://doi.org/10.1111/area.12515.

Larmour, Peter. 2005. "Corruption and Accountability in the Pacific Islands." Policy and Governance Discussion Paper 05-10. ANU Crawford School of Public Policy. Accessed November 29, 2019. https://openresearch-repository.anu.edu.au/handle/10440/1159.

Larzelere, Alex. 1988. *Castro's Ploy-America's Dilemma: The 1980 Cuban Boatlift.* Washington, DC: National Defense University.

Lasky, Kathryn. 2012. *Surtsey: The Newest Place on Earth.* Great Neck, NY: Seymour Science.

Lasserre, Frédéric. 2011. "The Geopolitics of Arctic Passages and Continental Shelves." Public Sector Digest. https://publicsectordigest.com/article/geopolitics-arctic-passages-and-continental-shelves.

Lauer, Matthew, Simon Albert, Shankar Aswani, Benjamin S. Halpern, Luke Campanella, and Douglas La Rose. 2013. "Globalization, Pacific Islands, and the Paradox of Resilience." *Global Environmental Change* 23 (1): 40-50.

Lawrence, David Herbert. 1928. *The Woman Who Rode Away, and Other Stories.* London, UK: Penguin Classics.

Lawson, Stephanie. 2013. "'Melanesia': The History and Politics of an Idea." *Journal of Pacific History* 48 (1): 1-22.

Lawton, Laura Jane, and Richard Butler. 1987. "Cruise Ship Industry – Patterns in the Caribbean 1880–1986." *Tourism Management* 8 (4): 329-343.

Lazrus, Heather. 2012. "Sea Change: Island Communities and Climate Change." *Annual Review of Anthropology* 41: 285-301.

Le Juez, Brigitte, and Olga Springer. 2015. "Introduction: Shipwrecks and Islands as Multilayered, Timeless Metaphors of Human Existence". In *Introduction: Shipwrecks and Islands as Multilayered, Timeless Metaphors of Human Existence*, edited by Brigitte Le Juez and Olga Springer Leiden, 1-13. Leiden, NLD: Brill.

Ledwell, Frank. 2002. *The North Shore of Home.* Charlottetown, PE: Acorn Press.

Ledwell, Jane. 2005. *Last Tomato.* Charlottetown, PE: Acorn Press.

Lee, Su-Hsin, Wen-Hua Huang, and Adam Grydehøj. 2017. "Relational Geography of a Border Island: Local Development and Compensatory Destruction on Lieyu, Taiwan." *Island Studies Journal* 12 (2): 97-112.

Lemelin, Harvey, Jackie Dawson, Emma J. Stewart, Pat Maher, and Michael Lueck. 2010. "Last-Chance Tourism: The Boom, Doom, and Gloom of Visiting Vanishing Destinations." *Current Issues in Tourism* 13 (5): 477-493.

Lenzen, Manfred, Ya-Yen Sun, Futu Faturay, Yuan-Peng Ting, Arne Geschke, and Arunima Malik. 2018. "The Carbon Footprint of Global Tourism." *Nature Climate Change* 8: 522-528.

Leonard, Dympna, Robyn McDermott, Kerin O'Dea, Kevin G. Rowley, Poi Pensio, Edna Sambo, Aletia Twist, Raima Toolis, Simone Lowson, and James D. Best. 2002. "Measuring Prevalence: Obesity, Diabetes and Associated Cardiovascular Risk Factors Among Torres Strait Islander People." *Australian and New Zealand Journal of Public Health* 26 (2): 144-149.

Leslie, Helen, and Gerard Prinsen. 2018. "French Territories in the Forum: Trojan Horse or Paddles for the Pacific Canoe?" *Asia Pacific Viewpoint* 59 (3): 384-390.

Levantis, Theo. 2010. "Is Tourism the Key to Pacific Prosperity?" Devpolicy Blog, December 15, 2010. https://devpolicy.org/is-tourism-the-key-to-pacific-prosperity20101215.

Levison, Michael, R. Gerard Ward, and John W. Webb. 1972. "The Settlement of Polynesia: A Report on a Computer Simulation." *Archaeology and Physical Anthropology in Oceania* 7 (3): 234-245.

Levison, Michael, R. Gerard Ward, and John W. Webb. 1973. *The Settlement of Polynesia: A Computer Simulation.* Canberra: ANU Press.

Lew, Alan A., Pin T. Ng, Chin-Cheng Ni, and Tsung-Chiung Wu. 2016. "Community Sustainability and Resilience: Similarities, Differences, and Indicators." *Tourism Geographies* 18 (1): 18-27.

Lewis, David. 1994. *We, the Navigators: The Ancient Art of Landfinding in the Pacific,* 2nd ed. Honolulu: University of Hawaii Press.

Lewis, James. 1999. *Development in Disaster-Prone Places: Studies of Vulnerability.* London, UK: Intermediate Technology.

Lewis, Nancy Davis, and Moshe Rapaport. 1995. "In a Sea of Change: Health Transitions in the Pacific." *Health & Place* 1 (4): 211-226.

Lewis, Simon L., and Mark A. Maslin. 2015. "Defining the Anthropocene." *Nature* 519 (7542): 171-180.

Lind, Niels C. 1992. "Some Thoughts on the Human Development Index." *Social Indicators Research* 27: 89-101.

Lindstrom, Lamont. 2007. "A Body of Postcards from Vanuatu." In *Embodying Modernity and Postmodernity: Ritual, Praxis, and Social Change in Melanesia,* edited by Sandra Bamford, 257-282. Durham, NC: Carolina Academic Press.

Lipo, Carl P., Robert J. DiNapoli, and Terry L. Hunt. 2018. "Commentary: Rain, Sun, Soil, and Sweat: A Consideration of Population Limits on Rapa Nui (Easter Island) before European Contact." *Frontiers in Ecology and Evolution* 6: 25.

Lockhart, Douglas G. 1997. "Islands and Tourism: An Overview." In *Island Tourism: Trends and Perspectives*, edited by Douglas G. Lockhart and David Drakakis-Smith, 3-21. London, UK: Pinter.

Lockhart, Douglas G., and David W. Drakakis-Smith, eds. 1997. *Island Tourism: Trends and Prospects.* London, UK: Pinter.

Lorenzo, Fely Marilyn E., Jaime Galvez-Tan, Kriselle Icamina, and Lara Javier. 2007. "Nurse Migration from a Source Country." *Health Services Research* 42 (3 Pt 2): 1406-1418.

Lowe, M. Kimberly. 2004. "The Status of Inshore Fisheries Ecosystems in the Main Hawaiian Islands at the Dawn of the Millennium: Cultural Impacts, Fisheries Trends and Management Challenges." In *Status of Hawaii's Coastal Fisheries in the New Millennium*, edited by Alan Friedlander, 12-107. Honolulu: American Fisheries Society, Hawai'i Chapter.

Loxley, Diana. 1990. *Problematic Shores: The Literature of Islands.* London, UK: Macmillan Press.

Lund, Katrín Anna, Kristín Loftsdóttir, and Michael Leonard. 2017. "More Than a Stopover: Analysing the Postcolonial Image of Iceland as a Gateway Destination." *Tourist Studies* 17 (2): 144-163.

MacArthur, Robert H., and Edward O. Wilson. 1967. *The Theory of Island Biogeography.* Princeton, NJ: Princeton University Press.

MacCannell, Dean. 1973. "Staged Authenticity: Arrangements of Social Space in Tourist Settings." *American Journal of Sociology* 79 (3): 589-603.

MacCarthy, Michelle. 2012. "'Before it Gets Spoiled by Tourists': Constructing Authenticity in the Trobriand Islands of Papua New Guinea." PhD thesis, University of Auckland. https://researchspace.auckland.ac.nz/bitstream/handle/2292/19623/whole.pdf.

MacDonald, Edward. 2011. "A Landscape… With Figures: Tourism and Environment on Prince Edward Island." *Acadiensis* 40 (1): 70-85.

MacDonald, Edward, and Alan MacEachern. 2016. "Rites of Passage: Tourism and the Crossing to Prince Edward Island." *Histoire Sociale/Social History* 49 (99): 289-306.

MacDonald, Fraser. 2006. "The Last Outpost of Empire: Rockall and the Cold War." *Journal of Historical Geography* 32: 627-647.

MacKinder, Halford. 1942. *Democratic Ideals and Reality.* Washington, DC: National Defense University Press.

Maclellan, Nic. 2005. "The Nuclear Age in the Pacific Islands." *The Contemporary Pacific* 17 (2): 363-372.

Maclellan, Nic. 2019. "Nuclear Testing and Racism in the Pacific Islands." In *The Palgrave Handbook of Ethnicity*, edited by Steven Ratuva, 885-905. Singapore: Palgrave Macmillan.

MacLeod, Alistair. 2000. *Island: The Collected Short Stories of Alistair MacLeod.* Toronto: McClelland & Stewart Ltd.

MacLeod, Nicola. 2006. "Cultural Tourism: Aspects of Authenticity and Commodification." In *Cultural Tourism in a Changing World: Politics, Participation, and (Re)Presentation*, edited by Melanie Smith and Mike Robinson, 177-190. Clevedon, UK: Channel View.

MacNeill, Timothy, and David Wozniak. 2018. "The Economic, Social, and Environmental Impacts of Cruise Tourism." *Tourism Management* 66 (June): 387-404.

MacPherson, Cluny. 1997. "The Polynesian Diaspora: New Communities and New Questions." In *Contemporary Migration in Oceania: Diaspora and Network (JCAS Symposium Series No. 3)*, edited by Ken'ichi Sudo and Shuji Yoshida, 77-100. Osaka: Japan Center for Area Studies.

Magnusson, Magnus, and Hermann Pálsson, eds. 1965. *The Vinland Sagas: The Norse Discovery of America.* London, UK: Penguin Classics.

Mainwaring, Cetta. 2014. "Small States and Nonmaterial Power: Creating Crises and Shaping Migration Policies in Malta, Cyprus, and the European Union." *Journal of Immigrant & Refugee Studies* 12 (2): 103-122.

Malinowski, Bronislaw. 1922. *Argonauts of the Western Pacific.* London, UK: Routledge.

Malm, Thomas. 2001. "The Tragedy of the Commoners: The Decline of the Customary Marine Tenure System of Tonga." *SPC Traditional Marine Resource Management and Knowledge Information Bulletin* 13: 3-13.

Malm, Thomas. 2007. "No Island is an 'Island': Some Perspectives on Human Ecology and Development in Oceania." In *The World System and the Earth System: Global Socioenvironmental Change and Sustainability Since the Neolithic Account*, edited by Carole L. Crumley and Alf Hornborg, 268-279. London, UK: Routledge.

Malthus, Thomas. (1798) 1986. *An Essay on the Principle of Population.* Reprint, London, UK: Pickering. Citations refer to the Pickering edition.

Mantz, Jeffrey W. 2003. "Caribbean." *Encyclopedia of Food and Culture.* Accessed on June 9, 2019. https://www.encyclopedia.com/sports-and-everyday-life/food-and-drink/food-and-cooking/caribbean#3403400114.

Marjavaara, Roger. 2007. "Route to Destruction? Second Home Tourism in Small Island Communities." *Island Studies Journal* 2 (1): 27-46.

Maron, Nicole, and John Connell. 2008. "Back to Nukunuku: Employment, Identity and Return Migration in Tonga." *Asia Pacific Viewpoint* 49 (2): 168-184.

Marsh, Laura K., Colin A. Chapman, Marilyn A. Norconk, Stephen F. Ferrari, Kellen A. Gilbert, Julio Cesar Bicca-Marques, and Janette Wallis. 2003. "Fragmentation: Specter of the Future or the Spirit of Conservation?" In *Primates in Fragments*, edited by Laura K. Marsh, 381-398. Boston, MA: Springer.

Marshall, Joan. 1999a. "Bitter Harvest: For the People of Grand Manan, Government-Approved Exploitation of Rockweed is Yet Another Outside Threat to Local Ecosystems and Economic Sustainability." *Alternatives Journal* 25 (4): 10-15.

Marshall, Joan. 1999b. "Insiders and Outsiders: The Role of Insularity, Migration, and Modernity on Grand Manan, New Brunswick." In *Small Worlds, Global Lives: Islands and Migration*, edited by Russell King and John Connell, 95-113. London, UK: Pinter.

Marshall, Joan. 2001. "Connectivity and Restructuring: Identity and Gender Relations in a Fishing Community." *Gender, Place, and Culture: A Journal of Feminist Geography* 8 (4): 391-409.

Marshall, Joan. 2008. *Tides of Change on Grand Manan Island: Culture and Belonging in a Fishing Community.* Montréal: McGill-Queens University Press.

Marsters, Evelyn, Nick Lewis, and Wardlow Friesen. 2006. "Pacific Flows: The Fluidity of Remittances in the Cook Islands." *Asia Pacific Viewpoint* 47 (1): 31-44.

Massey, Doreen B. 2005. *For Space.* London, UK: Sage.

Massey, Douglas S. 1990. "The Social and Economic Origins of Immigration." *The Annals of the American Academy of Political and Social Science* 510: 60-72.

Massey, Douglas S., Jaoquín Arango, Graeme Hugo, Ali Kouaouci, Adela Pellegrino, and J. Edward Taylor. 1993. "Theories of International Migration: A Review and Appraisal." *Population and Development Review* 19 (3): 431-466.

Mathison, Ymitri. 2016. "Maps, Pirates, and Treasure: The Commodification of Imperialism in Nineteenth-Century Boys' Adventure Fiction." In *The Nineteenth-Century Child and Consumer Culture*, edited by Dennis Denisoff, 185-198. London, UK: Routledge.

Mazer, Harry. 1981. *The Island Keeper.* New York: Delacorte.

McCartan, Paul. 2010. "Asia-Pacific: Are Pacific Islanders Eating Themselves to Death?" *Alternative Law Journal* 35 (4): 237-238.

McCusker, Maeve, and Anthony Soares, eds. 2011. *Islanded Identities: Constructions of Postcolonial Cultural Insularity.* Amsterdam & New York: Rodopi.

McDaniel, Carl N., and John M. Gowdy. 2000. *Paradise for Sale: A Parable of Nature.* Berkeley: University of California Press.

McElroy, Jerome L. 2003. "Tourism Development in Small Islands Across the World." *Geografiska Annaler: Series B, Human Geography* 85 (4): 231-242.

McElroy, Jerome L. 2006. "Small Island Tourist Economies Across the Life Cycle." *Asia Pacific Viewpoint* 47 (1): 61-77.

McElroy, Jerome L., and Klaus de Albuquerque. 1998. "Tourism Penetration Index in Small Caribbean Islands." *Annals of Tourism Research* 25 (1): 145-168.

McElroy, Jerome L., and Klaus de Albuquerque. 1999. "Measuring Tourism Penetration in Small Islands." *Pacific Tourism Review* 3 (2): 161-169.

McElroy, Jerome L., and Perri E. Hamma. 2010. "SITEs Revisited: Socioeconomic and Demographic Contours of Small Island Tourist Economies." *Asia Pacific Viewpoint* 51 (1): 36-46.

McElroy, Jerome L., and Mary Mahoney. 2000. "The Propensity for Political Dependence in Island Microstates." *INSULA – The International Journal of Island Affairs* 9 (1): 32-35.

McElroy, Jerome L., and Courtney E. Parry. 2010. "The Characteristics of Small Island Tourist Economies." *Tourism and Hospitality Research* 10 (4): 315-328.

McElroy, Jerome L., and Kara Pearce. 2006. "The Advantages of Political Affiliation: Dependent and Independent Small-Island Profiles." *The Round Table* 95 (386): 529-539.

McGarvey, Stephen T., James R. Bindon, Douglas E. Crews, and Diana E. Schendel. 1989. "Modernization and Adiposity: Causes and Consequences." In *Human Population Biology: A Transdisciplinary Science,* edited by Michael Little and Jere Haas, 263-279. New York: Oxford University Press USA.

McGillivray, Mark. 1991. "The Human Development Index: Yet Another Redundant Composite Development Indicator?" *World Development* 19 (10): 1461-1468.

McGinnity, Frances, and Mérove Gijsberts. 2018. "The Experience of Discrimination Among Newly Arrived Poles in Ireland and the Netherlands." *Ethnic and Racial Studies* 41 (5): 919-937.

McGlone, Matt S. 1989. "The Polynesian Settlement of New Zealand in Relation to Environmental and Biotic Changes." *New Zealand Journal of Ecology* 12: 115-129.

McLean, Sheldon, and Ava Jordon. 2017. "An Assessment of the Challenges to Caribbean Offshore Financial Centres: St. Kitts and Nevis." United Nations Economic Commission for Latin America and the Caribbean. Port of Spain, TT: United Nations. https://www.cepal.org/en/publications/42726-assessment-challenges-caribbean-offshore-financial-centres-saint-kitts-and-nevis.

McLennan, Amy K., and Stanley J. Ulijaszek. 2014. "Obesity Emergence in the Pacific Islands: Why Understanding Colonial History and Social Change is Important." *Public Health Nutrition* 18 (8): 1499-1505.

McLeod, Kari S. 2000. "Our Sense of Snow: The Myth of John Snow in Medical Geography." *Social Science & Medicine* 50 (7-8): 923-935.

McMahon, Elizabeth. 2010. "Australia, the Island Continent: How Contradictory Geography Shapes the National Imaginary." *Space and Culture* 13 (2): 178-187.

McMahon, Elizabeth, and Suvendrini Perera. 2009. *Australia and the Insular Imagination: Beaches, Borders, Boats, and Bodies.* New York: Palgrave Macmillan.

McMillen, Heather L., Tamara Ticktin, Alan Friedlander, Stacy D. Jupiter, Randolph Thaman, John Campbell, Joeli Veitayaki, Thomas Giambelluca, Salesa Nihmei, Etika Rupeni, et al. 2014. "Small Islands, Valuable Insights: Systems of Customary Resource Use and Resilience to Climate Change in the Pacific." *Ecology and Society* 19 (4): 44.

McNamara, Karen. 2015. "Cross-Border Migration with Dignity in Kiribati." *Forced Migration Review* 49: 62

McQuade, Walter. 1975. "The Smallest Richest Republic in the World." *Fortune* 92 (6): 132-140.

McSorley, Kevin, and Jerome L. McElroy. 2007. "Small Island Economic Strategies: Aid-Remittance Versus Tourism-Dependence." *E-Review of Tourism Research* 5 (6): 140-148.

McWethy, David B., Janet M. Wilmshurst, Cathy Whitlock, Jamie R. Wood, and Matt S. McGlone. 2014. "A High-Resolution Chronology of Rapid Forest Transitions Following Polynesian Arrival in New Zealand." *PLoS One* 9 (11): e111328.

Mead, Margaret. 1928. *Coming of Age in Samoa: A Psychological Study of Primitive Youth for Western Civilization.* New York: Harper Collins.

Mead, Margaret. 1957. "Introduction to Polynesia as a Laboratory for the Development of Models in the Study of Cultural Evolution." *Journal of the Polynesian Society* 66 (1): 145.

Meadows, Donella H., Dennis L. Meadows, Jørgen Randers, and William W. Behrens III. 1972. *The Limits to Growth: A Report for the Club of Rome's Project on the Predicament of Mankind.* New York: Universe.

Mebratu, Desta. 1998. "Sustainability and Sustainable Development: Historical and Conceptual Review." *Environmental Impact Assessment Review* 18: 493-520.

Medina-Muñoz, Diego R., and Rita D. Medina-Muñoz. 2012. "Determinants of Expenditures on Wellness Services: The Case of Gran Canaria." *Regional Studies* 46 (3): 309-319.

Meierhenrich, Jens. 2014. *Genocide: A Reader.* New York: Oxford University Press USA.

Merriam Webster. n.d. "Island." http://www.merriam-webster.com/dictionary/island.

Michelucci, Stefania. 2002. "The Violated Silence: D.H. Lawrence's 'The Man Who Loved Islands.'" In *Beyond the Floating Islands*, edited by Stephanos Stephanides and Susan Bassnett, 128-134. Bologna, IT: University of Bologna.

Middleton, Victor T. C., with Rebecca Hawkins. 1998. *Sustainable Tourism: A Marketing Perspective.* Oxford, UK: Butterworth Heinemann.

Mimura, Nobuo. 1999. "Vulnerability of Island Countries in the South Pacific to Sea Level Rise and Climate Change." *Climate Research* 12 (2-3): 137-143.

Mimura, Nobuo, Roger McLean, John Agard, Lino Briguglio, Penehuro Lefale, Rolph Payet, Graham Sem, Will Agricole, Kristie Ebi, Donald Forbes, et al. 2007. "Small Islands." In *Climate Change 2007: Impacts, Adaptation, and Vulnerability. Contribution of Working Group II to the Fourth Assessment Report of the Intergovernmental Panel on Climate Change,* edited by Martin Parry, Osvaldo Canziani, Jean Palutikof, Paul van der Linden, and Clair Hanson, 687-716. Cambridge, UK: Cambridge University Press.

Minooee, Arézou, and Leland S. Rickman. 1999. "Infectious diseases on cruise ships." Clinical infectious diseases 29 (4): 737-743. Mintz, Sidney. 1991. "Pleasure, Profit, and Satiation." In *Seeds of Change: A Quincentennial Commemoration,* edited by Herman Viola and Carolyn Margolis, 112-129. London, UK: Smithsonian Institution Press.

Mitchell, Wallace, J. Chittleborough, B. Ronai, and G. W. Lennon. 2001. "Sea Level Rise in Australia and the Pacific." In *Proceedings Science Component – Linking Science and Policy: Pacific Islands Conference on Climate Change, Climate Variability and Sea Level Rise, Rarotonga, Cook Islands, April 3-7, 2000*, 47-58. Adelaide, AU: National Tidal Facility Australia.

Mlachila, Montfort, Paul Cashin, and Cleary Haines. 2010. "Caribbean Bananas: The Macroeconomic Impact of Trade Preference Erosion." Working Paper #WP/10/59. International Monetary Fund. https://www.imf.org/external/pubs/ft/wp/2010/wp1059.pdf.

Mokyr, Joel, and Cormac Ó Gráda. 1984. "New Developments in Irish Population History, 1700-1850." *The Economic History Review* 37 (4): 473-488.

Montanez, Ana Maria Rios. 2019a. "Cayman Islands: Number of Tourist Arrivals 2005-2017." Statista. https://www.statista.com/statistics/816372/cayman-islands-number-of-tourist-arrivals.

Montanez, Ana Maria Rios. 2019b. "Turks & Caicos: Tourist Arrivals 2008-2017." Statista. https://www.statista.com/statistics/813540/number-tourist-arrivals-turks-caicos-islands.

Montgomery, Lucy Maud. 1908. *Anne of Green Gables*. Boston: L.C. Page & Co.

Morgan, Paula E. 2013. "Meet Me in the Islands: Sun, Sand, and Transactional Sex in Caribbean Discourse." *Anthurium: A Caribbean Studies Journal* 10 (1): 5.

Morowitz, Laura. 1998. "From Gauguin to Gilligan's Island." *Journal of Popular Film and Television* 26 (1): 2-10.

Morris, Julia C. 2019. "Violence and Extraction of a Human Commodity: From Phosphate to Refugees in the Republic of Nauru." *The Extractive Industries and Society*. Accessed on December 5, 2019. https://doi.org/10.1016/j.exis.2019.07.001.

Morrison, Keith. 2017. "The Role of Traditional Knowledge to Frame Understanding of Migration as Adaptation to the 'Slow Disaster' of Sea Level Rise in the South Pacific." In *Identifying Emerging Issues in Disaster Risk Reduction, Migration, Climate Change and Sustainable Development: Shaping Debates and Policies*, edited by Karen Sudmeier-Rieux, Manuela Fernández, Ivanna M. Penna, Michael Jaboyedoff, and J. C. Gaillard, 249-266. New York: Springer.

Morse, Sidney. 1863. *System of Geography for the Use of Schools*. New York: Harper & Brothers.

Mortimer, Nick, and Hamish Campbell. 2014. *Zealandia: Our Continent Revealed*. Auckland: Penguin Random House New Zealand Limited.

Morwood, Michael J., Paul B. O'Sullivan, Fachroel Aziz, and Asaf Raza. 1998. "Fission-Track Ages of Stone Tools and Fossils on the East Indonesian Island of Flores." *Nature* 392 (6672): 173-176.

Moseley, Malcolm. 1974. *Growth Centres in Spatial Planning*. Oxford, UK: Pergamon.

Mountz, Alison. 2011. "The Enforcement Archipelago: Detention, Haunting, and Asylum on Islands." *Political Geography* 30 (3): 118-128.

Mountz, Alison. 2015. "Political Geography II: Islands and Archipelagos." *Progress in Human Geography* 39 (5): 636-646.

Mountz, Alison. 2017. "Island Detention: Affective Eruption as Trauma's Disruption." *Emotion, Space, and Society* 24: 74-82.

Mullan, Fitzhugh. 2005. "The Metrics of the Physician Brain Drain." *New England Journal of Medicine* 353: 1810-1818.

Müller, Frank G. 2000. "Ecotourism: An Economic Concept for Ecological Sustainable Tourism." *The International Journal of Environmental Studies* 57 (3): 241-251.

Murray, Roy. 2010. "Notes on the Early History of the Caymans." *Journal of the University College of the Cayman Islands* 4: 109-124.

Myers, Norman. 1988. "Threatened Biotas: 'Hot Spots' in Tropical Forests." *The Environmentalist* 8 (3): 187-208.

Myers, Norman, Russell A. Mittermeier, Cristina G. Mittermeier, Gustavo A. B. da Fonesca, and Jennifer Kent. 2000. "Biodiversity Hotspots for Conservation Priorities." *Nature* 403: 853-858.

Nadarajah, Yaso, and Adam Grydehøj. 2016. "Island Studies as a Decolonial Project (Guest Editorial Introduction)." *Island Studies Journal* 11 (2): 437-446.

Nally, David. 2008. "'That Coming Storm': The Irish Poor Law, Colonial Biopolitics, and the Great Famine." *Annals of the Association of American Geographers* 98 (3): 714-741.

Nanditha, Arun, Ronald C.W. Ma, Ambady Ramachandran, Chamukuttan Snehalatha, Juliana C.N. Chan, Kee Seng Chia, Jonathan E. Shaw, and Paul Z. Zimmet. 2016. "Diabetes in Asia and the Pacific: Implications for the Global Epidemic." *Diabetes Care* 39 (3): 472-485.

National Geographic Society. 2012a. "Atoll." *Resource Library: Encyclopedia.* Accessed March 9, 2020. https://www.nationalgeographic.org/encyclopedia/atoll.

National Geographic Society. 2012b. "Island." *Resource Library: Encyclopedia.* Accessed March 9, 2020. https://www.nationalgeographic.org/encyclopedia/island.

Neel, James V. 1962. "Diabetes Mellitus: A 'Thrifty' Genotype Rendered Detrimental by 'Progress'?" *American Journal of Human Genetics* 14 (4): 353-362.

New Zealand Ministry for Culture and Heritage. 2017. "The Treaty in Brief: Treaty FAQs." *New Zealand History.* https://nzhistory.govt.nz/politics/treaty/treaty-faqs.

Ng, Marie, Tom Fleming, Margaret Robinson, Blake Thomson, Nicholas Graetz, Christopher Margono, Erin C. Mullany, et al. 2014. "Global, Regional, and National Prevalence of Overweight and Obesity in Children and Adults During 1980-2013: A Systematic Analysis for the Global Burden of Disease Study 2013." *The Lancet* 384 (9945): 766-781.

Ngata, Tina. 2014. "For Whom the Taika Roars (An Open Letter to Taika Waititi)." The Non-Plastic Maori. https://thenonplasticmaori.wordpress.com/2014/10/24/for-whom-the-taika-roars-an-open-letter-to-taika-waititi.

Nhan, Tu-Xuan, and Didier Musso. 2015. "The Burden of Chikungunya in the Pacific." *Clinical Microbiology and Infection* 21 (6): e47-8.

Nimführ, Sarah, Laura Otto, and Gabriel Samateh. 2020. "Denying, While Demanding Integration: An Analysis of the Integration Paradox in Malta and Refugees' Coping Strategies." In *Politics of (Dis)Integration,* edited by Sophie Hinger and Reinhard Schweitzer, 161-181. Cham, CH: Springer Open.

Niusulu, Anita Latai. 2018. "Challenging the Notion of 'Vulnerable Islands': A Review of Paradigms in the Climate Change Literature." *Journal of the Arts Faculty of the National University of Samoa* 4: 3-22.

Nunn, Patrick. 1992. "Human and Non-Human Impacts on Pacific Island Environments." Occasional Papers of the Program on Environment, paper no. 13. Honolulu: East-West Center.

Nunn, Patrick. 1994. *Oceanic Islands.* Oxford, UK: Blackwell.

Nunn, Patrick. 1999. "Geomorphology." In *The Pacific Islands: Environment and Society*, edited by Moshe Rapoport, 43-55. Hong Kong: Bess Press.

Nunn, Patrick. 2000. "Illuminating Sea-Level Fall Around AD 1220-1510 (730-440 cal yr BP) in the Pacific Islands: Implications for Environmental Change and Cultural Transformation." *New Zealand Geographer* 56 (1): 46-54.

Nunn, Patrick. 2001. "On the Convergence of Myth and Reality: Examples from the Pacific Islands." *The Geographical Journal* 167 (2): 125-138.

Nunn, Patrick. 2003. "Fished up or Thrown Down: The Geography of Pacific Island Origin Myths." *Annals of the Association of American Geographers* 93 (2): 350-364.

Nunn, Patrick. 2004a. "Myths and the Formation of Niue Island, Central South Pacific." *The Journal of Pacific History* 39 (1): 99-108.

Nunn, Patrick. 2004b. "Through a Mist on the Ocean: Human Understanding of Island Environments." *Tijdschrift voor Economische en Sociale Geografie* 95 (3): 311-325.

Nunn, Patrick. 2007. *Climate, Environment, and Society in the Pacific During the Last Millennium.* Oxford, UK: Elsevier.

Nunn, Patrick. 2009. *Vanished Islands and Hidden Continents of the Pacific.* Honolulu: University of Hawai'i Press.

Nunn, Patrick. 2012. "Understanding and Adapting to Sea-Level Rise." In *Global Environmental Issues,* 2nd ed., edited by Frances Harris, 87-104. Chichester, UK: John Wiley & Sons.

Nunn, Patrick. 2013. "The End of the Pacific? Effects of Sea Level Rise on Pacific Island Livelihoods." *Singapore Journal of Tropical Geography* 34: 143-171.

Nunn, Patrick, and James Britton. 2001. "Human-Environment Relationships in the Pacific Islands Around AD 1300." *Environment and History* 7 (1): 3-22.

Nunn, Patrick, William Aalbersberg, Shalini Lata, and Marion Gwilliam. 2014. "Beyond the Core: Community Governance for Climate-change Adaptation in Peripheral Parts of Pacific Island Countries." *Regional Environmental Change* 14: 221-235.

Nunn, Patrick, Joeli Veitayaki, Vina Ram-Bidesi, and Aliti Vunisea. 1999. "Coastal Issues for Oceanic Islands: Implications for Human Futures." *Natural Resources Forum* 23: 195-207.

Nunn, Patrick, Rosalind Hunter-Anderson, Mike Carson, Frank Thomas, Sean Ulm, and Michael Rowland. 2007. "Times of Plenty, Times of Less: Last-Millennium Societal Disruption in the Pacific Basin." *Human Ecology* 35: 385-401.

Nurse, Keith. 2007. "The Cultural Industries and Sustainable Development in Small Island Developing States." UNESCO Open Access Publications Portal. http://portal.unesco.org/en/files/24726/110805219811CLT3.doc/CLT3.doc.

Nurse, Keith, Danielle Edwards, and Denyse Dookie. 2018. "Climate Change Governance and Trade Policy: Challenges for Travel and Tourism in Small Island Developing States." In *Global Climate Change and Coastal Tourism: Recognizing Problems, Managing Solutions and Future Expectations,* edited by Andrew Jones and Michael Phillips, 74-91. Wallingford, UK: CABI.

Nurse, Leonard A., Roger F. McLean, and Avelino G. Suarez. 1997. "Small Island States." In *The Regional Impacts of Climate Change: An Assessment of Vulnerability – Special Report of IPCC Working Group II,* edited by Robert T. Watson, Marufu C. Zinyowera, and Richard H. Moss, 331-354. Cambridge, UK: Cambridge University Press.

Nurse, Leonard, Roger McLean, John Agard, Lino Briguglio, Virginie Duvat-Magnan, Netatua Pelesikoti, Emma Tompkins, and Arthur Webb. 2014. "Small Islands." In *Climate Change 2014: Impacts, Adaptation, and Vulnerability. Part B: Regional Aspects. Contribution of Working Group II to the Fifth Assessment Report of the Intergovernmental Panel on Climate Change,* edited by Vicente R. Barros, Christopher B. Field, David Jon Dokken, Michael D. Mastrandrea, Katharine J. Mach, T. Eren Bilir, Monalisa Chatterjee, et al., 1613-1654. Cambridge, UK: Cambridge University Press.

O'Neill, Dan. 2014. "Gross Domestic Product." In *Degrowth: A Vocabulary for a New Era,* edited by Giacomo D'Alisa, Federico Demaria, and Giorgios Kallis, 103-106. London, UK: Routledge.

Oberst, Ashley, and Jerome L. McElroy. 2007. "Contrasting Socioeconomic and Demographic Profiles of Two, Small Island, Economic Species: MIRAB versus PROFIT/SITE." *Island Studies Journal* 2 (2): 164-176.

Ogwang, Tomson. 1994. "The Choice of Principle Variables for Computing the Human Development Index." *World Development* 22 (12): 2011-2014.

Opeskin, Brian, and Therese MacDermott. 2009. "Resources, Population, and Migration in the Pacific: Connecting Islands and Rim." *Asia Pacific Viewpoint* 50 (3): 353-373.

Orange, Claudia. 2015. *The Treaty of Waitangi.* Wellington, NZ: Bridget Williams.

Organisation for Economic Cooperation and Development [OECD]. 1998. "Harmful Tax Competition: An Emerging Global Issue." Accessed December 5, 2019. https://web.archive.org/

web/20180624150446/http://www.oecd.org/tax/transparency/about-the-global-forum/publications/harmful-tax-competition-emerging-global-issue.pdf

Organisation for Economic Cooperation and Development [OECD]. 2012. *The Future of the Ocean Economy: Exploring the Prospects for Emerging Ocean Industries to 2030.* Paris: OECD Publishing.

Organisation for Economic Cooperation and Development [OECD]. 2019. "Development Aid Drops in 2018, Especially to Neediest Countries." https://www.oecd.org/newsroom/development-aid-drops-in-2018-especially-to-neediest-countries.htm.

Otto, Ton. 1993. "Empty Tins for Lost Traditions? The West's Material and Intellectual Involvement in the Pacific." In *Pacific Island Trajectories: Five Personal Views,* edited by Ton Otto, 1-28. Canberra: The Australian National University.

Ourbak, Timothée, and Alexandre K. Magnan. 2018. "The Paris Agreement and Climate Change Negotiations: Small Islands, Big Players." *Regional Environmental Change* 18 (8): 2201-2207.

Pacific Islands Forum Secretariat. 2007. "The Pacific Plan for Strengthening Regional Cooperation and Integration." https://www.adb.org/sites/default/files/linked-documents/robp-pac-2010-2013-oth01.pdf.

Palan, Ronen, Richard Murphy, and Christian Chavagneux. 2013. *Tax Havens: How Globalization Really Works.* Ithaca, NY: Cornell University Press.

Paneth, Nigel. 2004. "Assessing the Contributions of John Snow to Epidemiology: 150 Years After Removal of the Broad Street Pump Handle." *Epidemiology* 15 (5): 514-516.

Pantelescu, Andreea Marin. 2012. "Trends in International Tourism." *Cactus Tourism Journal* 3 (2): 31-35.

Panum, Peter. 1988. "Observations Made During the Epidemic of Measles on the Faroe Islands in the Year 1846." In *The Challenge of Epidemiology: Issues and Selected Readings*, edited by Carol Buck, Alvaro Llopis, Enrique Nájera, and Milton Terris, 37-41. Washington, DC: Pan American Health Organization.

Paraskevaidis, Pavlos, and Konstantinos Andriotis. 2015. "Values of Souvenirs as Commodities." *Tourism Management* 48: 1-10.

Pardini, Renata, Elizabeth Nichols, and Thomas Püttker. 2017. "Biodiversity Response to Habitat Loss and Fragmentation." In *Earth Systems And Environmental Sciences: Encyclopedia of the Anthropocene,* vol. 3, edited by Dominick DellaSala and Michael Goldstein, 229-239. Amsterdam: Elsevier.

Parry, John Horace. 1974. *The Discovery of the Sea.* New York: Dial Press.

Pascali, Luigi. 2017. "The Wind of Change: Maritime Technology, Trade, and Economic Development." *American Economic Review* 107 (9): 2821-54.

Pasifika Futures. 2017. "Pasifika People in New Zealand: How Are We Doing?" Auckland: Whānau Ora Commissioning Agency. http://pasifikafutures.co.nz/wp-content/uploads/2015/06/PF_How-AreWeDoing-RD2-WEB2.pdf.

Patke, Rajeev. 2004. "The Islands of Poetry; The Poetry of Islands." *Partial Answers: Journal of Literature and the History of Ideas* 2 (1): 177-194.

Pattullo, Polly. 1996. *Last Resorts: The Cost of Tourism in the Caribbean.* Kingston, JAM: Ian Randle.

Pauli, Gunter A. 2010. *The Blue Economy: 10 Years, 100 Innovations, 100 Million Jobs.* Taos, NM: Paradigm.

Peck, Deborah. 1998. *Teaching Culture: Beyond Language.* New Haven, CT: New Haven Teachers Institute. Accessed on September 20, 2019. http://teachersinstitute.yale.edu/curriculum/units/1984/3/84.03.06.x.html.

Peckham, Robert. 2002. "Coasting." In *Beyond the Floating Islands*, edited by Stephanos Stephanides and Susan Bassnett, 85-89. Bologna, IT: University of Bologna.

Peiser, Benny. 2005. "From Genocide to Ecocide: The Rape of Rapa Nui." *Energy & Environment* 16 (3-4): 513-539.

Pelling, Mark, and Juha Uitto. 2001. "Small Island Developing States: Natural Disaster Vulnerability and Global Change." *Global Environmental Change Part B: Environmental Hazards* 3 (2): 49-62.

Pérez, Louis A. 2002. "Fear and Loathing of Fidel Castro: Sources of US Policy Toward Cuba." *Journal of Latin American Studies* 34 (2): 227-254.

Péron, Françoise. 2004. "The Contemporary Lure of the Island." *Tijdschrift voor Economische en Sociale Geografie* 95 (3): 326-339.

Perrings, Charles, and Madhav Gadgil. 2003. "Conserving Biodiversity: Reconciling Local and Global Public Benefits." In *Providing Global Public Goods: Managing Globalization*, edited by Inge Kaul, Pedro Conceicao, Katell Le Goulven, and Ronald U. Mendoza, 532-555. New York: United Nations Development Program and Oxford University Press.

Peters, Everson J. 2010. "Impact of Hurricane Ivan on Grenada Water Supply." *Proceedings of the Institute of Civil Engineers: Water Management* 163 (2): 57-64.

Peterson, Ryan. 2011. "Screaming in Silence: Seeking Sustainable Tourism in Island Societies." Paper presented at the *1st International Conference on Governance for Sustainable Development of Caribbean Small Island Developing States, Curaçao, March 4-7, 2011.* http://sidsgg.webs.com/2012/proceedings/Peterson_Screaming%20in%20Silence%20Seeking%20Sustainable%20Tourism.pdf.

Petzold, Jan, and Beate Ratter. 2019. "More Than Just SIDS: Local Solutions for Global Problems on Small Islands." *Island Studies Journal* 14 (1): 3-8.

Philipsen, Dirk. 2015. *The Little Big Number: How GDP Came to Rule the World and What to Do About It.* Princeton, NJ: Princeton University Press.

Philpot, Dean, Tim S. Gray, and Selina M. Stead. 2015. "Seychelles, A Vulnerable or Resilient SIDS? A Local Perspective." *Island Studies Journal* 10 (1): 31-48.

Philpott, Stuart B. 1999. "The Breath of 'the Beast': Migration, Volcanic Disaster, Place and Identity in Montserrat." In *Small Worlds, Global Lives: Islands and Migration,* edited by Russell King and John Connell, 137-159. London, UK: Pinter.

Phoca-Cosmetatou, Nellie, ed. 2011. *The First Mediterranean Islanders: Initial Occupation and Survival Strategies.* University of Oxford School of Archaeology: Monograph 74. http://www.academia.edu/1138037/The_first_Mediterranean_islanders_initial_occupation_and_survival_strategies.

Pigou-Dennis, Elizabeth, and Adam Grydehøj. 2014. "Accidental and Ideal Island Cities: Islanding Processes and Urban Design in Belize City and the Urban Archipelagos of Europe." *Island Studies Journal* 9 (2): 259-276.

Pimm, Stuart L., Peter Raven, Alan Peterson, Cagan H. Sekercioglu, and Paul R. Ehrlich. 2006. "Human Impacts on Rates of Recent, Present and Future Bird Extinctions." *Proceedings of the National Academy of Sciences* 103 (29): 10941-10946.

Pitt, David. 1980. "Sociology, Islands, and Boundaries." *World Development* 8 (12): 1051-1059.

Platt, David. 2004. "Islandness." In *Holding Ground: The Best of the Island Journal 1984-2004,* edited by Philip Conkling and David Platt, 1. Rockland, ME: Island Institute.

Plecher, H. 2019. "Growth Rate of the Real Gross Domestic Product (GDP) in Madagascar from 2014 to 2024." *Statista.* Accessed November 10, 2019. https://www.statista.com/statistics/460320/gross-domestic-product-gdp-growth-rate-in-madagascar.

Ploch, Lauren, and Nicolas Cook. 2012. "Madagascar's Political Crisis." CRS Report for Congress R40448. Washington, DC: Congressional Research Service. http://www.fredsakademiet.dk/ordbog/mord/madagascar_policy.pdf.

Pocock, John Greville Agard. 2005. *The Discovery of Islands: Essays in British History.* Cambridge, UK: Cambridge University Press.

Pool, Ian, and Tahu Kukutai. 2011. "Taupori Māori – Māori population change." *Te Ara: The Encycolopedia of New Zealand.* https://teara.govt.nz/en/taupori-maori-maori-population-change.

Ponting, Clive. 1991. *A Green History of the World: The Environment and the Collapse of Great Civilizations.* New York: St. Martin's.

Portes, Alejandro, and József Böröcz. 1989. "Contemporary Immigration: Theoretical Perspectives on its Determinants and Modes of Incorporation." *International Migration Review* 23 (3): 606-630.

Post, Stephen G. 2005. "Altruism, Happiness, and Health: It's Good to be Good." *International Journal of Behavioral Medicine* 12 (2): 66-77.

Pratt, Godfrey A. 2002. "Sustainable Tourism Development in the Caribbean: The Role of Education." In *Tourism and Hospitality Education and Training in the Caribbean*, edited by Chandana Jayawardena, 301-327. Kingston, JAM: University of the West Indies Press.

Pratt, Stephen. 2015. "The Economic Impact of Tourism in SIDS." *Annals of Tourism Research* 52 (C): 148-160.

Premdas, Ralph. 1996. "Ethnicity and Identity in the Caribbean: Decentering a Myth." Working Paper no. 234. Notre Dame, IN: Kellogg Institute for International Studies. https://kellogg.nd.edu/sites/default/files/old_files/documents/234.pdf.

Prince, Solène. 2018. "Science and Culture in the Kerguelen Islands: A Relational Approach to the Spatial Formation of a Subantarctic Archipelago." *Island Studies Journal* 13 (2): 129-144.

Prinsen, Gerard, and Séverine Blaise. 2017. "An Emerging 'Islandian' Sovereignty of Non-Self-Governing Islands." *International Journal* 72 (1): 56-78.

Prinsen, Gerard, Yves Lafoy, and Julien Migozzi. 2017. "Showcasing the Sovereignty of Non-Self-Governing Islands: New Caledonia." *Asia Pacific Viewpoint* 58 (3): 331-346.

Přívara, Andrej. 2019. "Citizenship-for-Sale Schemes in Bulgaria, Cyprus, and Malta." *Migration Letters* 16 (2): 245-254.

Prosser, Gary. 1995. "Tourist Destination Life Cycles: Progress, Problems, and Prospects." In *CAUTHE 1995: Proceedings of the National Tourism and Hospitality Conference, February 14-17, 1995*, 318-328. Canberra: Australian Bureau of Tourism Research.

Pugh, Jonathan. 2013. "Island Movements: Thinking with the Archipelago." *Island Studies Journal* 8 (1): 9-24.

Pugh, Jonathan. 2018. "Relationality and Island Studies in the Anthropocene." *Island Studies Journal* 13 (2): 92-110.

Puig-Cabrera, Miguel, and Concepción Foronda-Robles. 2019. "Tourism, Smallness, and Insularity: A Suitable Combination for Quality of Life in Small Island Developing States (SIDS)?" *Island Studies Journal* 14 (2): 61-80.

Putnam, Robert. 2000. *Bowling Alone: The Collapse and Revival of American Community.* New York: Simon and Schuster.

Quammen, David. 1996. *The Song of the Dodo: Island Biogeography in an Age of Extinctions.* New York: Scribner.

Quinn, Ben. 2016. "Migrant Death Toll Passes 5,000 After Two Boats Capsize off Italy." *Guardian Weekly*, December 23, 2106. https://www.theguardian.com/world/2016/dec/23/record-migrant-death-toll-two-boats-capsize-italy-un-refugee.

Quirk, Genevieve, and Quentin Hanich. 2016. "Ocean Diplomacy: The Pacific Island Countries' Campaign to the UN for an Ocean Sustainable Development Goal." *Asia-Pacific Journal of Ocean Law and Policy* 1 (1): 68-95.

Raby, Peter. 2002. "Alfred Russel Wallace: A Life." *London Naturalist* 81: 86-107.

Rainbird, Paul. 1999. "Islands Out of Time: Towards a Critique of Island Archaeology." *Journal of Mediterranean Archaeology* 12 (2): 216-234.

Rainbird, Paul. 2007. *The Archaeology of Islands.* Cambridge, UK: Cambridge University.

Rallu, Jean-Louis, and Dennis Ahlburg. 1999. "Demography." In *The Pacific Islands: Environment and Society*, edited by Moshe Rapaport, 258-269. Honolulu: Bess Press.

Ramírez de Arellano, Annette B. 2011. "Medical Tourism in the Caribbean." *Signs: Journal of Women in Culture and Society* 36 (2): 289-297.

Randall, James E. 2014. "Immigrants, Islandness and Perceptions of Quality of Life on Prince Edward Island, Canada." *Island Studies Journal* 9 (2): 343-362.

Randall, James E. forthcoming. "Island Studies Inside (and Outside) of the Academy: The State of This Interdisciplinary Field." In *The Challenges of Island Studies*, edited by Ayano Ginoza, London, UK: Springer.

Rankey, Eugene C. 2011. "Nature and Stability of Atoll Island Shorelines: Gilbert Island Chain, Kiribati, Equatorial Pacific." *Sedimentology* 58 (7): 1831-1859.

Rapaport, Moshe. 2006. "Eden in Peril: Impact of Humans on Pacific Island Ecosystems." *Island Studies Journal* 1 (1): 109-124.

Ratter, Beate. 2018. *Geography of Small Islands: Outposts of Globalization.* Cham, CH: Springer.

Ravallion, Martin. 1997. "Good and Bad Growth: The Human Development Reports." *World Development* 25 (5): 631-638.

Ravenstein, Ernst Georg. 1885. "The Laws of Migration." *Journal of the Statistical Society of London* 48 (2): 167-235.

Razafindrakoto, Mireille, F. Roubaud, and Jean-Michel Wachsberger. 2018. "The Puzzle of Madagascar's Economic Collapse Through the Lens of Social Sciences." *Dialogue* 51. https://hal.archives-ouvertes.fr/hal-01921824/document.

Razak, Victoria. 1995. "Culture Under Construction: The Future of Native Arubian Identity." *Futures* 27 (4): 447-459.

Read, Robert. 2004. "The Implications of Increasing Globalization and Regionalism for the Economic Growth of Small Island States." *World Development* 32 (2): 365-378.

Reaser, Jamie K., Rafe Pomerance, and Peter O. Thomas. 2000. "Coral Bleaching and Global Climate Change: Scientific Findings and Policy Recommendations." *Conservation Biology* 14 (5): 1500-1511.

Redclift, Michael. 2005. "Sustainable Development (1987–2005): An Oxymoron Comes of Age." *Sustainable Development* 13 (4): 212-227.

Redclift, Michael, and Delyse Springett, eds. 2015. *Routledge International Handbook of Sustainable Development.* London, UK: Routledge.

Refugee Council of Australia. 2019. "Offshore Processing Statistics." Statistics. https://www.refugeecouncil.org.au/operation-sovereign-borders-offshore-detention-statistics.

Reuters. June 27, 2009. "Sarkozy to Offer Martinique Autonomy Vote." Boston.com. http://archive.boston.com/news/world/europe/articles/2009/06/27/sarkozy_will_offer_martinique_autonomy_vote.

Richards, Greg. 2009. "Creative Tourism and Local Development." In *Creative Tourism: A Global Conversation*, edited by Rebecca Wurzburger, Sabrina Pratt, and Alex Pattakos, 78-90. Santa Fe, NM: Sunstone.

Richardson, Harry W. 1976. "Growth Pole Spillovers: The Dynamics of Backwash and Spread." *Regional Studies* 10 (1): 1-9.

Rick, Torben C., Patrick V. Kirch, Jon M. Erlandson, and Scott M. Fitzpatrick. 2013. "Archeology, Deep History, and the Human Transformation of Island Ecosystems." *Anthropocene* 4: 33-45.

Ridgell, Reilly. 1995. *Pacific Nations and Territories: The Islands of Micronesia, Melanesia, and Polynesia.* Honolulu: Bess Press.

Risse, Mathias. 2009. "The Right to Relocation: Disappearing Island Nations and Common Ownership of the Earth." *Ethics & International Affairs* 23 (3): 281-300.

Roberts, Susan. 1995. "Small Place, Big Money: The Cayman Islands and the International Financial System." *Economic Geography* 71 (3): 237-256.

Robinson, Stacey-Ann. 2017. "Climate Change Adaptation Trends in Small Island Developing States." *Mitigation and Adaptation Strategies for Global Change* 22: 669-691.

Rodrigue, Jean-Paul, and Theo Notteboom. 2013. "The Geography of Cruises: Itineraries, Not Destinations." *Applied Geography* 38: 31-42.

Rodrigue, Jean-Paul, Claude Comtois, and Brian Slack. 2016. *The Geography of Transport Systems,* 4th ed. London, UK: Routledge.

Rodríguez, Juan Ramón Oreja, Eduardo Parra-López, and Vanessa Yanes-Estévez. 2008. "The Sustainability of Island Destinations: Tourism Area Life Cycle and Teleological Perspectives. The Case of Tenerife." *Tourism Management* 29 (1): 53-65.

Rogers, Robert. 1995. *Destiny's Landfall: A History of Guam.* Honolulu: University of Hawai'i Press.

Rogoziński, Jan. 1994. *A Brief History of the Caribbean: From the Arawak and the Carib to the Present.* New York: Meridian.

Rohrer, Judy. 2010. *Haoles in Hawaii.* Honolulu: University of Hawai'i Press.

Roig, Annasofia A. 2019. "No Way, USA: The Lack of a Repatriation Agreement with Cuba and its Effects on US Immigration Policies." *FIU Law Review* 13 (4): 875.

Ronström, Owe. 2009. "Island Words, Island Worlds: The Origins and Meanings of Words for 'Islands' in North-West Europe." *Island Studies Journal* 4 (2): 163-182.

Ronström, Owe. 2012. "Gute, Gotlander, Mainlander, Swede: Ethnonyms and Identifications in a Changing Island Society." Paper presented at *Travelling in Time: Islands of the Past, Islands of the Future: The 8ᵗʰ Annual Small Island Cultures Conference, Cape Breton, Canada, June 6-9, 2012.* Accessed on May 1, 2017. http://sicri-network.org/ISIC8/c.%20ISIC8P%20Ronstrom.pdf.

Rosenblat, Ángel. 1992. "The Population of Hispaniola at the Time of Columbus." In *The Native Population of the Americas in 1492,* 2nd ed., edited by William Denevan, 43-66. Madison, WI: University of Wisconsin Press.

Ross, David. 2002. *Ireland: History of a Nation.* New Lanark, UK: Geddes & Grosset.

Ross, Michael. 1999. "The Political Economy of the Resource Curse." *World Politics* 51(2): 297-322.

Ross, Michael. 2003. "The Natural Resource Curse: How Wealth Can Make You Poor." In *Natural Resources and Violent Conflict: Options and Actions,* edited by Ian Bannon and Paul Collier, 17-42. Washington, DC: World Bank.

Rostow, Walt Whitman. 1959. "The Stages of Economic Growth." *The Economic History Review* 12 (1): 1-16.

Royle, Stephen. 1995. "Health in Small Island Communities: The UK's South Atlantic Colonies." *Health & Place* 1 (4): 257-264.

Royle, Stephen. 1999. "Leaving the 'Dreadful Rocks': Irish Island Emigration and its Legacy." *History Ireland* 7 (2): 34-37.

Royle, Stephen. 2001. *A Geography of Islands: Small Island Insularity.* London, UK: Routledge.

Royle, Stephen. 2007. "Island Definitions and Typologies." In *A World of Islands,* edited by Godfrey Baldacchino, 33-56. Charlottetown, PE: Island Studies Press.

Royle, Stephen, and Laurie Brinklow. 2018. "Definition and Typologies." In *The Routledge International Handbook of Island Studies,* 3-20. London, UK: Routledge.

Ruhanen, Lisa. 2005. "Introduction to Australia." In *Oceania: A Tourism Handbook,* edited by Chris Cooper and C. Michael Hall, 7-16. Clevedon, UK: Channel View.

Rutz, Kerry. 2012. "Artificial Islands Versus Natural Reefs: The Environmental Cost of Development in Dubai." *International Journal of Islamic Architecture* 1 (2): 243-267.

Sacks, Oliver. 1996. *The Island of the Colourblind.* New York: Knopf.

Sagan, Carl. 1994. *Pale Blue Dot: A Vision of the Human Future in Space.* New York: Ballantine.

Sallabank, Julia. 2011. "Norman Languages of the Channel Islands: Current Situation, Language Maintenance and Revitalisation." *Shima: The International Journal of Research into Island Cultures* 5 (2): 19-44.

Sammler, Katherine Genevieve. 2016. "The Deep Pacific: Island Governance and Seabed Mineral Development." In *Island Geographies*, edited by Elaine Stratford, 24-45. London, UK: Routledge.

Sancha Pastor, Ana. 2019. "Past and Present of the Development Model of Small Island Developing States. Case study: Tourism as a Determinant of the Sustainable Development of the Maldives." Master's thesis, Comillas Universidad Pontificia. https://repositorio.comillas.edu/xmlui/bitstream/handle/11531/43706/TFM001300.pdf.

Sand, Peter H. 2011. "'Marine Protected Areas' of UK Overseas Territories: Comparing the South Orkneys Shelf and the Chagos Archipelago." *The Geographical Journal* 178 (3): 201-207.

Sanders, Ronald. 2002. "The Fight Against Fiscal Colonialism: The OECD and Small Jurisdictions." *The Round Table* 91 (365): 325-348.

Savory, Elaine. 2011. "Utopia, Dystopia, and Caribbean Heterotopias: Writing/Reading the Small Island." *New Literatures Review* 47-48: 35-56.

Sawyers, June. 2001. *Celtic Music: A Complete Guide*. Boston: Da Capo.

Scheier, Michael F., and Charles S. Carver. 1985. "Optimism, Coping, and Health: Assessment and Implications of Generalized Outcome Expectancies." *Health Psychology* 4 (3): 219-247.

Scheier, Michael F., and Charles S. Carver. 1987. "Dispositional Optimism and Physical Well-Being: The Influence of Generalized Outcome Expectancies on Health." *Journal of Personality* 55 (2): 169-210.

Scheyvens, Regina, and Janet Momsen. 2008. "Tourism in Small Island States: From Vulnerability to Strengths." *Journal of Sustainable Tourism* 16 (5): 491-510.

Schouten, Frans. 2007. "Cultural Tourism: Between Authenticity and Globalization." In *Cultural Tourism: Global and Local Perspectives,* edited by Greg Richards, 25-37. New York: Routledge.

Schroeder, Jonathan E., and Janet L. Borgerson. 1999. "Packaging Paradise: Consuming Hawaiian Music." In *North American Advances in Consumer Research Volume 26*, edited by Eric J. Arnould and Linda M. Scott, 46-50. Provo, UT: Association for Consumer Research.

Schroeder, Thomas. 2009. "Climate on Islands." In *Encyclopedia of Islands*, edited by Rosemary G. Gillespie and David A. Clague, 171-174. Berkeley: University of California Press.

Scott, Heidi. 2014. "Havens and Horrors: The Island Landscape." *Interdisciplinary Studies in Literature and Environment* 21 (3): 636-657.

Searle, Roger. 2009. "Plate Tectonics." In *Encyclopedia of Islands*, edited by Rosemary G. Gillespie and David A. Clague, 752-755. Berkeley: University of California Press.

Secretariat of the Pacific Community. 2019. "2018 Pocket Statistics Summary." Prism. http://www.spc.int/DigitalLibrary/Doc/SDD/Pocket_Summary/Pocket_Statistical_Summary_18.pdf.

Seeley, John Robert. 1883. *The Expansion of England: Two Courses of Lectures*. London, UK: Macmillan.

Seidel, Henrike, and Padma N. Lal. 2010. *Economic Value of the Pacific Ocean to the Pacific Island Countries and Territories*. Gland, CH: IUCN. https://www.iucn.org/sites/dev/files/import/downloads/economic_value_of_the_pacific_ocean_to_the_pacific_island_countries_and_territories_p.pdf.

Seitz, Sharon, and Stuart Miller. 2011. *The Other Islands of New York City: A History and Guide*, 3rd ed. Woodstock, VT: Countryman.

Shah, Saeeda. 2003. "The Researcher/Interviewer in Intercultural Context: A Social Intruder!" *British Educational Research Journal* 30 (4): 549-575.

Shankman, Paul. 2018. "Samoan Journeys: Migration, Remittances, and Traditional Gift Exchange." In *Change and Continuity in the Pacific*, edited by John Connell and Helen Lee, 87-101. New York: Routledge.

Shareef, Riaz. 2003. "Small Island Tourism Economies: A Bird's Eye View." *Proceedings of the International Conference on Modelling and Simulation: Socio-Economic Systems, Townsville, Australia, Volume III*: 1124-1129.

Shareef, Riaz, Suhejla Hoti, and Michael McAleer. 2008. *The Economics of Small Island Tourism: International Demand and Country Risk Analysis*. Cheltenham, UK: Edward Elgar.

Sharp, Hannah, Josefine Grundius, and Jukka Heinonen. 2016. "Carbon Footprint of Inbound Tourism to Iceland: A Consumption-Based Life-Cycle Assessment Including Direct and Indirect Emissions." *Sustainability* 8 (11): 1147.

Sharpley, Richard. 2004. "Islands in the Sun: Cyprus." In *Tourism Mobilities: Places to Play, Places in Play*, edited by Mimi Sheller and John Urry, 22-31. London. UK: Routledge.

Sharpley, Richard. 2012. "Island Tourism or Tourism on Islands?" *Tourism Recreation Research* 37 (2): 167-172.

Shaxson, Nicholas. 2012. *Treasure Islands: Tax Havens and the Men Who Stole the World*. London, UK: Palgrave Macmillan.

Shell, Marc. 2014. *Islandology: Geography, Rhetoric, Politics*. Stanford, CA: Stanford University Press.

Shetland Amenity Trust. n.d. "Geology." Accessed March 5, 2020. https://www.shetlandamenity.org/assets/files/Natural%20Heritage/Geopark%20Shetland/Geology.pdf

Sigurdsson, Gisli. 2004. "The Medieval Icelandic Saga and Oral Tradition." Translated by Nicholas Jones. Cambridge, MA: Milman Parry Collection of Oral Literature.

Skinner, Jonathan. 2002. "Introduction: Social, Economic, and Political Dimensions of Formality and Informality in 'Island' Communities." *Social Identities* 8 (2): 205-215.

Small, Cathy A., and David L. Dixon. 2004. "Tonga: Migration and the Homeland." Migration Information Source, The Online Journal of the Migration Policy Institute. http://www.migrationpolicy.org/article/tonga-migration-and-homeland.

Smiley, Jane, ed. 2001. *The Sagas of the Icelanders*. London, UK: Penguin Classics.

Smith, Anita, and Kevin Jones. 2007. *Cultural Landscapes of the Pacific Islands*. Paris: UNESCO, International Council on Monuments and Sites. https://whc.unesco.org/document/10061.

Smith, Barbara B. 1983. "Musics of Hawai'i and Samoa: Exemplar of Annotated Resources." *Music Educators Journal* 69 (9): 62-65.

Smith, Bernard W. 1985. *European Vision and the South Pacific*, 2nd ed. New Haven, CT: Yale University Press.

Smith, Joel, Hans-Joachim Schellnhuber, and Monirul Qader Mirza. 2001. "Vulnerability to Climate Change and Reasons for Concern: A Synthesis." In *Climate Change 2001: Impacts, Adaptation, and Vulnerability: Contribution of Working Group II to the Third Assessment Report of the Intergovernmental Panel on Climate Change* edited by James McCarthy, Osvaldo Canziani, Neil Leary, David Dokken and Kasey White, 913-970. Cambridge, UK: Cambridge University Press.

Smith-Godfrey, Simon. 2016. "Defining the Blue Economy." *Maritime Affairs: Journal of the National Maritime Foundation of India* 12 (1): 58-64.

Snowdon, Wendy, Mark Lawrence, Jimaima Schultz, Paula Vivili, and Boyd Swinburn. 2010. "Evidence-Informed Process to Identify Policies That Will Promote a Healthy Food Environment in the Pacific Islands." *Public Health Nutrition* 13 (6): 886-892.

Snoxell, David. 2008. "Expulsion from Chagos: Regaining Paradise." *The Journal of Imperial and Commonwealth History* 36 (1): 119-129.

Snoxell, David. 2009. "Anglo/American Complicity in the Removal of the Inhabitants of the Chagos Islands, 1964-73." *The Journal of Imperial and Commonwealth History* 37 (1): 127-134.

Spaiser, Viktoria, Shyam Ranganathan, Ranjula Bali Swain, and David J. T. Sumpter. 2017. "The Sustainable Development Oxymoron: Quantifying and Modelling the Incompatibility of Sustainable Development Goals." *International Journal of Sustainable Development & World Ecology* 24 (6): 457-470.

Spalding, Mark. 2016. "The New Blue Economy: The Future of Sustainability." *Journal of Ocean and Coastal Economics* 2 (2): Article 8.

Spalding, Mark, Corinna Ravilious, and Edmund Green. 2001. *The World Atlas of Coral Reefs*. Berkeley: University of California Press.

Squire, Shelagh. 1996. "Literary Tourism and Sustainable Tourism: Promoting 'Anne of Green Gables' in Prince Edward Island." *Journal of Sustainable Tourism* 4 (3): 119-134.

Srinivasan, T. N. 1994. "Data Base for Development Analysis: An Overview." *Journal of Development Economics* 44 (1): 3-27.

Stannard, David. 1993. "Disease, Human Migration, and History." In *The Cambridge World History of Human Disease,* edited by Kenneth F. Kiple, 35-42. Cambridge, UK: Cambridge University Press.

Stanton, Elizabeth A. 2007. "The Human Development Index: A History." Political Economy Research Institute Working Paper Series no. 127. Amherst, MA: University of Massachusetts. http://scholarworks.umass.edu/cgi/viewcontent.cgi?article=1101&context=peri_workingpapers.

Statistics Iceland. 2009. "Return Migration 1986-2008." https://hagstofa.is/en/publications/news-archive/population/return-migration-1986-2008.

Stats NZ. 2006. "Demographics of New Zealand's Pacific Population." http://archive.stats.govt.nz/browse_for_stats/people_and_communities/pacific_peoples/pacific-progress-demography/population-growth.aspx.

Steffler, John. 1985. *The Grey Islands: A Journey*. Toronto: McClelland & Stewart.

Steig, William. 1976. *Abel's Island*. Toronto: McGraw-Hill Ryerson.

Steinberg, Philip. 2005. "Insularity, Sovereignty, and Statehood: The Representation of Islands on Portolan Charts and the Construction of the Territorial State." *Geografiska Annaler: Series B, Human Geography* 87 (4): 253-265.

Stephanides, Stephanos, and Susan Bassnett. 2008. "Islands, Literature, and Cultural Translatability." *Transtext(e)s Transcultures: Journal of Global Cultural Studies* Occasional Series 8: 5-21.

Stevens-Arroyo, Anthony M. 1993. "The Inter-Atlantic Paradigm: The Failure of Spanish Medieval Colonization of the Canary and Caribbean Islands." *Comparative Studies in Society and History* 35 (3): 515-543.

Stevenson, Robert Louis. 1883 (2009). *Treasure Island*. Reprint, New York: Gramercy.

Stoddart, David R., and James Alfred Steers. 1977. "The Nature and Origin of Coral Reef Islands." In *Biology and Geology of Coral Reefs Volume 4: Geology 2*, edited by Owen A. Jones and Robert Endean, 60-106. London, UK: Academic Press.

Storey, Donovan, and Vanessa Steinmayer. 2013. "Mobility as Development Strategy: The Case of the Pacific Islands." *Asia-Pacific Population Journal* 26 (4): 57-72.

Storr, Cait. 2016. "Islands and the South: Framing the Relationship Between International Law and Environmental Crisis." *The European Journal of International Law* 27 (2): 519-540.

Stoutenburg, Jenny Grote. 2015. *Disappearing Island States in International Law*. Boston, MA: Brill Nijhoff.

Stratford, Elaine. 2003. "Flows and Boundaries: Small Island Discourses and the Challenge of Sustainability, Community and Local Environments." *Local Environment* 8 (5): 495-499.

Stratford, Elaine. 2008. "Islandness and Struggles over Development: A Tasmanian Case Study." *Political Geography* 27 (2): 160-175.

Stratford, Elaine. 2013. "The Idea of the Archipelago: Contemplating Island Relations." *Island Studies Journal* 8 (1): 3-8.

Stratford, Elaine, Godfrey Baldacchino, Elizabeth McMahon, Carol Farbotko, and Andrew Harwood. 2011. "Envisioning the Archipelago." *Island Studies Journal* 6 (2): 113-130.

Stuart, Kathleen. 2008. "A Listing of the World's Populated Sub-National Island Jurisdictions." In *Pulling Strings: Policy Insights for Prince Edward Island From Other Sub-National Island Jurisdictions*, edited by Godfrey Baldacchino and David Milne, 174-185. Charlottetown, PE: Island Studies Press.

Stuart, Kathleen. 2009. "A Listing of the World's Sub-National Island Jurisdictions." In *The Case for Non-Sovereignty: Lessons from Sub-National Island Jurisdictions*, edited by Godfrey Baldacchino and David Milne, 11-20. London, UK: Routledge.

Stylidis, Dimitrios, Matina Terzidou, and Konstantinos Terzidis. 2008. "Islands and Destination Image: The Case of Ios." *Tourismos: An International Multidisciplinary Journal of Tourism* 3 (1): 180-189.

Sutherland, Douglas, and Jane Stacey. 2017. "Sustaining Nature-Based Tourism in Iceland." OECD Economics Department Working Paper no. 1422. Organization for Economic Cooperation and Development, Paris. http://dx.doi.org/10.1787/f28250d9-en.

Suwa, Jun'ichiro. 2007. "The Space of Shima." *Shima: The International Journal of Research into Island Cultures* 1 (1): 6-14.

Swift, Jonathan. 1726. *Gulliver's Travels into Several Remote Nations of the World*. Transcribed by David Price from the 1892 George Bell & Sons Edition. https://www.gutenberg.org/files/829/829-h/829-h.htm.

Tabarelli, Marcello, and Claude Gascon. 2005. "Lessons from Fragmentation Research: Improving Management and Policy Guidelines for Biodiversity Conservation." *Conservation Biology* 19 (3): 734-739.

Tagawa, Hideo, Eizi Suzuki, Tukirin Partomihardjo, and Ade Suriadarma. 1985. "Vegetation and Succession on the Krakatau Islands, Indonesia." *Vegetatio* 60: 131-145.

Taglioni, François. 2011. "Insularity, Political Status, and Small Insular Spaces." *Shima: The International Journal of Research into Island Cultures* 5 (2): 45-67.

Tamaira, A. Mārata Ketekiri, and Dionne Fonoti. 2018. "Beyond Paradise? Retelling Pacific Stories in Disney's Moana." *The Contemporary Pacific* 30 (2): 297-327.

Tavares, Rodrigo. 2016. *Paradiplomacy: Cities and States as Global Players*. Oxford, UK: Oxford University Press.

Tax Justice Network. 2019. "Estimates of Tax Avoidance and Evasion." Accessed November 28, 2019. https://www.taxjustice.net/topics/more/estimates-of-tax-avoidance-and-evasion.

Taylor, David G. P. 2000. "British Colonial Policy in the Caribbean the Insoluble Dilemma – The Case of Montserrat." *The Round Table* 89 (355): 337-344.

Taylor, Sam. 2009. *The Island at the End of the World*. New York: Penguin.

Taylor, Theodore. 1969. *The Cay*. New York: Avon.

Teaiwa, Teresia, and Selena T. Marsh. 2010. "Albert Wendt's Critical and Creative Legacy in Oceania: An Introduction." *The Contemporary Pacific* 22 (2): 233-248.

Teaiwa, Teresia. 2010. "What Remains to Be Seen: Reclaiming the Visual Roots of Pacific Literature." *Publications of the Modern Language Association [PMLA]* 125 (3): 730-736.

The Island Review. July 19, 2013. "Island Sounds: Roddy Woomble." http://www.theislandreview.com/island-music-roddy-woomble-2.

The Star Online. 2018. "Number of Injured in Indonesia Tsunami Surges to Over 14,000." https://www.thestar.com.my/news/regional/2018/12/31/number-of-injured-in-indonesia-tsunami-surges-to-over-14000.

The Free Dictionary. 2019a. "Diaspora." Accessed November 5, 2019. http://www.thefreedictionary.com/Diaspora.

The Free Dictionary. 2019b. "Geopolitics." Accessed October 30, 2019. http://www.thefreedictionary. com/geopolitics.

The Free Dictionary. n.d. "Island." Accessed September 17, 2019. http://www.thefreedictionary.com/ island.

Thomas, Frank R. 2019. "Atoll Archaeology in the Pacific." In *Encyclopedia of Global Archaeology*, edited by Claire Smith, 1-12. Basel, CH: Springer Nature.

Thompson, Krista. 2007. *An Eye for the Tropics: Tourism, Photography, and Framing the Caribbean*. Durham, NC: Duke University Press.

Thompson, Maddy, and Margaret Walton-Roberts. 2019. "International Nurse Migration from India and the Philippines: The Challenge of Meeting the Sustainable Development Goals in Training, Orderly Migration, and Healthcare Worker Retention." *Journal of Ethnic and Migration Studies* 45 (14): 2583-2599.

Thornton, Ian. 1996. *Krakatau: The Destruction and Reassembly of an Island Ecosystem*. Cambridge, MA: Harvard University Press.

Thornton, Ian, Tim New, David McLaren, Sudarman Husen Kartodihardjo, and Patrick Vaughan. 1988. "Air-Borne Arthropod Fall-Out on Anak Krakatau and a Possible Pre-Vegetation Pioneer Community." *Philosophical Transactions of the Royal Society of London* 322: 471-479.

Toatu, Teuea. 2004. "Keeping the Nauru Economy Afloat." *Pacific Economic Bulletin* 19 (2): 123-128.

Tognotti, Eugenia. 2013. "Lessons from the History of Quarantine, From Plague to Influenza A." *Emerging Infectious Diseases* 19 (2): 254-259.

Togolo, Mel. 2006. "The 'Resource Curse' and Governance: A Papua New Guinean Perspective." In *Globalisation and Governance in the Pacific Islands*, edited by Stewart Firth, 275-285. Canberra: Australian National University ePress.

Toh, Rex S., Habibullah Khan, and Ai-Jin Koh. 2001. "A Travel Balance Approach for Examining Tourism Area Life Cycles: The Case of Singapore." *Journal of Travel Research* 39 (4): 426-432.

Tolley, Hillary, Wendy Snowdon, Jillian Wate, A. Mark Durand, Paula Vivili, Judith McCool, Rachel Novotny, et al. 2016. "Monitoring and Accountability for the Pacific Response to the Non-Communicable Diseases Crisis." *BMC Public Health* 16: 958.

Tompkins, Emma L., and Lisa-Ann Hurlston. 2005. "Natural Hazards and Climate Change: What Knowledge is Transferable." Tyndall Centre for Climate Change Research Working Paper no. 69. Tyndall Centre for Climate Change Research, University of East Anglia, Norwich, UK. http://citeseerx.ist.psu.edu/viewdoc/download?doi=10.1.1.503.3701&rep=rep1&type =pdf.

Torres-Delgado, Anna, and Jarkko Saarinen. 2014. "Using Indicators to Assess Sustainable Tourism Development: A Review." *Tourism Geographies* 16 (1): 31-47.

Tortella, Bartolomé Deyà, and Dolores Tirado. 2011. "Hotel Water Consumption at a Seasonal Mass Tourist Destination: The Case of the Island of Mallorca." *Journal of Environmental Management* 92 (10): 2568-2579.

Trautman, Lawrence J. 2017. "Following the Money: Lessons from the Panama Papers – Part 1: Tip of the Iceberg." *Penn State Law Review* 121 (3): 807-874.

Tuan, Yi-Fu. 1974. *Topophilia: A Study of Environmental Perception, Attitudes, and Values*. Upper Saddle River, NJ: Prentice-Hall.

Turks and Caicos Statistics Department. no date. "Latest Statistics". Accessed November 13, 2019. https://www.gov.tc/stats.

Turner, Louise, and John Ash. 1975. *The Golden Hordes: International Tourism and the Pleasure Periphery*. London, UK: Constable.

Twain, Mark. 2018. "Roughing It, Complete." *Project Gutenberg eBook #3177*. http://www.gutenberg. org/files/3177/3177-h/3177-h.htm.

Uekusa, Shinya, and Steve Matthewman. 2017. "Vulnerable and Resilient? Immigrants and Refugees in the 2010-2011 Canterbury and Tohoku Disasters." *International Journal of Disaster Risk Reduction* 22: 355-361.

Ulijaszek, Stanley J., ed. 2006. *Population, Reproduction, and Fertility in Melanesia*. New York: Berghahn.

Underwood, Jane Hainline. 1969. "Preliminary Investigations of Demographic Features and Ecological Variables of a Micronesian Island Population." *Micronesica* 5 (1): 1-24.

United Nations. 1994. *Report of the Global Conference on the Sustainable Development of Small Island Developing States*. New York: United Nations. https://www.un.org/en/events/pastevents/SIDS_1994.shtml.

United Nations. 2015. "Transforming Our World: The 2030 Agenda for Sustainable Development." https://sustainabledevelopment.un.org/post2015/transformingourworld/publication.

United Nations. 2018. "Convention on the Law of the Sea, Part VII, Article 121, Regime of Islands." https://www.un.org/depts/los/convention_agreements/texts/unclos/part8.htm.

United Nations Conference on Trade and Development [UNCTAD]. 2014. "The Oceans Economy: Opportunities and Challenges for Small Island Developing States." New York: United Nations. http://unctad.org/en/PublicationsLibrary/ditcted2014d5_en.pdf.

United Nations Development Program [UNDP]. 2018. "Human Development Indices and Indicators: 2018 Statistical Update." New York: United Nations Development Program. http://hdr.undp.org/sites/default/files/2018_human_development_statistical_update.pdf.

United Nations Education Scientific and Cultural Organization [UNESCO]. 2010. *Atlas of the World's Languages in Danger*. Paris: UNESCO

United Nations High Commissioner for Refugees [UNCHR]. 2019. "UNHCR Refugee and Migrant Arrivals to Europe in 2019 (Mediterranean) (January-March 2019)." Accessed November 7, 2019. https://data2.unhcr.org/en/documents/details/69500.

United Nations Office of the High Representative for the Least Developed Countries, Landlocked Developing Countries, and Small Island Developing States [UN-OHRLLS]. 2019. "Country Profiles." Accessed October 31, 2019. http://unohrlls.org/about-sids/country-profiles.

United Nations Secretariat. 2010. *Trends in Sustainable Development: Small Island Developing States (SIDS)*. New York: United Nations Secretariat.

United Nations World Commission on Environment and Development [UNWCED]. 1987. *Our Common Future*. Oxford, UK: Oxford University Press. https://sustainabledevelopment.un.org/content/documents/5987our-common-future.pdf.

Urry, John. 2010. "Consuming the Planet to Excess." *Theory, Culture & Society* 27 (2-3): 191-212.

US State Department. 2001. "Background Note: Papua New Guinea." *Archive: Bureau of East Asian and Pacific Affairs*. http://1997-2001.state.gov/background_notes/papua_new_guinea_1098_bgn.html.

USA International Business Publications. 2011. *Island States: Small Island States Handbook - Volume 1: Development Strategies and Programs*. Washington, DC: International Business Publications.

Uyarra, Maria C., Isabelle M. Cote, Jennifer A. Gill, Rob R. T. Tinch, David Viner, and Andrew R. Watkinson. 2005. "Island-Specific Preferences of Tourists for Environmental Features: Implications of Climate Change for Tourism-Dependent States." *Environmental Conservation* 32 (1): 11-19.

Valencia, Mark J. 1997. "Asia, the Law of the Sea, and International Relations." *International Affairs* 73 (2): 263-282.

Valencia, Mark J. 2007. "The East China Sea Dispute: Context, Claims, Issues, and Possible Solutions." *Asian Perspective* 31 (1): 127-167.

Valentine, Nadine Angelita. 2016. "Wellness Tourism: Using Tourists' Preferences to Evaluate the Wellness Tourism Market in Jamaica." *Review of Social Sciences* 1 (3): 25-44.

Vallandingham, Christopher. 2013. "Tracking Down Legal Sources on Prestatehood Florida." In *Prestatehood Legal Matters: A Fifty-state Legal Guide*, edited by Michael Chiorazzi and Marguerite Most, 249-275. New York: Routledge.

van den Bergh, Jeroen. 2007. "Abolishing GDP." Tinbergen Institute Discussion Paper TI 2007-019/3. Amsterdam: Vrije Universiteit and Tinbergen Institute. https://research.vu.nl/ws/portalfiles/portal/73350121/07019.

van den Bergh, Jeroen. 2009. "The GDP Paradox." *Journal of Economic Psychology* 30 (2): 117-135.

van Dyke, Jon, and Robert Brooks. 1983. "Uninhabited Islands: Their Impact on the Ownership of the Oceans' Resources." *Ocean Development & International Law* 12 (3-4): 265-300.

van Fossen, Anthony. 2018. "Passport Sales: How Island Microstates Use Strategic Management to Organise the New Economic Citizenship Industry." *Island Studies Journal* 13 (1): 285-300.

van Rekom, Johan, and Frank Go. 2006. "Cultural Identities in a Globalizing World: Conditions for Sustainability of Intercultural Tourism." In *Tourism and Social Identities*, edited by Peter Burns and Marina Novelli, 95-106. London, UK: Routledge.

Vannini, Phillip. 2011. "Constellations of (In-)Convenience: Disentangling the Assemblages of Canada's West Coast Island Mobilities." *Social & Cultural Geography* 12(5): 471-492.

Vannini, Phillip, and Jonathan Taggart. 2013. "Doing Islandness: A Non-Representational Approach to an Island's Sense of Place." *Cultural Geographies* 20 (2): 225-242.

Veenendaal, Wouter. 2016. "Smallness and Status Debates in Overseas Territories: Evidence from the Dutch Caribbean." *Geopolitics* 21 (1): 148-170.

Verne, Jules. 1874. *The Mysterious Island*. http://www.gutenberg.org/files/1268/1268-h/1268-h.htm.

Vine, David. 2011. *Island of Shame: The Secret History of the US Military Base on Diego Garcia*. Princeton, NJ: Princeton University Press.

Vine, David. 2019. "No Bases? Assessing the Impact of Social Movements Challenging US Foreign Military Bases." *Current Anthropology* 60 (S19): s158-s172.

Vitaliano, Dorothy B. 2007. "Geomythology: Geological Origins of Myths and Legends." *Geological Society London – Special Publications* 273 (1): 1-7.

Vitousek, Peter M., Henning Adersen, and Lloyd L. Loope. 1995. "Introduction: Why Focus on Islands?" In *Islands: Biological Diversity and Ecosystem Function*, edited by Peter M. Vitousek, Henning Adersen, and Lloyd L. Loope, 1-4. Berlin: Springer.

Voigt, Cornelia, and Christof Pforr. 2014. *Wellness Tourism: A Destination Perspective*. Abingdon, UK: Routledge.

von Mossner, Alexa Weik. 2015. "Small Islands in Documentary Film." *Global Environment* 8 (1): 178-195.

Vourdoubas, John. 2019. "Estimation of Carbon Emissions Due to Tourism in the Island of Crete, Greece." *Journal of Tourism Hospitality Management* 7 (2): 24-32.

Walker, Lawrence R., and Peter Bellingham. 2011. *Island Environments in a Changing World*. Cambridge, UK: Cambridge University Press.

Wallace, Alfred R. 1902. *Island Life: or, The Phenomena and Causes of Insular Faunas and Floras: Including a Revision and Attempted Solution of the Problem of Geological Climates*. London, UK: Macmillan.

Ward, R. Gerard, John W. Webb, and Michael Levison. 1973. "The Settlement of the Polynesian Outliers: A Computer Simulation." *The Journal of the Polynesian Society* 82 (4): 330-342.

Ward, Trevor J., and Alan Butler. 2006. "Coasts and Oceans: Theme Commentary Prepared for the 2006 Australia State of the Environment Committee." *Department of Environment and Heritage, Canberra*. https://webarchive.nla.gov.au/wayback/20120319030224/http://www.environment.gov.au/soe/2006/publications/commentaries/coasts/pubs/coasts.pdf.

Warrington, Edward, and David Milne. 2007. "Island Governance." In *A World of Islands: An Island Studies Reader*, edited by Godfrey Baldacchino, 379-428. Charlottetown, PE: Island Studies Press.

Warrington, Edward, and David Milne. 2018. "Governance." In *The Routledge International Handbook of Island Studies: A World of Islands,* edited by Godfrey Baldacchino, 173-201. New York: Routledge.

Wasem, Ruth Ellen. 2009. "Cuban Migration to the United States: Policy and Trends." CRS Report for Congress 7-5700. Washington, DC: Congressional Research Service. https://www.fas.org/sgp/crs/row/R40566.pdf.

Watts, Ronald. 2009. "Island Jurisdictions in Comparative Constitutional Perspective." In *The Case for Non-Sovereignty: Lessons from Sub-National Island Jurisdictions*, edited by Godfrey Baldacchino and David Milne, 21-39. London, UK: Routledge.

Weale, David. 1991. "Islandness." *Island Journal* 8: 81-82.

Weale, David. 1992. *Them Times.* Charlottetown, PE: Island Studies Press.

Weaver, David B. 1993. "Ecotourism in the Small Island Caribbean." *GeoJournal* 31 (4): 457-465.

Weaver, David B. 2006. "The 'Plantation' Variant of the TALC in the Small-Island Caribbean." In The Tourism Area Life Cycle Volume 2: Conceptual and Theoretical Issues, edited by Richard W. Butler, 185-197. Clevedon, UK: Channel View.

Weber, Eberhard. 2014. "Environmental Change and (Im)Mobility in the South." In *A New Perspective on Human Mobility in the South*, edited by Rudolf Anich, Jonathan Crush, Susanne Melde, and John Oucho, 119-148. New York: Springer.

Weir, Tony, Liz Dovey, and Dan Orcherton. 2016. "Social and Cultural Issues Raised by Climate Change in Pacific Island Countries: An Overview." *Regional Environmental Change* 17 (4): 1017-1028.

Weiskel, Tim. 1989. "Lessons of the Past: An Anthropology of Environmental Decline." *The Ecologist* 19: 104-119.

Wendt, Albert. 1976. "Towards a New Oceania." *Mana Review* 1 (1): 49-60.

Wendt, Albert. 1982. "Towards a New Oceania." In *Writers in East-West Encounter: New Cultural Bearings*, edited by Guy Amirthanayagam, 202-215. London, UK: Palgrave Macmillan.

Whelan, Harry T., Heather Annis, and Phillip Guajard. 2014. "From Land to Sea; Embracing a Renewable Future." Supplement, *Journal of Biotechnology & Biomaterials* S6.

Whittaker, Robert J. 2009. "Krakatau." In *Encyclopedia of Islands*, edited by Rosemary G. Gillespie and David A. Clague, 517-520. Berkeley: University of California Press.

Whittaker, Robert J., and Jose Maria Fernandez-Palacios. 2007. *Island Biogeography: Ecology, Evolution, and Conservation*, 2nd ed. Oxford, UK: Oxford University Press.

Whittaker, Robert J., José María Fernández-Palacios, Thomas J. Matthews, Michael K. Borregaard, and Kostas A. Triantis. 2017. "Island Biogeography: Taking the Long View of Nature's Laboratories." *Science* 357 (6354): eaam8326.

Wikipedia. n.d. "Social Determinants of Health." Accessed November 11, 2019. https://en.wikipedia.org/wiki/Social_determinants_of_health.

Wilkinson, Paul F. 1987. "Tourism in Small Island Nations: A Fragile Dependence." *Leisure Studies* 6 (2): 127-146.

Wilkinson, Paul F. 1989. "Strategies for Tourism in Island Microstates." *Annals of Tourism Research* 16 (2): 153-177.

Wilkinson, Paul F. 1999. "Caribbean Cruise Tourism: Delusion? Illusion?" *Tourism Geographies* 1 (3): 261-282.

Williams, Maslyn, and Barrie MacDonald. 1985. *The Phosphateers. A History of the British Phosphate Commissioners and the Christmas Island Phosphate Commission.* Melbourne: Melbourne University Press.

Williams, Raymond. 2006. "The Analysis of Culture." In *Cultural Theory and Popular Culture*, 3rd ed., edited by John Storey, 32-40. Harlow, UK: Pearson Education.

Williams, Rhodri. 2018. "One Hundred Years of Solitude: The Significance of Land Rights for Cultural Protection in the Åland Islands." *Journal of Autonomy and Security Studies* 2 (1): 50-81.

Williamson, Ian, and Michael D. Sabath. 1982. "Island Population, Land Area, and Climate: A Case Study of the Marshall Islands." *Human Ecology* 10 (1): 71-84.

Williamson, Jeffrey G. 1986. "Migration and Urbanization in the Third World." Discussion Paper no. 1245. Harvard Institute of Economic Research. Cambridge, MA: Harvard University.

Williamson, Robert W. 2013. *Religious and Cosmic Beliefs of Central Polynesia*, vol. 2. Cambridge, UK: Cambridge University Press.

Wilmshurst, Janet M., Atholl J. Anderson, Thomas F. G. Higham, and Trevor H. Worthy. 2008. "Dating the Late Prehistoric Dispersal of Polynesians to New Zealand Using the Commensal Pacific Rat." *Proceedings of the National Academy of Sciences* 105 (22): 7676-7680.

Wilson, Janet M. 2018. "Deconstructing Home: 'The Return' in Pasifika Writing of Aotearoa New Zealand." *Journal of Postcolonial Writing* 54 (5): 641-654.

Wilson, John. 2005. "European Discovery of New Zealand – Abel Tasman." *Te Ara: The Encyclopedia of New Zealand.* http://www.TeAra.govt.nz/en/european-discovery-of-new-zealand/page-2.

Wilson, Samuel, ed. 1997 (1999). *The Indigenous People of the Caribbean.* Gainesville, FL: University Press of Florida.

Winters, L. Alan, and Pedro M. G. Martins. 2004. "When Comparative Advantage is Not Enough: Business Costs in Small Remote Economies." *World Trade Review* 3 (3): 347-383.

Wodzicki, Kazimierz A. 1950. *Introduced Mammals of New Zealand: An Ecological and Economic Survey.* Wellington, NZ: CABI.

Wood, Robert E. 2000. "Caribbean Cruise Tourism: Globalization at Sea." *Annals of Tourism Research* 27 (2): 345-370.

Woodham-Smith, Cecil. 1991. *The Great Hunger: Ireland 1845-1849.* London, UK: Penguin.

World Bank. 2000. "Chapter 1: The Ocean to Pacific Island People." *Cities, Seas, and Storms: Managing Change in Pacific Island Economies,* 1-4. Accessed September 17, 2019. https://pacific-data.sprep.org/system/files/180.pdf.

World Bank. 2016. *Migration and Remittances Factbook 2016,* 3rd ed. Washington, DC: International Bank for Reconstruction and Development. Accessed November 30, 2019. http://siteresources.worldbank.org/INTPROSPECTS/Resources/334934-1199807908806/4549025-1450455807487/Factbookpart1.pdf.

World Bank. 2019. "Air Transport, Passengers Carried." Accessed November 11, 2019. https://data.worldbank.org/indicator/is.air.psgr.

World Bank Group. 2019a. "Migration and Remittances: Recent Developments and Outlook." *Migration and Development Brief 31.* https://www.knomad.org/publication/migration-and-development-brief-31.

World Bank Group. 2019b. "Personal Remittances, Received (% of GDP)." Accessed November 8, 2019. https://data.worldbank.org/indicator/BX.TRF.PWKR.DT.GD.ZS.

World Health Organization, Regional Office for South-East Asia. 2008. *Health in Asia and the Pacific.* Manila: World Health Organization Regional Offices for South-East Asia and the Western Pacific. https://apps.who.int/iris/handle/10665/205227.

World Tourism Organization. 2016. *UNWTO Tourism Highlights, 2016 Edition.* Madrid: UNWTO. http://www.e-unwto.org/doi/pdf/10.18111/9789284418145.

World Tourism Organization. 2019. *International Tourism Highlights, 2019 Edition.* Madrid: UNWRO. https://doi.org/10.18111/9789284421152.

Wynne, Barbara Groome. 2007. "Social Capital and Social Economy in Sub-National Island Jurisdictions." *Island Studies Journal* 2 (1): 115-132.

Wyss, Johann David. 1812. *The Swiss Family Robinson; or Adventures on a Desert Island.* http://www. gutenberg.org/files/11703/11703-h/11703-h.htm.

Yiallourides, Constantinos. 2019. "First Chagos, Then Cyprus? Cyprus Gains Legal Tool in ICJ Ruling on Chagos Islands." *SSRN Electronic Journal.* https://ssrn.com/abstract=3351969.

Yoshinaga, Ida. 2019. "Disney's Moana, the Colonial Screenplay, and Indigenous Labor Extraction in Hollywood Fantasy Films." *Narrative Culture* 6 (2): 188-215.

Younger, Stephen M. 2009. "Violence and Warfare in the Pre-Contact Caroline Islands." *Journal of the Polynesian Society* 118 (2): 135-164.

Zafar, Ali. 2011. "Mauritius: An Economic Success Story." In *Yes Africa Can: Success Stories from a Dynamic Continent,* edited by Punam Chuhan-Pole and Manka Angwafo, 91-106. Washington, DC: World Bank.

Zamel, Noe. 1995. "In Search of the Genes of Asthma on the Island of Tristan da Cunha." *Canadian Respiratory Journal* 2 (1): 18-22.

Zimmet, Paul, Gary Dowse, Caroline Finch, Sue Serjeantson, and Hilary King. 1990. "The Epidemiology and Natural History of NIDDM: Lessons from the South Pacific." *Diabetes/Metabolism Reviews* 6 (2): 91-124.

Zoromé, Ahmed. 2007. "Concept of Offshore Financial Centers: In Search of an Operational Definition." IMF Working Paper WP/07/87. International Monetary Fund, Washington, DC. https:// www.imf.org/external/pubs/ft/wp/2007/wp0787.pdf.

Index